Sustainable Alternatives for Poverty Reduction and Eco-Justice

Sustainable Alternatives for Poverty Reduction and Eco-Justice: Volume 1 2nd Edition

Edited by

Lucas Andrianos, Jan-Willem Sneep, Guillermo Kerber and Robin Attfield

CAMBRIDGE SCHOLARS
PUBLISHING

Sustainable Alternatives for Poverty Reduction and Eco-Justice:
Volume 1 2nd Edition,
Edited by Lucas Andrianos, Jan-Willem Sneep, Guillermo Kerber and Robin Attfield

This book first published 2013
Institute of Theology and Ecology, Orthodox Academy of Crete,
Chania, Greece
This edition published 2014

Cambridge Scholars Publishing

12 Back Chapman Street, Newcastle upon Tyne, NE6 2XX, UK

British Library Cataloguing in Publication Data
A catalogue record for this book is available from the British Library

ISBN (10): 1-4438-6689-X, ISBN (13): 978-1-4438-6689-7

CONTENTS

Poverty, Financial Crisis and Human Rights

Green Economy and Food Security

Global Crisis and Case Studies

PATRIARCHAL GREETINGS

H.A.H. BARTHOLOMEW

BY THE GRACE OF GOD
ARCHBISHOP OF CONSTANTINOPLE,
NEW ROME AND ECUMENICAL PATRIARCH
SPIRITUAL MENTOR OF THE OAC[1]

Your Eminence Metropolitan of Kissamos and Selinon and Exarch of Western Crete, Our beloved brother and concelebrant in the Holy Spirit, Mgr Amphilochios, President of the Orthodox Academy of Crete, may the grace and peace of God be with your Eminence.

The Orthodox Church has always attributed the greatest importance to the purpose of the creation of humankind, which has been succinctly defined in the book of Genesis: "And the Lord God took the man whom he had formed, and placed him in the garden of Delight, to cultivate and keep it" (Gen. 2.15 Septuagint); the Church understood as paradise the whole creation made by God for the sake of humankind.

Unfortunately, the rupture in the relation between Man and God that occurred through the disobedience of Adam and Eve broke the relation of humankind and the natural environment, leading to irreparable distortions. This, because fallen man, by appropriating the creation, considered himself as its master, instead of steward who gained the right to enjoy the fruits and goods of nature by divine dispensation, and subjected the

[1] Translated by Dr. Vasilios Meichanetsidis, scientific staff of the Orthodox Academy of Crete.

creation to the "vanity" of his greed and arrogance, as St. Paul accurately states (Rom 8:20). This illogical and selfish use of creation by a few not only damaged it, but deprived a great number of people of the necessary resources for their survival, and continues to do so.

The Mother Church of Constantinople, within the framework of her sacred mission, labours with all Her power for the awakening of the conscience of man, especially so of the leaders, and for the sensitization towards the pursuit of new solutions and ways, capable of stopping the painful consequences of the continuing and ever increasing human greed and the destruction of the natural environment resulting from this. For that reason and with great joy and satisfaction, We place under Our aegis the conference, entitled "Sustainable Alternatives for Poverty Reduction and Eco-Justice", organized by the Institute of Theology and Ecology of the Orthodox Academy of Crete with the active participation of various Institutes and Associations, in order to exchange views and to undertake a series of actions, combining the protection of the environment with the just distribution of its goods to all people, and aiming at the resolving of phenomena of poverty and wretchedness. Being convinced that any solution to this most important issue, that is of the relation of humankind with the natural environment, cannot dissociate the scientific dimension from the ethical one, We are pleased and delighted about this common consultation between representatives of the Church and the scientific world under the auspices of our Holy Church and of the Orthodox Academy of Crete, looking forward with great interest to the results and conclusions of this international conference. Thus, We pray to our God and Creator of the whole universe for the success of the deliberations of this first international ecological and theological conference to be of benefit for humankind and for the protection of the natural environment.

May the grace of God and His infinite mercy be with you all.

24th of September, 2012
† Bartholomew of Constantinople, your beloved brother in Christ

MESSAGE OF WELCOME

JAN-WILLEM SNEEP,
CHAIRMAN OF THE PLANTA EUROPA FOUNDATION

Planta Europa Foundation, Laan van Meerdervoort 1030,
2564 AW The Hague, The Netherlands
E-mail: jwsneep@live.nl

Distinguished guests, ladies and gentlemen, friends,

It is a great honour and pleasure for me to be here. I would like to thank the Orthodox Academy of Crete for organising this conference to share our experiences and knowledge on sustainable alternatives for poverty reduction and ecological justice.

With more than 7 billion people, we are living in a fascinating world with a very rich biodiversity, a world with millions of living organisms. This world provides the raw materials, goods and services we need to be able to live, work, produce and consume. To make the strengthening and sustainable use of biodiversity possible, changes are needed in current consumption and production patterns, the distribution of wealth, access to technology and the effectiveness of international institutions. This is not only an ecological task, but also a socio-economic one.

The concept of sustainability rests in fact on three pillars: ecological, economic and socio-cultural, which together must remain in balance at all times. There can be no sustainability if growth in one of the pillars proceeds at the expense of another pillar. Sustainable development also excludes passing on problems to other parts of the world, or to future generations. The three areas of transition necessary for sustainable development are: renewable energy, sustainable use of biodiversity and natural resources, and sustainable agriculture.

Poor populations are often dependent for their survival on the use of natural resources without the space and means necessary to achieve sustainable use. As a result of extreme income inequality and a skewed balance of ownership of valuable natural resources, poor populations are forced into marginal areas. Not only do the poor benefit from biodiversity; biodiversity also benefits from poverty reduction measures. After all, the lack of basic security, ambiguous land rights, and restricted access to health care, credit and education lead to overpopulation, migration, and non-sustainable use of natural resources. Isolation of poor populations deepens their dependence on natural resources. The elimination of poverty and the strengthening and conservation of biodiversity therefore go hand-in-hand.

The downward spiral of degradation and impoverishment can be broken by stimulating participative management of natural resources which focuses on sustainability. Giving the population a voice should be coupled with a transfer of knowledge and skills. This gives the population a feeling of responsibility for resources management with a longer-term perspective. If the natural habitat remains ecologically intact and the local population can survive there, this will create the necessary conditions to discourage migration to the cities.

Development processes are often coupled with the non-sustainable use of natural resources which has a negative impact on natural forests, water reservoirs and wetlands. Unfortunately, many poverty reduction strategies devote little attention to sustainable development in relation to the environment and biodiversity.

Macro-economic processes and international trade relations, which may alleviate poverty in the short term, may be very negative for biodiversity in the long term. Examples include the trade in soybeans for animal feed in tropical forests and dry areas in Latin America and Asia, the investments in palm oil plantations in the tropical rain forests of Indonesia, and the exploitation of wood in central Africa and its effect on large mammal populations. Trade liberalisation can have positive impacts on poverty reduction and biodiversity provided these clearly take place in the context of sustainability.

The disproportionate claim made on biodiversity and natural resources in development countries by rich countries and the fact that biodiversity is not reflected in prices serve to promote inequalities. Sustainable trade relations and the integration of the environmental factor into prices, the abolition of unfair subsidies and the prevention of new trade barriers have the potential to reduce poverty as well as to promote the conservation and sustainable use of biodiversity and natural resources.

The strengthening and sustainable use of biological diversity has to be seen as a general concern of humanity.

Due to the increasing use of social media we will be better informed on people who are living in other areas. Social disruptions like war, crime and corruption divert resources from the area of greatest human need, damage the capacity of societies to plan for the future, and generally threaten human well-being and the environment.

This reminds me of the first children's book of the United Nations: Pepito's speech at the United Nations. One evening at dinner Pepito made a decision. On the plate before him were Spanish rice, Danish rolls, Swiss cheese and salad with French dressing and Greek olives. The dessert would be American apple pie. He said: the United Nations should be like this dinner. All parts go well together. I am going to make a speech in the General Assembly Hall to help solve the problems of the world. With his parents he visited the United Nations. During the guided tour he felt asleep. In his dream he was guest speaker and poised to speak to all the nations. But he could not speak. Than he smiled and then (wonder of wonders), his big sunny smile was reflected on the faces of all delegates from all the countries of the world, and everyone smiled back at him. Pepito had not given the speech he had prepared, but his real speech was his smile, and it said more than any words of any language. It said to everyone who gave it: I am a friend and I offer you friendship. Everybody was telling him his smile was the best speech ever given. So, the little boy Pepito brings the world kindness, a single smile and a dream of peace understood by all the people of the world.

The Earth is our home and all Earth's species, including Homo sapiens erectus, are our family, our sisters and brothers. It is our duty to take care of all habitats and their living inhabitants, as we also take care on our own children. To take actions together now for life on Earth we must learn to live together.

I hope this conference will show us a multitude of sustainable alternatives for poverty reduction and ecological justice, allowing us to continue to believe in a wonderful world for all people in future, And I invite you all to give your smile, just as Pepito did, to the world, so that everyone in it may work together on poverty reduction and ecological justice for the world as a whole.

I wish you a very pleasant and fruitful conference.

Thank you for your attention.

Dr Jan-Willem Sneep,
Chairman,
PlantaEuropa Network

FOREWORD

GUILLERMO KERBER
PROGRAMME EXECUTIVE ON CARE FOR CREATION AND CLIMATE JUSTICE WCC

World Council of Churches
Geneva,
Switzerland
Email: gkm@wcc-coe.org

Climate, sustainability and ecological justice

As part of the long lasting collaboration between the World Conference of Churches and the Orthodox Academy of Crete, in 2012 the International Conference on Sustainable Alternatives for Poverty Reduction and Eco-Justice was organized. On this occasion about twenty members of the WCC Working Group on Climate Change participated in the Conference, bringing their various experiences and contributing to the objective of the conference of engaging multidisciplinary and ecumenical dialogues on key social, economic and ecological concerns.

When wc look at sustainability and ecological justice, two of the core concepts of the conference, one of the most challenging threats of the world today is climate change. The scientific consensus on the topic shows that global warming is impacting various regions of the world and if a cut on CO2 emissions doesn't take place immediately, this will have dramatic consequences for the whole world and future generations. And those who are especially affected are the vulnerable and poor communities in various regions of the world.

In the last thirty years climate change has changed from being a matter for a few specialists to become a matter of discussion in various public spheres. At the international arena, for instance, the adoption of the United Nations Framework Convention on Climate Change (UNFCCC) in 1992 during the UN Conference on Environment and Development in Rio de

Janeiro, Brazil, was a benchmark in a process which had started twenty years before with the UN Conference on the Human Environment, which took place in Stockholm in 1972. The UNFCCC aims at "stabilizing greenhouse gas concentrations in the atmosphere at a level that would prevent dangerous anthropogenic interference with the climate system" (Article 2). In 1995, the UNFCCC opened its Conference of Parties (COPs) in Berlin, and in 1997 the COP adopted the Kyoto Protocol, the binding instrument of the Convention, which sets binding obligations on industrialised countries to reduce emissions of greenhouse gases.

The World Council of Churches (WCC) has a long history dealing with climate issues. It began referring to climate change (then called "global warming")[1] as early as 1988 in the Justice, Peace and Integrity of Creation (JPIC) conciliar process which led to the JPIC Convocation in Seoul in 1990. It is at this time that the Working Group was created. The group helped to analyse the scientific data, deepen the ethical and theological reflection and advocate at the UN, having participated at all UNFCCC annual conferences (called COPs – Conference of Parties). The COPs have had in recent years an extraordinary development. From a few hundred participating at the first COPs, tens of thousands participated in the COPs in Copenhagen in 2009, Cancun in 2010 and Durban in 2011.

In parallel to international negotiations, the mobilization of civil society in recent years has been impressive all over the world with for instance the 350 campaigns or the Global Call for Climate Action, best known as "tcktcktck"[2]. Churches and the ecumenical movement at large have been involved in this process, with their own contribution, the Time for Climate Justice campaign[3].

One of the key components of the ecumenical advocacy work has been to link climate change with justice. An example of the rationale for this linkage is expressed in the joint declaration from the Lutheran World Federation and WCC delegates to a UN Climate Conference in Cancún, México in 2010 (COP 16)[4]: "In the churches' perspective, justice must be

---

[1] See e.g.WCC, "An evaluation of the UNCED Conventions" in *Searching for the new heavens and the new earth: An Ecumenical response to UNCED*, where under the section titled Climate Change, the term global warming is frequently used as a synonym of climate change (1992: 13 -14).

[2] Cf. www.350.org and tcktcktck.org.

[3] Cf. www.climatejusticeonline.org

[4] See the Declaration "Why are the churches at the UN Conference on Climate

the basic criterion of applied ethics in all decisions concerning the measures to cope with climate change. Although climate change is a global issue affecting all peoples and nations, those who are and will increasingly be affected by negative climate change consequences are the vulnerable communities who have contributed the least to global emissions. These include women and children, indigenous peoples, poorest communities, people with disabilities and inhabitants of coastal low lying areas. Vulnerable communities and states are also much more dependent on natural resources for their subsistence and do not have the means to mitigate emissions and to adapt to climate change. Their survival is at risk, and justice requires that the nations most responsible historically for the adverse ecological conditions should take the greatest responsibility towards the adaptation of these vulnerable communities and nations."

Advocacy for climate justice is rooted in the churches' affirmation that climate change has to be addressed holistically. This has meant recognizing the interconnectedness of the various dimensions of the climate crisis—including environmental, economic, social, and cultural. Furthermore, churches have been stressing that the climate change crisis, as part of the ecological crisis, is also an ethical and spiritual crisis. We are quite aware of the environmental, political, economic, social and cultural aspects of the crisis. But sometimes, the spiritual dimension is overlooked. When we look at sustainability and ecological justice, we need to bring in the spiritual dimension as well.

This has been one of the achievements of the Orthodox Academy of Crete when organizing various conferences: bridging the gaps between the sciences, ethics, theology and spirituality. The various contributions to the conference, collected in this book, show the interlinks between these different aspects and deepen reflection on various topics.

Participating at SAPREJ 2012 was, for the WCC Working Group on Climate Change and for me personally, a profound experience which allowed us to discuss matters of concern with experts from all over the world and to build a community that lasts beyond the period of the conference. For this we are thankful to the authorities of the OAC and

---

Change in Cancun" at: http://www.oikoumene.org/en/resources/documents/wcc-programmes/justice-diakonia-and-responsibility-for-creation/climate-change-water/why-are-the-churches-at-the-un-conference-on-climate-change-in-cancun (accessed 12 August 2013)

especially to Dr Lucas Andrianos, who, putting together all the building blocks, guaranteed the meaningful flow of the conference.

Dr. Guillermo Kerber
Programme Executive
World Council of Churches

EDITORIAL

LUCAS ANDRIANOS
HEAD OF THE INSTITUTE OF THEOLOGY AND ECOLOGY, ORTHODOX ACADEMY OF CRETE

Kolympari, Chania, Crete, Greece
E-mail: laoac@otenet.gr

"A common search for solutions to global crisis and injustice"

The first international conference on Sustainable Alternatives for Poverty Reduction and Eco-Justice (SAPREJ-12) took place in Kolympari, Chania, Crete, at the Orthodox Academy of Crete (OAC) in September 27th – 30th, 2012. This first conference was organized by the Institute of Theology and Ecology - OAC and as an outcome of this initiative, the current book 'Sustainable Alternatives for Poverty Reduction and Eco-Justice-12, Vol.1" is published and distributed worldwide.

The editorial of this SAPREJ-12 book began in the same way as the previous book series of the Orthodox academy publications. The reason is simple and self-explanatory, as SAPREJ-12 is the natural continuation of the efforts of the Institute of Theology and Ecology OAC in publishing ecological texts related to ecological concerns and the search for common sustainable strategies to the preservation of the Divine Creation.

Consequently, the content of this book addresses, once more, the need for human response to the ecological crisis and human sufferings – poverty – from socio-economic problems. It affirms human responsibility for the well-being of the Earth and the vast diversity of Life. More than forty five (45) abstracts were submitted from about twenty four countries worldwide: *Algeria, Australia, Austria, Belgium, Bangladesh, Canada, Czech Republic, Germany, Ghana, Greece, India, Kenya, Korea, Madagascar, Romania, South Africa, Sweden, Switzerland, Tanzania, Thailand, The Netherlands, Turkey, United Kingdom and the United*

States of America. From these papers, about thirty (30) were selected for presentation and publication in this book.

The aim of the SAPREJ-12 conference is to engage multidisciplinary and ecumenical dialogues on key social, economic and ecological concerns from a variety of perspectives. Among others, the goal of the conference is to build a permanent partnership and long life learning process to promote social and ecological justice.

Participants were interested in multi-disciplinary exchanges and insights, with a focus on religious-based and scientific approaches to sustainability problems and injustice. The conference was a blend of learning and discussion, while attending to the magnificent Earth and cultural context of this region of Crete.

The selected papers are classified in six thematic sections: (1) Biodiversity, sustainable policy, religion and ethics; (2) Financial crisis, the global market and poverty; (3) Ecological crisis, climate change and eco-justice; (4) Economical ethics and ecological economics; (5) Production, distribution and consumption patterns; (6) Permaculture and case studies.

The expected perspective for the publication of this book is to promote collaboration between theology and applied science in order to find solutions to poverty and environmental problems. This step was initiated since the foundation of the Institute of Theology and Ecology at the OAC and it is reinforced by the recent publication of the ECOTHEE books. Now we expect more in-depth progress so it is essential that we develop the SAPREJ book series so that our skills to combine science and religion achieve a better management of our planet.

I would like to thank all our sponsors and supporting institutions for the SAPREJ-12 conference. Special thanks are due to the World Council of Churches – Care for Creation on Climate Justice, the Holy Convent of Chrysopigi and to all SAPREJ-12 participants for their valuable contributions and unforgettable fellowship.

Last but not least, we would like to thank the “PlantaEuropa” Foundation, the local Television Channel “NEA TV”, all the primary and secondary “schools” from Chania and “Flowers of Crete” for their support in organizing the environmental exhibition and aspects that made the

SAPREJ-12 a successful event, promoting ecological awareness and awareness of the need to protect the earth and all Creation.

Blessings and Peace to All in Christ,

Dr. Lucas Andrianos
Head of Institute and SAPREJ founder
Institute of Theology and Ecology
Orthodox Academy of Crete

Biodiversity and Ecological Crisis

NATURE CONSERVATION AND ECOLOGICAL ETHICS

JAN-WILLEM SNEEP

Planta Europa Foundation, Laan van Meerdervoort 1030,
2564 AW The Hague, The Netherlands
E-mail: j.w.sneep@live.nl

Abstract

Our planet has a very rich biodiversity. The number of species of plants, animals and micro organisms, the enormous diversity of genes in these species, the different ecosystems on the planet, such as deserts, rainforests and coral reefs are all part of a biologically diverse Earth. The biodiversity we see today is the fruit of billions of years of evolution, shaped by natural processes and, increasingly, by the influence of humans. It forms the web of life of which we are an integral part and upon which we so fully depend. Healthy ecosystems provide vital goods and services to human and other organisms. The ongoing growth of human populations has an enormous impact on the use of land and the increasing use of our natural resources. Our soils, freshwater, oceans, forests and biodiversity are being rapidly degraded. Reducing negative human impact and enhancing ecosystem services needs good management of oceans, freshwater systems, land and atmosphere. The human impact is called the Ecological Footprint, which is defined as the area of productive land and water ecosystems required to produce the resources that the population consumes and assimilate the wastes that the population produces, wherever on Earth the land or water is located.

Biodiversity and development are closely linked. Biodiversity sustains development and development has an impact on biodiversity, either positive or negative. Some key development sectors and related activities depend directly on biodiversity, and contribute to poverty alleviation. These sectors include agriculture and livestock, forestry, fishing, tourism and, in many areas of the world, housing construction.

Sustainability is the route towards economic and social development. The Convention on Biological Diversity (CBD) invites the Parties to integrate the conservation and sustainable use of biological diversity into relevant sectoral or cross-sectoral plans, programmes and policies. The underlying driver of direct human impact on the environment is human consumption. This impact is reduced by not only diminished consumption but by also making the full cycle of production, use and disposal more sustainable. It is time to rethink how we grow, share and consume our food. The "reduce, reuse and recycle" shoppers are using their purchasing power for ethical consumerism.

1. Introduction

Today we are living with 7 billion people on this planet. There are 21 cities with more than 10 million citizens (in 1975 there were only 3 such cities). In 2050 more than 9 billion people will be expected. Most of the increase will be in developing countries whose population is projected to rise from 5.6 billion in 2009 to 7.9 billion in 2050. Europe has become highly urbanised in recent decades. Two out of three of the more than 800 million Europeans live in cities now, which accounts for 1% of the continent's surface area.

Our planet has a very rich biodiversity. (*See YouTube 'Earth': this you need to see.*). This means: the variety of all animals and plant species on our planet, together with all the places where these species are found (the ecosystems).

The number of species of plants, animals and micro-organisms, the enormous diversity of genes in these species, the different ecosystems on the planet, such as deserts, rainforests and coral reefs are all part of a biologically diverse Earth.

Most nature conservation organisations are only focussed on the conservation of priority animals. The focus of WWF is on very attractive, good looking species such as Polar Bear, Panda, Elephants, Tigers, Rhinoceros, Sea Turtles and Whales. These species are more attractive than (for example) threatened snakes and spiders. They call up positive emotions.

Scientists estimate there are between 5 and 15 million species of plants, animals and micro-organisms existing on Earth. We know there must be at least 12 million living species. Tropical forests represent some 5 percent of the earth's surface but harbour 50 percent of all living species. There are 29,375 plant species in Indonesia, compared to 3,270 in Canada. At some places in the upper Amazon basin you can find 1,200 kinds of butterflies, or about 7 percent of the world's species. There are over 50,000 tree species in tropical areas but only around 2,000 in temperate regions. (700 species of tree are found in all of Canada and the United States.)

The total number of species is unknown. Only about 1.8 million species of plants, animals and micro-organisms have been identified (with numbers which are only a soft guide (2010).

The biodiversity we see today is the fruit of billions of years of evolution, shaped by natural processes and, increasingly, by the influence of humans. It forms the web of life of which we are an integral part and upon which we so fully depend.

2. Why Is Biodiversity Important?

Biodiversity fosters ecosystem productivity, where each species, no matter how small, has an important role to play. Examples:

- A larger number of plant species means a greater variety of crops.
- Greater species diversity ensures natural sustainability for all life forms.
- Healthy ecosystems can better withstand and recover from a variety of disasters.

And so, while we dominate this planet, we still need to preserve the diversity in wildlife. The tremendous growth in human population is the root of biodiversity crises.

A healthy biodiversity provides a number of natural services for everyone: Ecosystem services, such as

- Protection of water resources
- Soils formation and protection
- Nutrient storage and recycling
- Pollution breakdown and absorption
- Contribution to climate stability
- Maintenance of ecosystems
- Recovery from unpredictable events

Biological services, such as:

- Food
- Medicinal resources and pharmaceutical drugs
- Wood products
- Ornamental plants
- Breeding stocks, population reservoirs
- Future resources
- Diversity in genes, species and ecosystems

Social benefits, such as:

- Research, education and monitoring
- Recreation and tourism
- Cultural values (aesthetic, spiritual)

That is quite a lot of services we get for free. But at the same time we fully depend on the biodiversity on our planet. We simply cannot live without it.

Also each species depends on the services provided by other species to ensure survival. It is a type of cooperation based on mutual survival and is often what a “balanced ecosystem” refers to. One third of all our food – fruit and vegetables – would not exist without pollinators visiting flowers. But honeybees, the primary species that fertilizes food-producing plants, have suffered dramatic declines in recent years, mostly from afflictions introduced by humans. Bees are vital to biodiversity.
(See YouTube ‘Biodiversity starts with a B’.)

3. Impact on Biodiversity

Healthy ecosystems provide vital goods and services to human and other organisms. Through the provision of biological resources and ecosystem services, biodiversity is an essential component of human development. The ongoing growth of human populations has an enormous impact on the use of land and the increasing use of our natural resources. Our soils, freshwater, oceans, forests and biodiversity are being rapidly degraded. Climate change is putting even more pressure on the resources we depend on.

Today’s biodiversity crisis is being caused by ourselves. Reducing negative human impact and enhancing ecosystem services need good management of oceans, freshwater systems, land and atmosphere.

Loss of biodiversity stems largely from the habitat loss and fragmentation produced by the human appropriation of land for development (urbanisation and mobility), forestry and agriculture as natural capital is progressively converted to man-made capital. Ten thousand years ago humans began to cultivate crops, drastically changing the human ecological niche. People around the world utilize over 40.000 species every day. All the crops that we cultivate (corn, wheat, potatoes, tomatos, apples, oranges and so on) are domesticated from wild species. Species diversity is also important to agriculture: approximately 7000 plant species are cultivated worldwide. Genetic diversity allows adaptation to climate change, pests and diseases and is therefore important to global food security.

We are destroying thousands of acres of forest a year, not only in the tropics, but everywhere. Over 1.6 billion people rely on forests and non-timber forest products for their livelihoods. Communities around the world depend on forests for goods such as food, medicines, firewood and building materials. Forests also maintain important ecosystem services including carbon sequestration, erosion control, watershed protection and nutrient cycling, and provide habitat for approximately 80% of the remaining terrestrial biodiversity.

The majority of biofuels (sourced from plants and extracted by a variety of methods), including bio-ethanol, bio-diesel and celluloid ethanol, require an increasing amount of land for biofuel crop production.

Food is essential to life. Feeding more than 7 billion human bodies takes a heavy toll on the Earth's resources. This begins with the appropriation of about 38% of the Earth's land surface and about 20% of its net primary productivity. Added to this are the resource-hungry activities of industrial agribusiness – everything from the crop need for irrigation water, synthetic fertilizers and pesticides to the resource costs of food packaging, transport (now a major part of global trade) and retail.

A profound change of the global food and agriculture system is needed if we are to nourish today's 925 million hungry people and the additional 2 billion people expected by 2050. The food and agriculture sector offers key solutions for development, and is central for hunger and poverty eradication.

The world's oceans – their temperature, chemistry, currents and life – drive global systems that make the Earth habitable for humankind. Our rainwater, drinking water, weather, climate, coastlines, much of our food, and even the oxygen in the air we breathe, are all ultimately provided and regulated by the sea. Careful management of this essential global resource is a key feature of a sustainable future. We are overharvesting the world's oceans. Many people depend on marine and coastal biodiversity for their livelihoods, including many people in the developing countries for whom fishing is a main subsistence and commercial activity. 25% of the marine species are found in coral reefs, which provide livelihoods for over 100 million people.

Water covers 71% of the Earth's surface. Of this is 97.5% is the salty water of the oceans and only 2.5% freshwater, most of which is locked up in the Antarctic ice sheet. The remaining freshwater is found in glaciers, lakes, rivers, wetlands, the soil, aquifers and the atmosphere. Due to the water cycle, fresh water supply is continually replenished by precipitation; however there is still a limited amount, necessitating management of this resource. During the 20th century, more than half the world's wetlands have been lost along with their valuable environmental services. Increasing urbanisation pollutes clean water supplies and much of the world still does not have access to clean, safe water.

Currently towards 35% of human water use is unsustainable. From 1961 to 2001 water demand doubled – agriculture use increased by 75%, industrial use by more than 200% and domestic use by more than 400%. In the 1990s it was estimated that humans were using 40-50% of the globally available freshwater in the approximate proportion of 70% for agriculture, 22% for industry, and 8% for domestic purposes, with total use progressively increasing.

Clean, accessible water for all is an essential part of the world we want to live in. There is sufficient fresh water on the planet to achieve this dream. But due to bad economics or poor infrastructure, every year millions of people, most of them children, die from diseases associated with inadequate water supply, sanitation and hygiene. Water scarcity, poor water quality and inadequate sanitation negatively impact food security, livelihood choices and educational opportunities for poor families across the world. Drought afflicts some of the world's poorest countries, worsening hunger and malnutrition.

We depend on the living world for the oxygen that we breathe. Green plants produce oxygen through photosynthesis (sun energy). Since the industrial revolution the concentrated energy of the Sun stored in fossilized plants as fossil fuels has been a major driver of technology which, in turn, has been the source of both economic and political power. Fossil fuel (oil, coal, natural gases) is used for energy. The atmospheric increase in CO2 was human-induced, mostly as a result of fossil fuel emissions.

Human impacts on the atmosphere also include the air pollution in cities, the pollutants including toxic chemicals like nitrogen oxides, sulphur oxides, volatile organic compounds and particulate matter that produce photochemical smog and acid rain, and the chlorofluorocarbons that degrade the ozone layer. Stabilizing the world's climate will require high–income countries to reduce their emissions by 60-90% (Kyoto protocol).

4. Ecological Footprint

Biodiversity and development are closely linked. Biodiversity sustains development and development has an impact on biodiversity, either positive or negative. Some key development sectors and related activities depend directly on biodiversity, and contribute to poverty alleviation. These sectors include agriculture and livestock, forestry, fishing, tourism and, in many areas of the world, housing construction..

In this way we claim a great deal of the natural resources and biodiversity of other countries. This impact is called the Ecological Footprint. An ecological footprint is a measurement of human impact on the environment. It calculates the amount of land and water required to maintain a particular population.

The import of animal feed, wood, meat and other (often agricultural) products from all over the world makes up an ecological footprint. Wherever these imports are based on non-sustainable production, nature is impacted and we see a loss of biodiversity.

Mass tourism, the largest economic sector in the world, has a major impact on the use of space, the use of water and energy and other natural resources, and on biodiversity. There is also a lost of biodiversity whenever financial institutions invest on a large scale in non-sustainable production processses of foreign governments and enterprises. Some palm-oil plantations in Africa and Asia are conspicious examples of this.

Land clearance, mining activities, land reclamation, dredging and similar water-related activities usually involve large-scale interventions which can have negative impacts on ecosystems.

In 2006, the Living Planet Report ranked the United Arabic Emirates as having the highest per capita Ecological Footprint in the world. This movie shows an initiative to work together to better understand the Ecological Footprint of the UAE, and monitor consumption patterns in the country. (See You Tube ‘Ecological Footprint Animation’.)

For millennia, mankind’s use of biodiversity and ecosystem services has contributed to human well-being and economic development. The world’s poor (70% of the world’s poor live in rural areas) depend directly on biological resources for as much as 90% of their needs, including food, fuel, medicine, shelter and transportation.

Habitat degradation and the loss of biodiversity are threatening the livelihoods of more than 1 billion people living in dry and sub-humid lands. Livestock is essential to many aspects of rural livelihood strategies, such as the subsistence consumption of home-produced milk, eggs or meat, the use of manure as fertilizer and the provision of labour for agricultural tasks.

Biodiversity is crucial to the reduction of poverty, due to the basic goods and ecosystem services it provides. Strategies to protect biodiversity must therefore be developed for achieving poverty reduction and sustainable development.

A sustainable food system encourages local production and distribution infrastructures and makes nutritious food available, accessible, and affortable to all. Futher it is humane and just, protecting farmers and other workers, consumers, and communities.

If there is growing awareness on the value and importance of biodiversity, there is a lack of consensus on how it can be measured. Biodiversity offers the potential to place unique products on the market and to generate income for local communities. Many of these products are very valuable; yet their sale seldom benefits the people who harvest them.

Biodiversity’s capacity to deliver benefits to the poor is diminishing around the world. In fact, poverty and hunger are worsening as ecosystems unravel and water and soil resources continue suffering from disruption and over-exploitation.

Reversing ecosystems' degradation while meeting an increasing demand for their services is a challenge. It can be met through significant changes to policies, regulations and practices.

Reducing poverty in developing economies requires:

- Policies, investments and reforms to enhance the sustainable and efficient use of natural resources and the production processes on which they depend.
- Ensuring that the financial returns from more sustainable activities are re-invested in the industrial activities, infrastructure, health services, and the education and skills necessary for long-term economic development.
- Targeting investments and other policy measures to improve the livelihoods of the rural poor, especially those living in fragile environments.
- Protecting and improving the provision of ecosystem services on which the extreme poor depend.

5. Convention on Biological Diversity

Sustainability is the route towards economic and development. The Convention on Biological Diversity (1992) is key to sustainable development. However, its practical implementation (integration of biodiversity considerations into development policies and programmes) still requires much work.

But what is sustainability? Sustainability is the capacity to endure. For humans, sustainability is the long-term maintenance of well being, which has environmental, economic and social dimensions, and encompasses the concept of union, an interdependent relationship and mutually responsible position with all living and non-living things on earth. In ecology, sustainability describes how biological systems remain diverse and productive over time, a necessary precondition for human well-being. Long-lived and healthy wetlands and forests are examples of sustainable biological systems.

The Earth Charter Initiative (2000) speaks of "a sustainable global society founded on respect for nature, universal human rights, economic justice, and a culture of peace".

Through *Article 6 (b)* the Convention invites the Parties to "integrate, as far as possible and as appropriate, the conservation and sustainable use of biological diversity into relevant sectoral or cross-sectoral plans, programmes and policies".

The important links between biodiversity and development are now reflected in *numerous decisions* of the Convention on Biological Diversity's Conference of the Parties. (*See YouTube: 'CBD What does it mean?')*

The need to mainstream the conservation and sustainable use of biological resources across all sectors of the national economy, society and through policy-making frameworks was recognized by the Parties of CBD a long time ago.

Actions to be undertaken by the international community should be:

- Maintain ecosystems' health and productivity to provide ecosystem services to cover basic needs in the long term;
- Instigate proper legislation for a fair and equitable access to ecosystems;
- Institute market regulations and economic incentives at all levels to build up a green economy focusing on pro-poor growth;
- Combine scientific and traditional knowledge in order to create the capacities for local communities to sustainably manage biodiversity;
- Overcome national and institutional individualism and build up the required international framework for effective global environmental governance.

6. Human Consumption

From the production perspective, it found that fossil-fuel combusting processes, agriculture and fisheries have the most important impacts. Meanwhile, from the final consumption perspective, it found that household consumption related to mobility, shelter, food and energy-using products cause the majority of the life-cycle impacts of consumption.

The challenge historically has been a close correlation between economic growth and environmental degradation; as communities grow, so the environment declines.

The underlying driver of direct human impact on the environment is human consumption. This impact is reduced by not only diminished consumption but by also making the full cycle of production, use and disposal more sustainable.

Today the consumption levels of the developed world population are unsustainable. So, it is time to rethink how we grow, share and consume our food.

If done right, agriculture, forestry and fisheries can provide nutritious food for all and generate decent incomes, while supporting people-centred rural development and protecting the environment.

The challenge for sustainability is to curb and manage Western consumption while raising the standard of living of the developing world without increasing its resource use and environmental impact. This must be done by using strategies and technology that break the link between, on the one hand, economic growth and on the other, environmental damage and resource depletion.

Fair Trade Certification empowers farmers and farm workers to lift themselves out of poverty by investing in their farms and communities, protecting the environment, and developing the business skills necessary to compete in the global marketplace. Fair Trade is not just about earning a higher price for the harvest; it's about learning techniques for sustainable agriculture that help protect the environment and make the world a better place for future generations.

To become Fair Trade Certified farms must adhere to a strict set of environmental standards, including but not limited to: no use of GMOs, minimal and safe use of chemicals, elimination of slash and burn techniques, and responsible disposal of waste and water runoff. These standards not only ensure sustainable production practices, but also worker health and safety. Additionally, while Fair Trade does not require organic certification, it does encourage organic conversion to further enhance environmental commitment. (*See YouTube: 'Fair trade: the story', an example on fair trade in Costa Rica on the production of coffeebeans*).

What more can we do? Every economic activity produces material that can be classified as waste. To reduce waste, industry, business and government are now mimicking nature by turning the waste produced by industrial metabolism into resources. Dematerialization is being encouraged through the idea of industrial ecology, eco-design and eco-labelling. In addition to the well-established "reduce, reuse and recycle", shoppers are using their purchasing power for ethical consumerism. William Mc Donough introduced the concept "Cradle to cradle". (*See YouTube: 'Cradle to cradle/ waste is food'.)*

7. Conclusion

We must learn to live together. We have been the enemy; now we must become the friend! We have to take care of the goods and services of our planet and therefore also of each other. Now, it is time for actions for life on Earth.

HAH Ecumenical Patriarch Bartholomew: The Green Patriarch and his Care for Creation

Lic. Elías Crisóstomo Abramides

Greek Orthodox Metropolis of Buenos Aires and South America
Ecumenical Patriarchate
WCC Working Group on Climate Change
World Council of Churches
Buenos Aires – Argentina

"The Earth, having no tongue, cries out sighing:
Why are you people polluting me with many evil things?"
Troparion – 9th ode

"The heart that has learnt to love has pity for all creation"
Writings of St. Siluan the Athonite

Abstract: *Action is needed now to rescue our planet with God's help from the effects of human greed.*

1. The Ecumenical Patriarchate – Introduction

In the first half of the 1st Century the Apostle St. Andrew, the "*First to be called*" by Jesus Christ, the "*Protóklitos*", established the Christian Church in Byzantium, an ancient town by the Strait of Bosphorus (Bosporus) at the confluence of Europe and Asia.

After the death of the Apostle St. Andrew, the first bishop of the city of Byzantium was St. Stachys, one of his beloved disciples. In 330 the Roman Emperor Constantine rebuilt the city and renamed it Constantinople - the city of Constantine - becoming the new capital of the Roman Empire. It was also called New Rome. Thus, St. Andrew was named the Patron Saint of the Empire and of the Church known today as the Ecumenical Patriarchate.

During the 6th Century the Patriarch of Constantinople received the title of Ecumenical Patriarch. The Apostolic Succession has continued up to the present. In October 1991 His All Holiness Bartholomew, Archbishop of Constantinople-New Rome and Ecumenical Patriarch, became the 270th successor of Apostle St. Andrew.

The Ecumenical Patriarchate is an Eastern Orthodox Church that fully participates in the life and activities of the World Council of Churches (WCC) since its inception in 1948. Ecumenical Patriarch Bartholomew served on the WCC Executive and Central Committees and in the Faith and Order Commission. He has worked to advance reconciliation with the Roman Catholic Church, the Anglican Communion and other Christian confessions through theological dialogue. Being a member of the Christian World Communions the Ecumenical Patriarchate also takes part in the international meetings of this group. It participates in local ecumenical bodies around the world, and joins in dialogue with monotheistic faiths: Islam and Judaism.

Patriarch Bartholomew strives earnestly to prepare the Orthodox Church for its role as mediator between the East and the West. In his role as "*Primus Inter Pares*" – First among Equals - among the Orthodox Churches he is the spiritual leader of the almost 300 million Orthodox faithful around the world.

A significant and central activity of Patriarch Bartholomew is his worldwide well-known efforts to increase environmental protection awareness around the world, caring for the good creation of God in order to preserve life on planet Earth, our home, in an environment in peace with its Creator. His constant achievements and positive impacts have earned him the honorific title of "Green Patriarch".

2. The Time for Creation – Encyclical Letters

On 1st September 1989 and marking the New Ecclesiastical Year in the Orthodox Church, late Ecumenical Patriarch Dimitrios sent a message to the entire Christian world to offer every year on that day, prayers and supplications to the Creator and Maker of all, both as thanksgiving for the great gift of creation, and as petitions for its protection and salvation. And

to request all those who are entrusted with the responsibility of governing the world's nations, to act without delay taking all necessary measures for the protection and preservation of the natural creation.
In the words of the late Ecumenical Patriarch Dimitrios, it should be the day "*when prayers and supplications should be offered for all Creation and for the reparation of the impairment caused to the natural environment.*"[1]

After his consecration in 1991, Ecumenical Patriarch Bartholomew continued and gave a strong impulse to this vital matter, making it one of the central issues for him to develop in his position as the spiritual head of Orthodoxy.

Today 1st September is known as "*Day for the Protection of All Creation*" and became established and adopted in full agreement with all the Orthodox Churches. It gave birth to the "*Time for Creation*" in the church calendar, starting on 1st September and ending on 4th October of every year, on the day when the Roman Catholic Church commemorates St. Francis of Assisi, their patron saint of ecology. Among others, it was adopted by the World Council of Churches (WCC), by the Conference of European Churches (CEC), and by the Council of European Bishops' Conferences (CCEE) during the Third European Ecumenical Assembly (EE3A) which met in Sibiu, Roumania, in 2007.

Every 1st September the Ecumenical Patriarch issues an Encyclical Letter on the environment. Some of his key concepts expressed across these many years are the following:

> "… Humanity has failed in what was its noble vocation: to participate in God's creative action in the world. It has succumbed to a theory of development that values production over human dignity and wealth over human integrity… This is why nature 'groans and travails' in all its parts… [Rom. 8:22]"[2]

We have disrupted the original natural balance, the perfect original equilibrium of creation. The Ecumenical Patriarch Bartholomew signalled

---

[1] Ecumenical Patriarch Dimitrios: "Encyclical Letter on the Environment, 1st September 1989". The Ecumenical Patriarchate and the World Wide Fund for Nature (WWF): "Orthodoxy and the Ecological Crisis". Phanar, Istanbul, Turkey, and Gland, Switzerland, 1990.

[2] Message of Ecumenical Patriarch Bartholomew on the "Day of Prayer for the Protection of All Creation", 1st September 1994.

our present attitude of constant abuse: "*The beauty of creation and our tragic abuse of it*".[3] He affirmed that "*...according to our current experiences and the concrete efforts of many, few positive steps have been made on the arduous path toward true and stable reconciliation of humankind with our surrounding physical world*", adding that: "*The so imperatively needed reconciliation, whenever and wherever it is accomplished, represents a 'par excellence' spiritual event*".[4]

In November 1997 Ecumenical Patriarch Bartholomew declared that all aggressions to God's good creation was a **sin**:

> "... to commit a crime against the natural world is a sin. For humans to cause species to become extinct and to destroy the biological diversity of God's creation; for humans to degrade the integrity of the Earth by causing changes in its climate, by stripping the Earth of its natural forests, or destroying its wetlands; for humans to injure other humans with disease; for humans to contaminate the Earth's waters, its land, its air, and its life with poisonous substances: these are sins."[5]

3. Water – Source of Life

Patriarch Bartholomew addressed the issue of the vital importance of water. He clearly declared: "*Just as water is the essence of all life, water is also the primary element in the life of a Christian, where the sacrament of Baptism marks the sacred source of the spiritual life... Water, then, signifies the depth of life and the calling to cosmic transfiguration. It can never be regarded or treated as private property or become the means and end of individual interest. Indifference towards the vitality of water constitutes both a blasphemy to God the Creator and a crime against humanity*".[6]

---

[3] Id., ibid.

[4] Message of Ecumenical Patriarch Bartholomew on the "Day of Prayer for the Protection of All Creation", 1st September 1996.

[5] Ecumenical Patriarch Bartholomew: "Address at the Environmental Symposium", Santa Barbara, California, USA, November 1997.

[6] Ecumenical Patriarch Bartholomew: "Statement for the WCC Working Group on Water", August 2005. Published by the WCC - Geneva, Switzerland, January 2006.

4. Climate Change

Climate change is an unprecedented threat to the integrity and diversity of life on earth. Ecumenical Patriarch Bartholomew has proclaimed that ecological abuse is a sin against God and has recognised it is also a crime against humanity. The resources within the human spirit must be discovered in order to develop a sense of urgency and action.

With reference to climate change he declared: "... *Climate change constitutes a matter of social and economic justice. For those who will most directly and severely be affected by climate change will be the poorer and most vulnerable nations – what Christian Scriptures refer to as our 'neighbours' – as well as younger and future generations, the world of our children, and of our children's children... There is a close link between the economy of the poor and the warming of our planet... We must serve our neighbour and preserve our world with both humility and generosity, in a perspective of frugality and solidarity alike*".[7]

In view of the current financial and economic crisis, on 1st September 2010 he stated: *"It is important to note that the current grievous financial crisis may spark the much-reported and absolutely essential shift to environmentally viable development; i.e., to a standard of economic and social policy whose priority will be the environment and not unbridled financial gain..."*[8]

Conferences and travels around the globe, prizes and honours, bilateral declarations with religious leaders of the world, and his permanent dedication to the protection of life on the planet and to what gives us our daily sustenance: clean water, clean air, nutritious food grown on fertile soils, nurturing our body and spirit, all continuously increase his fame as the "Green Patriarch" of the current times.

5. Halki Environmental Seminars

From 1994 to 1998, Ecumenical Patriarch Bartholomew organised five international environmental summer seminars at the Theological School of

---

[7] Ecumenical Patriarch Bartholomew: "Statement for the WCC Working Group on Climate Change", August 2005. Published by the WCC - Geneva, Switzerland, January 2006.

[8] Message of Ecumenical Patriarch Bartholomew on the "Day of the Protection of the Environment", 1st September 2010.

Halki on the grounds of the historical Monastery of the Holy Trinity in the island of Heybeliada, Halki in Greek, the second largest of the Princes' Islands in the Sea of Marmara just off the coast of Istanbul. The seminars were centred on the following key issues: Religious Education (1994); Ethics (1995); Communications (1996); Justice (1997); and Poverty (1998).[9]

In June 2012 under the auspices of Ecumenical Patriarch Bartholomew, the Ecumenical Patriarchate and Southern New Hampshire University convened as co-sponsors the "Halki Summit 2012" at the Theological School of Halki. The summit was envisaged as a conversation on environment, ethics and innovation. It stressed the importance of the carbon-neutral campus of Southern New Hampshire University, and the purpose all the participants have to pass along to the next generation a clean world with clean water and clean air. This pledge also mobilized the grassroots "350.org Campaign" which in a few years' span has emerged as the largest green movement in history.[10]

In his keynote address, Ecumenical Patriarch Bartholomew said. *"... Each of us is called to draw a distinction between* what we want *and* what we need, *or – more importantly – what the* world *needs. Greed and gratification reduce the world to a survival of the fittest; whereas generosity and gratitude transform the world into a community of sharing."*

He added: *"... We know what needs to be done and we know how it must be done... It is a long journey from the head to the heart...and an even longer journey from the heart to the hand... There is no way of endlessly manipulating our environment that comes without cost or consequence... this is a movement as critically urgent and as morally imperative as any campaign for fundamental human and civil rights."*

---

[9] Halki International Environmental Seminars: Five Seminars from 1994-1998.

[10] Halki Summit 2012. Sponsored by the Ecumenical Patriarchate and co-sponsored by Southern New Hampshire University, on Global Responsibility and Environmental Sustainability – 18th -20th June 2012.

6. Scientific International Symposia on World Rivers and Seas

Well-known are the eight international, inter-disciplinary and inter-religious symposia convened from 1995 to 2009 by the Ecumenical Patriarch, bringing together scientists, environmentalists, policy makers and religious leaders, drawing world attention to the degradation of the following bodies of water:

- *Symposium I*: The Aegean Sea (Revelation and the Environment - AD 95-1995) - Convened to commemorate the 1900th anniversary of the Book of Revelation. **1995**
- *Symposium II*: The Black Sea (The Black Sea in Crisis – An Encounter of Beliefs: A Single Objective). **1997**
- *Symposium III*: The Danube River (The Danube River – A River of Life: Down the Danube to the Black Sea). **1999**
- *Symposium IV*: The Adriatic Sea (The Adriatic Sea – A Sea at Risk: a Unity of Purpose - 2002) - In Venice, Ecumenical Patriarch Bartholomew and Pope John Paul II signed the "**Venice Common Declaration**" on ethical principles for the protection of creation, committing both sides to "*steer the earth towards our children's future*". In their words: *"Almighty God envisioned a world of beauty and harmony, and He created it, making every part an expression of His freedom, wisdom and love (Gen. 1:1-25). At the centre of the whole creation, He placed us, human beings, with our inalienable human dignity"*. June **2002**[11]
- *Symposium V*: The Baltic Sea (The Baltic Sea – Heritage and Responsibility). **2003**
- *Symposium VI*: The Amazon River (The Amazon River – Source of Life). **2006** - The destruction of the Amazon rain forests would disrupt the whole global circulation of air and water, and thus the climate of the world.
- *Symposium VII*: The Arctic Ocean (The Arctic Sea – Mirror of Life). **2007** - In the words of Ecumenical Patriarch Bartholomew: "*Time is short. The Ice of the Arctic is shrinking at a frightening pace according to scientists and the Greenlanders. If all of the ice in Greenland melts*

---

[11] Ecumenical Patriarch Bartholomew and Pope John Paul II Common Declaration, "Venice Common Declaration on Environmental Ethics". Symposium IV: "The Adriatic Sea: A Sea at Risk: a Unity of Purpose". Venice, Italy, 10th June 2002.

the consequences for Greenland and the rest of the world could be devastating. We must act together and act now".[12]

- *Symposium VIII*: The Mississippi River (Restoring Balance: The Great Mississippi River). **2009** - "*We shall also continue to pray and work for the protection of the natural environment*", the Ecumenical Patriarch affirmed in New Orleans.[13]

7. Honours – Environmental Awards

In October 1997, Ecumenical Patriarch Bartholomew was awarded the Global Medal of the United States Congress, the highest civilian honour bestowed by the United States Congress, based on his efforts to promote religious freedom and human rights, his initiatives on religious tolerance among the world's religions, and his work for international peace and environmental protection, which have placed him as an apostle of love, peace and reconciliation.[14]

He earned the title "Green Patriarch" and received such distinctive environmental awards as the "Scenic Hudson Visionary Award" in the United States of America in 2000,[15] and the international environment and development "Sophie Prize" by the Government of Norway in 2002.[16]

In May 2012, in order to keep the legacy of President Franklin Roosevelt alive, Ecumenical Patriarch Bartholomew was awarded the "Freedom of Worship Medal", presented by HM Queen Beatrix of The Netherlands and the Dutch Prime Minister, on behalf of the Roosevelt Institute. He was nominated for his extraordinary efforts in favour of religious tolerance among the world's religions, on his initiatives on religious freedom and human rights, and his dedication to a dialogue of reconciliation among Christians, Muslims and Jews, which he started together with HH Pope John Paul II. The "Green patriarch", who gained worldwide respect for his efforts towards the protection of the natural environment and life on the

---

[12] Symposium VII: "The Arctic Sea – Mirror of Life". Greenland, North Pole. September 2007.

[13] Symposium VIII: "Restoring Balance: The Great Mississippi River". Symposium convened in New Orleans, USA. October 2009.

[14] Global Medal of the United States Congress – 6th October 1997.

[15] Scenic Hudson Visionary Award – United States of America – 2000.

[16] Environment and development "Sophie Prize", awarded by the Government of Norway – June 2002.

planet, is one of the outstanding figures dedicated also to the accomplishment of freedom of worship.[17]

8. Messages to United Nations International Meetings

Addressing environmental crises either natural or induced by human activities, and environmental and development international conferences and events, Ecumenical Patriarch Bartholomew issues specific messages stressing the ethical, theological and spiritual aspects in search of solutions.

Every year on 5th June, World Environment Day as established by the United Nations, he issues a message stressing the importance this date has for present and future human communities and ecosystems around the world.

In the message to the United Nations Conference on Climate Change COP17-CMP7[18] which met in Durban, South Africa in 2011, he conveyed that climate change is a profoundly moral and spiritual issue. We human beings tend to ignore the lives of our neighbours, men and women, who are and will be directly affected by the conditions and consequences of climate change. We all are responsible for the future of our planet and for human life; therefore we cannot remain as spectators: climate change affects and will more intensely affect all peoples and nations.

In order to sensitise and energise 'the people of good will': policy makers, politicians, civil society and faith community members participating in COP17-CMP7, he shared three challenges to be met: i) the struggle to surmount national and regional considerations and to contemplate the whole picture; ii) to remember that sacrifice is needed to arrive at a successful conclusion; iii) and the priority in securing moral leadership.

In his message to the United Nations Conference on Sustainable Development "Rio+20"[19] which met in Rio de Janeiro in June 2012, he

---

[17] To keep the legacy of President Franklin D. Roosevelt alive, Ecumenical Patriarch Bartholomew was awarded the "Freedom of Worship Medal", presented by HM Queen Beatrix of The Netherlands and the Dutch Prime Minister, on behalf of the Roosevelt Institute. May 2012.

[18] United Nations Conference on Climate Change COP17-CMP7, Durban, South Africa, 28th November to 9th December 2011.

[19] United Nations Conference on Sustainable Development: UNCSD "Rio+20",

stressed the urgent need to search beyond the surface of our contemporary problems in order to have access to their deep foundations, which are rooted in the minds and the hearts of human beings. Otherwise no satisfactory conclusions will be reached. The searching of a sustainable development for human societies will need great personal sacrifice, while the current systems of greed and ambition will have to be replaced by systems of love and compassion.

9. Reflections

For over the last two centuries nature has been impaired by the challenges and aggressions it constantly receives. Humanity seems to forget we share one world and the same resources, one atmosphere and the same habitat.

To pursue the perfect original natural order to be at peace with creation we must commit ourselves to a change of paradigm. The pursuit of peace requires a deep and sincere *metanoia,* repentance, a conversion. As the human heart is where the crises are primarily centred, to attain peace in all fullness the conversion must be permanent, thus becoming a metamorphosis of the heart followed by a metamorphosis of the mind, bringing peace, dignity, justice and love to the lives of all human beings and a deep care for God's good creation.

Peace and justice are central in the Holy Scriptures: peace with oneself, peace with one's neighbour, peace with the environment, peace with nature, peace with the earth. Peace and justice are essential and integral components of freedom and these qualities are good to the eyes of our Creator. Peace is unthinkable without justice and also without freedom.

Greed and pride and an exacerbated consumerism imposed by the established economic order have discarded the role of human beings as priests of creation, turning them into its exploiters and abusers. Natural resources, water, air, land and even human beings are treated as commodities.

A concept frequently referred to by Ecumenical Patriarch Bartholomew is **greed**. The destruction of the environment, the irrational exploitation of

---

Rio de Janeiro, Brazil, 20th to 22nd June 2012, commemorating the 20th anniversary of the United Nations Conference on Environment and Development: UNCED, Rio de Janeiro, 1992.

nature, climate change and extreme poverty, are direct consequences of human greed and insatiable ambition. In his words during the Halki Summit of 2012 he affirmed: "*…any real hope of reversing climate change and addressing environmental pollution requires a radical transformation of the way we perceive and treat our planet… **greed** and gratification reduce the world to a survival of the fittest, whereas generosity and gratitude transform the world into a community of sharing…"*[20] We should be mature enough to overcome our spiritual crisis, and find the difference between what we want and what we need, or better what our neighbour, our community and we ourselves need.

The concept of **greed** is also found in the Ecumenical Patriarch's message to the recent UNCSD "Rio+20" meeting in Rio de Janeiro: *"… The present times take us to distinguish between our **greed** and the needs of our neighbour. Moderation and frugality require personal sacrifice and a sense of temperance in favour of the well-being of our planet"*. [19]

In the Encyclical Letter of 1st September 2012 we read: *"… dominion over the earth and its environment implies rational use and enjoyment of its benefits, and not destructive acquisition of its resources out of a sense of **greed**…"*

10. Conclusion

Despite the apparently negative signs of the times, we are still at a point where humankind may take a decision that could, even now, save the good creation of God from the otherwise inevitable desecration caused by human activities.

Following this light of optimism and Christian hope is why we are gathered today in the Orthodox Academy of Crete, in Kolympari, near Chania, Crete, Greece. We deeply thank His All Holiness Ecumenical Patriarch Bartholomew and through him the Ecumenical Patriarchate, and the authorities of the Orthodox Academy of Crete for giving us the great opportunity of meeting here today.

Ecumenical Patriarch Bartholomew has always stressed the deep and continuous commitment of the Ecumenical Patriarchate to the vision and

---

[20] Message of Ecumenical Patriarch Bartholomew on the "Day of the Protection of the Environment", 1st September 2012.

mission of the World Council of Churches, from the very early days up to the present time.

In 2008 in Geneva he addressed the WCC Central Committee meeting celebrating the 60th anniversary of its foundation. In 2009 he addressed the plenary of the Faith and Order Commission here in the Orthodox Academy of Crete. He communicated a message via video to the International Ecumenical Peace Convocation: IEPC, held in Jamaica in 2011. And less than one month ago he addressed the WCC Central Committee meeting here in the Orthodox Academy of Crete, the last WCC Central Committee meeting before the WCC 10th Assembly, to be convened in Busan, Republic of Korea, in 2013.

The insightful meaning of our supplications on 1st September to God the Almighty, the Creator, is a petition for help and support for our act of "*metanoia*" for our actions in destroying the value, the splendour, and the beauty of nature, our only source of material sustenance of life: the most precious gift given to us by God.

The survival of humanity and even more, the survival of life on the planet are based on the safeguarding and preservation of the marvellous constituent parts of the good Creation of God.

To attain our ethical and moral obligations and achievements we need to act now.

May God in his mercy grant us the wisdom to act in time!

ECA-JAT: Buenos Aires, 18th September 2012

ECUADOR'S YASUNI-ITT INITIATIVE: A CRITICAL ASSESSMENT[1]

CARLOS LARREA[2]

Universidad Andina Simón Bolivar
E-mail: clarrea_2000@yahoo.com

Abstract

The Yasuni-ITT Initiative is an innovative proposal presented by Ecuador with UNDP participation. It offers new ways of promoting climate change mitigation by not exploiting certain fossil fuel reserves in biologically sensitive areas in developing countries, while simultaneously promoting biodiversity conservation and equitable human development. The Yasuni-ITT Initiative forms part of a policy aimed to promote sustainable and equitable development in Ecuador, although the country still lacks a coherent environmental strategy. This article summarizes the contributions of the Initiative, not only towards global mitigation and biodiversity conservation but also to sustainable economic development in Ecuador. It also discusses its economic feasibility and possibilities of international replication.

1. Introduction

As Jeffrey Sachs points out, the global economy is not only following an unsustainable path, but it is also affected by interrelated social, environmental and financial crises (Sachs, 2011). New ideas are required for a shift towards an equitable and sustainable global economy. Ecuador's Yasuni-ITT Initiative is an innovative and feasible proposal that aims to do this by simultaneously addressing global climate change mitigation,

---

[1] An initial version of this paper was presented at the ENGOV (Environmental Governance in Latin America and the Caribbean) Project Conference, held in Brasilia between June 13 and 16, 2012.

[2] Professor, Universidad Andina Simón Bolivar, and former Technical Adviser of the Yasuni-ITT Initiative.

biodiversity preservation, and equitable social development for Ecuador.

As a large heavy oil reserve was confirmed in the Yasuni National Park, considered the most biodiverse hotspot in the Western Hemisphere (Bass, Finer, Jenkins, et al. 2010), in 2007 President Rafael Correa announced to the United Nations Ecuador's decision to keep the crude oil indefinitely underground, provided that the international community cooperates with Ecuador by contributing at least half of the revenue that the State would earn by extracting the oil (Correa, 2007). The UNDP now administers the Yasuni Fund, guaranteeing its efficiency and transparency. The Fund's capital will be invested exclusively on renewable energy in Ecuador, and the interest earned will be devoted to effectively preserving the Yasuni Park and the 35% of Ecuador's land area that is still covered by undisturbed rainforests and original ecosystems, thereby promoting sustainable social development in the Amazon, fostering energy efficiency, and sponsoring scientific research aimed towards ensuring sustainability.

2. Oil and Development in Ecuador

Ecuador, a small South American country, ranks 83rd among the 187 countries on the UN Human Development Index. Within Latin America, it is clearly a less developed country, with a per capita income below the regional average (UNDP, 2011)[3]. Ecuador's economic diversification remains at a low level, and according to ECLAC, primary products still represent 92% of exports, mostly composed of crude oil, bananas, shrimps, coffee, cacao, fish and flowers (ECLAC, 2009). Petroleum, the single most important product in the economy, accounted for 57% of total export revenue between 2004 and 2010, and oil revenues made up on average 26% of the government's revenues between 2000 and 2010 (Banco Central, 2011).

In 1967 large oil reserves were discovered in the Amazon region, and from 1972 onwards Ecuador has been an oil exporter. Almost four decades later, it can be concluded that oil has contributed little to equitable and sustainable development. Economic growth has remained elusive, with an average growth rate of 1.6% in per capita income between 1971 and 2009 (Chart 1). Despite important social achievements from 2006 onwards,

[3] Ecuador's per capita GDP was 8,268 PPP Dollars, compared with the Latin American average of 10,739 dollars in 2011.

Chart 1: Per capita income in Ecuador: 1950-2009

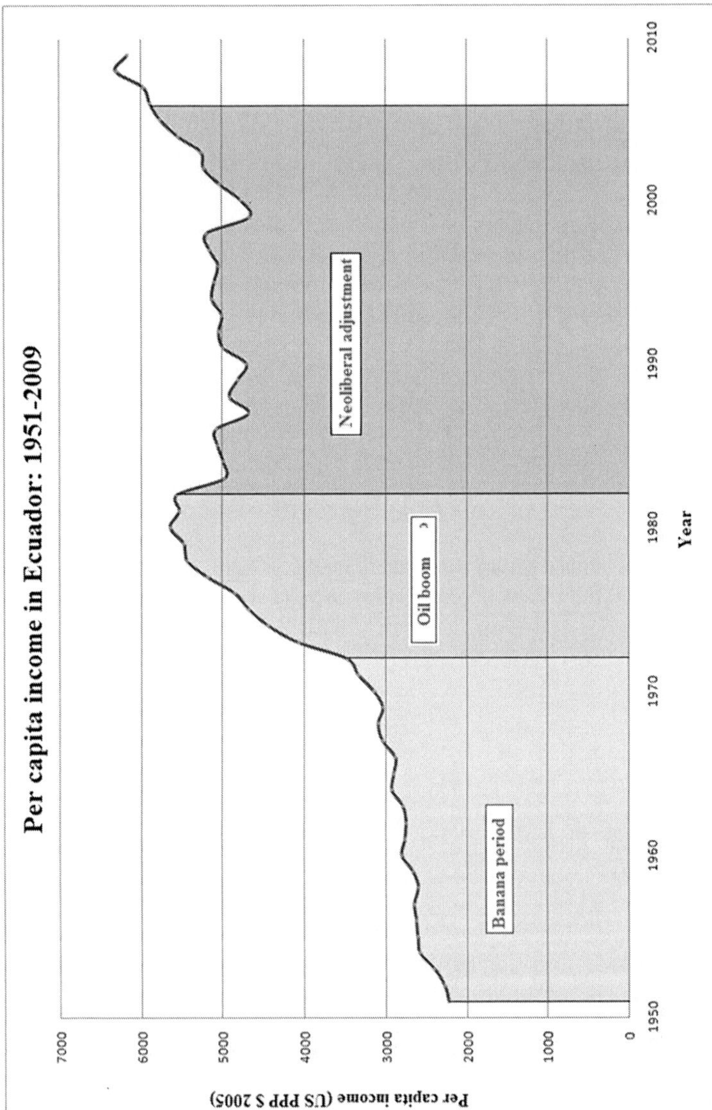

Source: University of Pennsylvania, PENN world table (http://pwt.econ.upenn.edu/php_site/pwt_index.php), 2011.

social, ethnic, and regional disparities that have historically affected the country have remained pervasive, as 39% of the population were living below the poverty line in 2010, underemployment affected 47% of the urban labour force (INEC, 2011), and social inequality had barely declined, as the Gini coefficient remained at 0.5 in 2009 (CEPALSTAT, 2011, Larrea, 2010, Falconi, Vallejo, Larrea, and Burbano, 2011).

Since oil extraction in Ecuador is located in a formerly undisturbed region in the Amazon basin, the environmental effects of oil activity have been severe, particularly regarding deforestation, loss of biodiversity, pollution, and human health hazards (Herbert, 2010, Amazon Defense Coalition, 2012).

Most oil exporter developing countries share difficulties in reaching sustained and equitable growth. Several studies have found that oil exports have had negative impacts on development. A comparative World Bank investigation concluded that most oil-exporting countries failed to efficiently channel oil revenues to development during the 1970s. In general, the economic results for national development were disappointing, as "Dutch Disease"[4] and other shared problems reduced the possibilities of economic diversification and stability (Gelb et al, 1988).

Jeffrey Sachs, adducing a sample of 97 developing countries between 1971 and 1989, found a negative and significant correlation between natural resources exports and economic growth (Sachs, 1995). Albert Berry, relying on a comparative analysis of Indonesia, Venezuela, Chile, and Nigeria, found poor outcomes in job creation and income distribution in oil and mineral exporting countries. Rosemary Thorp points out that mining and oil producer countries have serious long-term institutional development problems. In general, countries that are dependent on oil or mineral exports are vulnerable and fragile, and they share poor records in economic growth, diversification, institutional development, job creation, and equity (Berry, 2008; Thorp, 2009, Larrea and Warnars, 2009).

[4] The "Dutch Disease" theory refers to the negative effects of primary export booms on long-term development prospects for industrialization and economic diversification. Booming export activities generate effects on the exchange rate and domestic demand, which over-expand both the booming traded and shielded sectors, making other traded and import competing activities less competitive. Once the boom is over, the economy is affected by low diversification and deindustrialization. The term originated in Holland after the discovery of North Sea gas (Gelb, 1988).

Chart 2: Oil Extraction by fields in Ecuador: 1972-2026

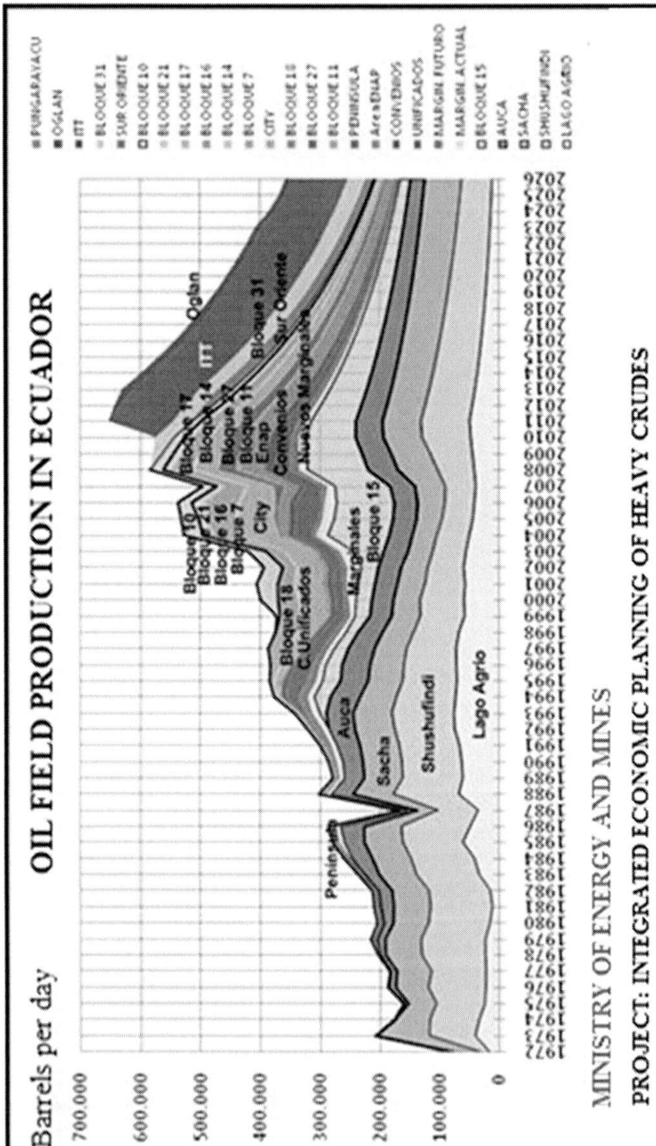

Source: Ministerio de Energía y Minas, Ecuador, 2007.

Table 1: Oil extraction, exports and imports in Ecuador: 2000-2011

		Exports					
Year	Extraction	Crude	Derivatives	Total (X)	Derivative Imports (I)	Internal consumption	Net Exports (X-I)
2000	146209	86197	15802	101999	5832	50042	96167
2001	148746	89907	14332	104239	8693	53200	95546
2002	143759	84263	13268	97531	6153	52381	91378
2003	153518	92442	11632	104074	15759	65203	88315
2004	192315	129409	13556	142965	17348	66698	125617
2005	194172	131595	12799	144394	22173	71951	122221
2006	195523	136634	13615	150249	25932	71206	124317
2007	186547	124,098	15,160	139258	29329	76618	109929
2008	184706	127,395	15,416	142812	27859	69753	114953
2009	177408	119,558	12,334	131892	32179	77696	99713
2010	177422	124,464	10,259	134723	41004	83702	93719
2011	182357	121,732	11,527	133259	37435	86533	95824

Source: Banco Central del Ecuador (March 2012 and previous issues). Información Estadística Mensual (www.be.fin.ec).

Future oil exports in Ecuador are also constrained by limited reserves. Currently, proven reserves estimates vary between 3.65 (Ecuador's Government estimate)[1], 6.5 (Energy Information Administration, 2011) and 7.2 billion barrels (OPEC, 2011), which in any case will permit less than 30 years of continued net exports, depending on future discoveries (Chart 2). Net oil exports have already declined by 23% since 2004 (Table 1). Therefore, a turn to alternative development strategies is required.

3. Ecuador's Biodiversity

Ecuador has one of the most diverse natural and cultural endowments in the word, with the highest number of vertebrates per square kilometre on earth. Additionally, Ecuador ranks among the first ten most abundant countries in the absolute number of amphibians, birds, and butterflies. Ecuador also has a rich cultural diversity, with 14 indigenous nationalities and 13 spoken languages.

In all, Ecuador has 17 different ecosystems, and about 35% of its land still remains covered by undisturbed ecosystems, mostly in the Amazon region. Within this region, the Yasuni National Park is the most important biological reserve.

According to recent research, the Yasuni National Park is the most biologically diverse hotspot in the Western Hemisphere (Bass, Finer, Jenkins, et al., 2010). It was created in 1979 and declared a UNESCO World Biosphere Reserve in 1989. The Park is located in the upper Napo basin in the western Amazon region, and it has an area of 928,000 ha. Its strategic position, close to the Equator and the Andean mountains, provides unique climatic conditions in the Amazon basin, with relatively high and uniform temperatures and rainfall levels.

Scientists agree on the park's unique value due to its extraordinary biodiversity, state of conservation, and cultural heritage. The reserve has an estimated 2,274 tree and bush species, and 655 species have been counted in just one hectare; this is similar to the total number of native tree species in the United States and Canada combined. The park has 593 recorded bird species, making it one of the world's most diverse avian sites. There are 80 bat, 150 amphibian and 121 reptile species as well as

---

[1] El Comercio, July 16, 2012. The official source is the National Hydrocarbon Direction.

4,000 vascular plant species per million hectares. The number of insect species is estimated to be 100,000 per hectare, the highest concentration on the planet. Furthermore, the species found in the park have a high level of endemism.

The park has the highest density of amphibian, mammal, bird, and plant species in the Amazon region. In addition to high biodiversity, the projected temperature rise in the park due to climate change will be comparatively moderate, which makes the region strategically important for the future conservation of species (Bass, Finer, Jenkins, et al., 2010; Hoorn, 2006).

According to a well-known theory, the upper Napo territory, where the Yasuni Park is located, was the largest Amazon refuge in the Pleistocene era, when glaciers drastically cooled the earth several times, converting most of the Amazon basin into grassland. Species concentrated in a few places like Yasuni – "the Pleistocene refuges" – where jungle still flourished, leading to a process of speciation (Haffer, 1969, Bush and de Oliveira, 2006, Haffer and Ptance, 2002, Solomon, Bacci, Martins, Gonçalves and Mueller, 2008).

4. The Yasuni-Itt Initiative

Large deposits of heavy crude petroleum have been recently confirmed in the ITT (Ishpingo-Tambococha-Tiputini) field, located in the Yasuni National Park. President Rafael Correa has announced to the United Nations that Ecuador had decided to maintain the crude petroleum in the ITT field indefinitely underground, in order to place social and environmental values first, and that it was exploring other ways to benefit economically from the Amazon and its conservation. If the international community cooperates with Ecuador by contributing at least half of the revenue that the State would receive by extracting the petroleum, the State would initially assume up to half of the opportunity cost of keeping the petroleum in the ground (Correa, 2007).

This original initiative proposes:

a) An innovative option for combating global warming, by avoiding the production of fossil fuels in areas which are biologically and culturally highly sensitive in developing countries;

b) Protecting the biodiversity of Ecuador and supporting the voluntary isolation of uncontacted indigenous peoples that live in the Yasuni Park (the Tagaeri and Taromenane);

c) Social development, nature conservation, and turning to the use of renewable energy sources as part of a strategy aimed to consolidate a new model of sustainable human development in the country.

Ecuador commits itself to indefinitely refrain from extracting the 846 million barrels of petroleum reserves in the ITT field, within the Yasuni National Park. The international community helps by providing a financial contribution, creating a capital fund to be administered by UNDP, with the participation of the Ecuadorian government, Ecuadorian civil society, and international contributors.

The **Fund's capital** will be invested in renewable energy projects in Ecuador that can promise stable and safe returns, taking advantage of the country's vast hydroelectric, geothermal, wind, and solar potential, in order to overcome its current dependence on fossil fuels, which currently account for 47% of all power generation. The **interest** earned from this fund will be invested by the state in the following objectives, within the guidelines of the National Development Plan:

1. The effective conservation and prevented deforestation of the Yasuni Park and 44 other protected areas, which account for 4.8 million hectares, along with other remaining ecosystems. The total area protected amounts to at least 19% of Ecuador's territory, one of the highest percentages in the world, and the total area covered by pristine forests to be conserved reaches 35% of the national territory. Properly conserving Yasuní Park would also allow the Tagaeri and Taromenane peoples to remain in voluntary isolation. A substantial reduction in the deforestation rate, currently regarded as one of the highest in South America, will be pursued.

2. The natural regeneration, reforestation, and afforestation of one million hectares of forest–managed by small landholders, on land currently threatened with degradation.

3. An increase in national energy efficiency and energy savings.

4. Social development in the zones of influence of the three preceding

objectives, with programs in education and training, health, technical assistance, and productive job creation in sustainable businesses, such as ecotourism and agro-forestry.

5. Research and development in science and technology aimed to ensure sustainability.

The Yasuni-ITT fund will promote the transition from the current development model, based on petroleum extraction, to a new strategy based on equality and sustainability (Larrea, 2008).

The project has received the official support of various internationally recognized individuals, including; Nobel Peace laureates Muhammad Yunus, Desmond Tutu, Jody Williams and Rigoberta Menchú; Rita Levi Montalcini, Nobel laureate in Medicine; Ban Ki-Moon, Secretary General of the United Nations and intellectuals like Vandana Shiva and Enrique Iglesias; ex-presidents Mikhail Gorbachev (USSR), Felipe González (Spain), Fernando Henrique Cardoso (Brazil), Ricardo Lagos and Michelle Bachelett (Chile), Helen Clark, former Prime Minister of New Zealand; Prince Charles of Great Britain and Danielle Mitterrand, President of the France Libertés Foundation, among others. It has also received formal backing from the German Parliament, with unanimous support from all political parties, as well as the European Union, and other international bodies such as OPEC (Organization of Petroleum Exporting Countries), CAN (Andean Community of Nations), CAF (Andean Development Corporation), the OAS (Organization of American States), numerous international organizations, like the IUCN (International Union for Conservation of Nature and Natural Resources), and various indigenous organizations and ecological groups in Ecuador. The active support from UNDP, which administrates the capital fund, is an important achievement. Chile became the first country in the world to contribute financially to the Initiative, followed by Spain. Other countries, such as Italy, Colombia, Georgia, Turkey, and Peru have also contributed to the Fund, as well as the Wallonia regional government in Belgium and several local governments in France. Non-governmental organizations, such as AVINA, also participate in the Initiative.

5. Contributions to the Yasuni Fund

The Fund is expected to reach at least 3.6 billion dollars, equivalent to half of the present value of oil revenues, in a 13 year period. Current contributions are voluntary, and they come from national governments, regions and cities, private corporations, NGOs, and individuals. Future funds may come from multilateral sources, based on a Post-Kyoto binding agreement on climate change, or biodiversity conservation resources. If the UNFCCC accepts the concept of avoided emissions from keeping unexploited fossil fuel reserves in sensitive areas in developing countries, financial recourses can be provided by multilateral institutions or new mitigation mechanisms.

6. The Yasuni-Itt Initiative as a Mitigation Tool

The ITT field contains recoverable proven reserves of at least 846 million barrels of heavy oil (Beicip Franlab, 2004). By avoiding the extraction and burning of this oil, 407 million tonnes of CO_2 emissions will be prevented. This amount is large and significant, surpassing the annual emissions of Brazil (332 million tonnes) and France (373 million tonnes), and the equivalent of Ecuador's emissions (29 million) over 13 years (UNDP, 2011).

Using a representative value of the European Emission Allowances (EUA) on the European market, i.e. US$ 19.81 per tonne of CO_2-eq[2], the economic value of the emissions prevented by the Initiative would amount to US$ 8.067 billion[3].

Avoided deforestation and achieved reforestation. Ecuador has a high proportion of undisturbed rainforests in its territory. According to ECLAC, the figure is 35%. Moreover, protected areas account for 19% of national territory, one of the highest proportions in Latin America. Most of the remaining forests are located in the Amazon region. However, Ecuador is also affected by a very high deforestation rate (1.4% per year), the highest in South America and the third highest in Latin America (ECLAC, 2010, FAO, 2010), that results largely from oil extraction in the Amazon region.

The goal is to significantly reduce deforestation in Ecuador, eliminating it in a 30 year period. The total area of avoided deforestation over 30 years will reach 1.35 million hectares, with avoided emissions of 791 million

---

[2] September 14, 2010.

[3] If the emissions prevented are distributed over a 13-year period, their current net value would be US$ 5.37 billion, using a social discount rate of 6% per annum. Taking the EUA price, as of February 18, 2011, of US$ 20.31 per tonne, the present value is 5.51 billion dollars.

tons of CO_2. The reforestation of a million hectares adds 68 million tons of reduced emissions. The recent experience of Brazil, which reached a substantial reduction in deforestation after 2005 with significant Norwegian support, demonstrates that the goal is feasible, given political will and economic support (Charts 3 and 4).

Additional mitigation from renewable energy sources. The Yasuni Fund will be invested exclusively in renewable energy facilities in Ecuador (hydroelectric, solar, wind, and geothermal). The country's electricity demand is growing at 6% per year and, given weak investment between 1990 and 2006, fossil fuel generation accounted for 47% of the electricity supply in 2006. The current government fostered investment in hydroelectric projects, increasing the share of renewable sources to 59% in 2008.

Ecuador´s hydroelectric facilities tap only 10% of the country's potential hydro capacity. Additionally, Ecuador has very large untapped renewable energy resources. The areas that have received attention include geothermal and solar energy and, more recently, wind energy. The Yasuni Fund will accelerate the total conversion of power supply to renewable sources. The contribution of energy conversion to mitigation has been estimated to reduce 43 million tons of CO_2 emissions, from which at least 30% will be the direct result of the Initiative.

In conclusion, the direct mitigation of 407 million tons from keeping the ITT reserves unexploited will be complemented with indirect mitigation of 791 million tons from avoided deforestation, 68 million tons from reforestation, and at least 12 million tons from building renewable energy facilities, over a 30 year period. The indirect mitigation adds up to 871 million tons of CO_2, bringing the total mitigation to 1,207 million tons of CO_2, about three times higher than the direct mitigation.

Chart 3: Deforestation in Brazilian Amazon: 1960-2011

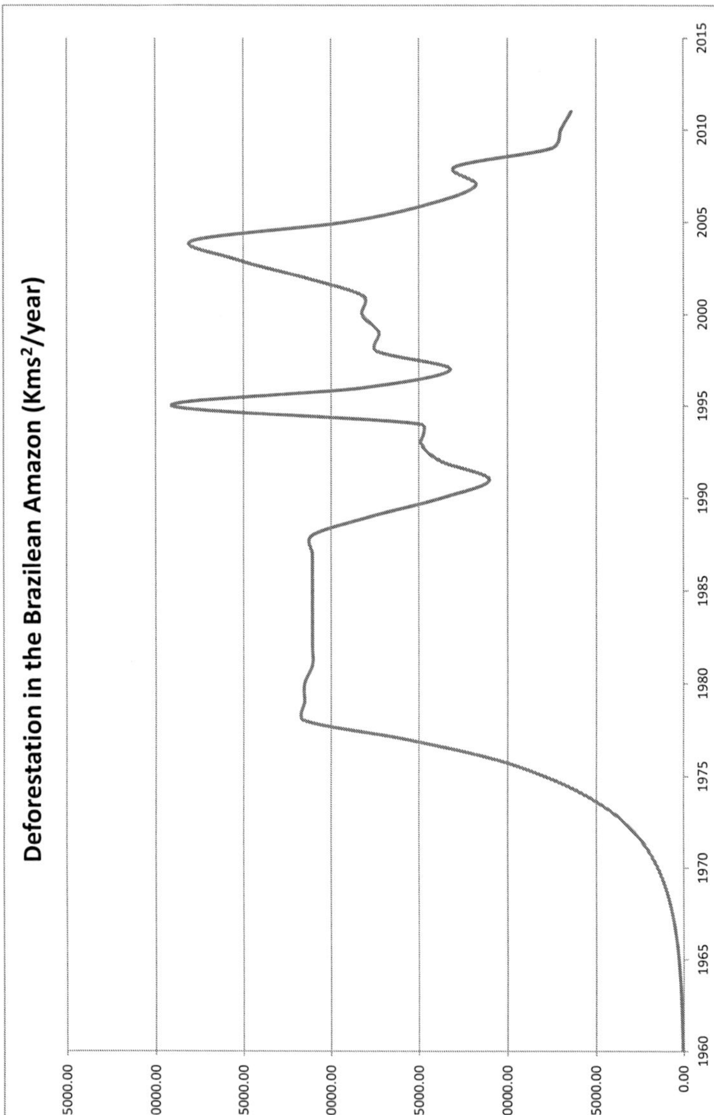

Source: INPE (Instituto Nacional de Pesquisas Espaciais, Sao Paulo, http://www.inpe.br. August 2012.

Chart 4: Contribution of the Yasuni-ITT Initiative to avoided deforestation

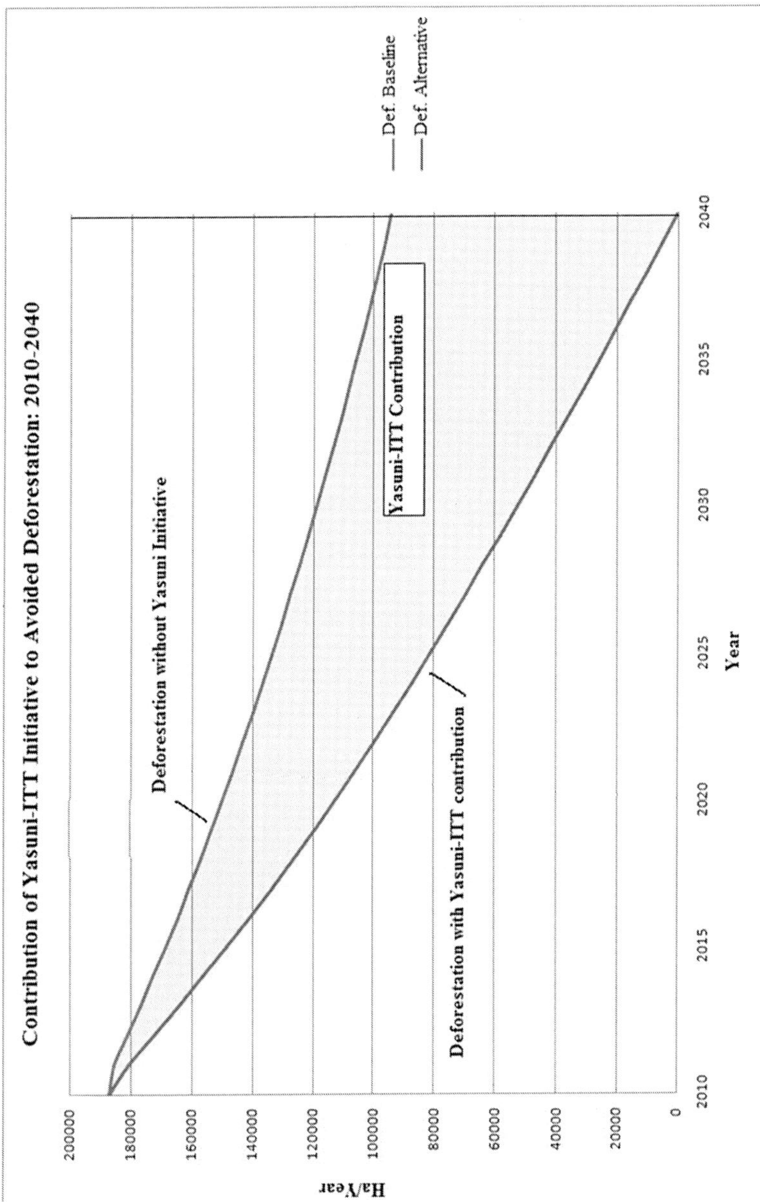

It has been argued that mitigation from not extracting fossil fuel reserves may be prevented by an international leakage, given that a reduction in extraction from one country can be easily replaced by increasing extraction from other countries.

It is true that, in the short term, the oil not supplied by one producer may be replaced by another, if the internationally installed capacity allows the substitution[4]. Yet, over the long term, the reduction in CO_2 emissions will be real because oil is a non-renewable natural resource and therefore finite. Estimates of world oil reserves show that, at the current rate of extraction, world oil production will last for only 40 years[5]. In this period, the non-extraction of oil reserves is a net contribution. Even in the short term, recent experience has shown that reductions in the supply of oil have not been easily compensated by the additional extraction of other oil-producing countries, as happened in the oil crisis of the 1970s and during the recent Libyan conflict.

Furthermore, the Yasuní Initiative envisages that the financial resources to be obtained will be invested in new projects that will absorb or reduce additional CO_2 emissions, in programs for deforestation prevention, reforestation and the development of clean energy sources. The joint potential for emission reduction of these projects is estimated to reach at least 820 million tonnes. The 407 million tonnes from the non-extraction of oil are additional to this figure.

7. Yasuni-Itt Initiative: An Idea to be Replicated

The Yasuni-ITT initiative is pioneering a mitigation activity in a developing country. It involves keeping fossil fuel reserves underground indefinitely in areas of high environmental and/or cultural fragility.

The Initiative is applicable in countries that meet the following conditions:

1. Developing countries. A critical element of the Initiative is that it seeks to simultaneously achieve three aims: to combat climate

[4] According to the widely accepted theory of the Hubbert peak, the world production capacity of oil is currently approaching its maximum and then will begin to decline below demand due to the limitation of world reserves. See: Deffeyes, Kenneth. *Hubbert's Peak. The Impending World Oil Shortage. Princeton: Princeton University Press,* 2001.

[5] BP (2008). *Statistical Review of World Energy*. London.

change, maintain biodiversity, and reduce poverty and inequality in a developing country. The Initiative promotes sustainable development.

2. Megadiverse countries. These countries hold most of the planet's biodiversity.

3. Countries with significant fossil fuel reserves in areas of high biological and cultural sensitivity.

Countries that meet all these criteria include Brazil, Colombia, Costa Rica, Democratic Republic of Congo, Ecuador, India, Indonesia, Madagascar, Malaysia, Papua New Guinea, Peru, Bolivia, The Philippines, and Venezuela.

According to recent research, the goal of keeping global warming within the acceptable low risk level of 2°C by the end of the century can only be achieved if a substantial fraction of current fossil fuel reserves will remain unexploited. If all reserves are extracted in the current century, global warming will increase to a range between 4°C and 7°C (Meinhausen et al, 2009).

World reserves of oil, natural gas and coal are limited and extraction rates will peak within the following decades, creating a physical limit to CO_2 emissions (Aleklett, 2012, Aleklett, Höök, Jakobsson, Lardelli, Snowden, Söderbergh and Bengt, 2010). Nevertheless, as mentioned, the maximum acceptable limit to GHG emissions is lower than the total amount generated by burning all currently known proven reserves. The latter has been estimated at 2800 Gt of CO_2 and the former in 1437 Gt of CO_2 (Chart 5).

If a significant fraction of world fossil fuel reserves must remain unexploited in order to keep global warming within safe limits, the unexploited reserves must be those where the environmental cost of exploiting them is the highest. Therefore, keeping the oil underground in the Yasuni National Park, and generally, in undisturbed rainforests, is a priority to achieve a global sustainable future.

8. Protecting Biodiversity

The Yasuni-ITT Initiative will not only allow the protection of the Yasuni Park but also the effective conservation of Ecuador's Amazon rainforests, currently jeopardized by oil extraction, illegal logging, hunting, and new settlements. The Amazon is the largest remaining rainforest in the planet.

Chart 5: Cumulative CO_2 Emissions, Projected Global Warming and Fossil Fuel Reserves

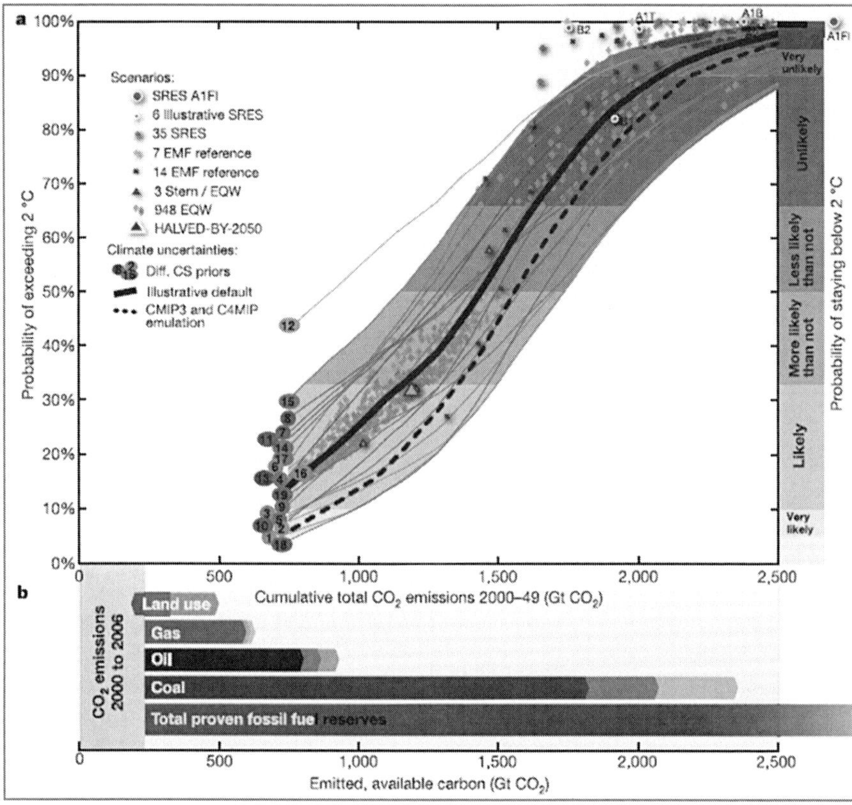

Source: Meinhausen et al (2009), *Greenhouse emission targets for limiting global warming to 2°C,* Nature, 458, April.

Biodiversity does not only have an intrinsic value; it also constitutes the very origin of our existence as species. The benefits of ecosystems in terms of regulating the weather, providing water, food, wood, pharmaceutical resources, and other renewable goods directly affect 1.6 billion people worldwide, mainly in developing countries (World Bank, 2003, Chivian and Bernstein, 2010).

Biodiversity, and particularly rainforests, provide several critical services to humankind. Most new medicines have been derived from properties of living organisms. Among the best known are "morphine from the Opium Poppy (*Papaver somniferum*), aspirin from the White Willow Tree (*Salix alba vulgaris*), and the anticoagulant coumadin from spoiled sweet clover (*Melilotus* species). Tropical plants such as the Madagascar, or Rosy, Periwinkle (*Catharanthus roseus*) have yielded vinblastine (which has revolutionized the treatment of Hodgkin's lymphoma, turning a disease that was once uniformly fatal into one that can now be totally cured in many patients) and vincristine (which has done the same for acute childhood leukemia). Medicines from animals include: the ACE inhibitors (which are among the most effective medicines known for treating high blood pressure) from the Pit Viper (*Bothrops jararaca*), and AZT (azidothymidine) used in the treatment of HIV-AIDS, patterned after compounds made by the marine sponge *Cryptotethya crypta*. Microbes have given us nearly all of our antibiotics such as penicillin, as well as the cholesterol lowering statins, and rapamycin (also called sirolimus), which is used to coat arterial stents, so that the cells lining the arteries opened by the stents do not divide and re-clog them." (Chivian, 2010, p. 11).

As infectious agents evolve and new diseases are transmitted from domestic animals, conventional medicines become less effective, resistance to antibiotics increases and new products are required. The potential value of rainforests for medical research becomes critical for the future of humankind.

The future of food security, which is increasingly recognized as a critical situation[6], also depends on biodiversity. Although the number of species used in agriculture is small, productivity depends on complex phenomena such as pollination, the stability of climate, soil fertility, water supply and genetic diversity of crops. All these processes in turn are upheld by biodiversity. In particular, tropical ecosystems play an indispensable role through services such as climate regulation and moisture, food supply and water purification, regulation of droughts and floods, storing carbon and the formation and detoxification of fertile soil. In summary, the conservation and survival of biodiversity, particularly in tropical ecosystems, is essential for the future of humankind.

---

[6] Future food security is constrained, among other factors, by limited or decreasing available arable land, water scarcity, population growth, and a mainstream agricultural technology intensive in energy and fossil fuels, which is not sustainable.

Tropical rainforests make up the greatest reserve of biodiversity on the planet, harbouring 28% of all land vertebrate species and an even greater percentage of invertebrates and other living species (UNEP, 2005). Human activity in the last 50 years has dealt a severe blow to biodiversity, in particular in tropical rainforests. The current rates of species extinction are up to 1,000 times higher than those from natural causes (UNEP, 2005), posing the greatest threat to planetary biodiversity since the extinction of dinosaurs 65 million years ago. The worldwide deterioration of biodiversity between 1970 and 2005 has been estimated at 30%, based on population counts of a high number of representative species. This problem is even more serious in tropical ecosystems, where the reduction reaches 51% (WWF, 2010).

9. Towards Equitable and Sustainable Development in Ecuador

The Yasuni-ITT Initiative may foster the transition to a sustainable development model in Ecuador, in a post-petroleum phase. Shifting energy supply to renewable sources, preserving the most important asset of Ecuador, its biodiversity, effectively conserving undisturbed ecosystems, and encouraging sustainable employment and social development are the specific means toward this goal.

The shift towards equitable and sustainable development may be consolidated as an integrated and coherent strategy in Ecuador. The new constitution, approved in a referendum with the support of two thirds of voters in 2008, became the first in the world to recognize rights for nature, guaranteeing ecosystems the right to exist and thrive. The new Constitution further establishes the concept of *buen vivir* (the possibility of living well) as the goal for participatory, intercultural, equitable, and sustainable development. *Buen vivir* has been a central tenet of indigenous cultures in Ecuador, entailing living in harmony with nature, recognizing environmental limits, and fostering community participation.

The National Development Plan (SENPLADES, 2008) is the principal instrument for implementing the concept of *buen vivir* in Ecuador. According to ECLAC, poverty in Ecuador declined from 48% to 42% between 2005 and 2009, as did income inequality (CEPALSTAT, 2011). Significant achievements have been made as well in education, health, employment, and housing. The plan also set up a goal for a 30% reduction in the deforestation rate by 2013, and includes a policy towards expanding renewable energy sources and promoting energy efficiency.

Unfortunately, environmental policies in Ecuador still lack coherence and effectiveness. There is no evidence of a significant reduction in deforestation rates, as illegal logging and hunting remain mostly uncontrolled in most sensitive areas in the country, including the cloud forests, national parks and protected areas. Road construction in areas associated with oil and mining exploration within remaining ecosystems continues, and programs aimed to protect rainforests, such as "Socio Bosque", are mostly ineffective in areas of expansion of the agricultural frontier.[7] The state has even provided credit for expanding palm oil plantations in areas formerly covered by forests.[8]

Therefore, an opportune and coherent transition to a more equitable and sustainable post–petroleum society in Ecuador requires both consistent political will and significant investment. The Yasuni-ITT Initiative may contribute to this goal.

The post-petroleum economy will be based on tourism, ecotourism, scientific research, and other services linked to the conservation of biodiversity, subject to a significant participation of megadiverse countries and indigenous communities in the results of future scientific research, which is not currently recognized by the international legislation on intellectual property rights. Most of these activities generate sustainable employment and are responsive to investments in education, health, and human development. Economic diversification, biodiversity conservation, and social equity will complement each other in a new path towards a sustainable society.

---

[7] The "Socio Bosque" Project provides an economic incentive to land holders and communities to preserve undisturbed forests. As the incentive per hectare is low and flat, and its value is substantially lower than the short term revenues from logging, deforestation continues to be a profitable short-term option to new settlers in areas of expansion of the agricultural frontier (Interview with Eugenio Pappalardo, 2011).

[8] As an example, the Secoya community received a credit from the Corporación Financiera Nacional to plant palm oil in the rainforest area (http://radioequinoccio.com/component/content/article/1-ultimas-noticias/2222-ecuador-multan-a-comunidad-secoya-con-375048-por-sembrar.html, 2011)

10. Required Resources

The Initiative aims to collect a capital of at least 3.6 billion dollars over a period of 13 years of fundraising. The international support for the Initiative and the increasing availability of international resources for climate change mitigation provide feasibility to the Initiative. The Fund would be fully funded with an annual contribution of one dollar per capita in industrialized countries.

Despite the impasse of negotiations toward a new internationally binding agreement to mitigate climate change, there is increasing availability of international resources for mitigation in developing countries. Norway's International Climate and Forests Initiative (NICFI), launched in 2007, has $500 million annually for mitigation in developing countries, and has included funding for $1000 million to reduce deforestation in the Amazon rainforest in Brazil, in a multiannual project, a similar one in Guyana for $250 million and another for Tanzania for $83 million, among others.[9] *Fast Start* funds for adaptation and mitigation of climate change in developing countries were committed by industrialized countries for the period 2010-2012 at the Copenhagen summit of 2009, reaching $29,224 million dollars, although a substantial fraction of these funds has not been disbursed or effectively committed. The European Union decision to auction permits for greenhouse gas emissions from 2012 onwards can generate resources of the same order of magnitude.

Given the availability of resources, the enthusiastic international acceptance of the Initiative and its holistic, pioneering and unique nature, the proposal may be regarded as financially feasible, assuming an adequate and consistent international promotion. While the global financial crisis has reduced the availability of resources in the short term, it should be noted that the Initiative has a period of 13 years for fundraising, enough to overcome the immediate effects of current restrictions. In the future, there will also be a growing awareness of global ecological problems.

---

[9] Ref.: www.regjeringen.no/en/dep/md/Selected-topics/klima/the-government-of-norways-international-.html?id=548491

11. Perspectives

The Initiative has established itself nationally and internationally as a possible option for mitigating climate change and preserving biodiversity. The road ahead is complex and difficult; it requires strong and continuous political support by the Ecuadorian state and an effective promotion campaign at an international level. If the initiative is consolidated, climate change mitigation mechanisms may be extended for developing countries, including emissions avoided through the non-extraction of fossil fuels. For Ecuador, the advantages of conserving the Yasuní National Park and the biodiversity of the Amazon are critical to future generations.

The obstacles to overcome, however, are complex. The international financial crisis, sparked in 2008 and deepened by the current Euro situation, may continue, seriously affecting the availability of resources for international cooperation towards climate change and biodiversity.

Since the failure of the Copenhagen Summit in 2009, international climate change negotiations have stalled. While this situation makes the Ecuadorian proposal more visible as a concrete, replicable and effective option, the lack of an internationally binding agreement reduces the available resources for the mitigation of climate change and heralds difficult negotiations for new mechanisms for developing countries, under the framework of a post-Kyoto agreement.

Finally, the Republican opposition in the United States has eliminated from the Senate agenda discussion on climate change legislation with binding targets for reducing emissions. In the absence of a federal policy on climate change, the support of the Initiative in North America is confined to the states and civil society.

This unfavourable short-term international context has hampered the Initiative. The negotiations may have also been weakened by the perception from several international groups of a conditional support of the Ecuadorian government to the Initiative, following public references to Plan B, oil extraction, or deadlines to the international negotiations of the Initiative. The Initiative has also been adversely affected by the frequent changes in its institutional structure, which has seen five different commissions in charge of its promotion since 2007. During these last three years, there have been critical problems regarding its dissemination and promotion.

These unfavorable factors can be neutralized or reversed by a strong and unequivocal support for the Initiative from the Ecuadorian government, accompanied by an extensive international promotion campaign. In the medium term, the growing international awareness of the urgent need for effective action on climate change and biodiversity loss may create favorable conditions to the Initiative. The extended period of fundraising will overcome the adverse cyclical elements that have been mentioned. The recently developed multiple criteria analysis on the Yasuní-ITT (Vallejo, Falconí, Larrea y Burbano, 2011), which integrates economic, social, environmental and political dimensions, concludes that the Initiative is definitely far more favourable to the country than extracting oil, the benefits of which are limited to financial gains in the short term, with serious environmental damage, whilst on the other hand, the Initiative possesses a high potential towards sustainability and equity. The Initiative is not only favourable to the country, preventing oil extraction in the ITT field, but also opens new alternatives for international cooperation on climate change and the irreversible loss of biodiversity.

Source: http://es.mongabay.com/news/2011/es1201-brazil_deforestation_2010.html, August 2012.

Editorial Note

Unfortunately President Correa is reported as having abandoned his government's support for this initiative in August 2013, on the ground that the donations received up to that point had been insufficient. The government later offered donors a choice between the return of their donations and investing them in other projects in Ecuador. It is also reported as having been involved in negotiations with a Chinese bank from 2009 onwards with a view to a $1 billion deal to drill for oil under the Yasuni national park. The decision to abandon the project could itself be rejected through a possible referendum, but such a vote would be opposed by the government. (See David Hill, 'Ecuador in drilling talks with China while soliciting donations to keep oil in ground', *The Guardian* (London), 20 February 2014, p. 20) However, the Yasuni initiative continues to illustrate what could be achieved (and might still be achieved elsewhere), given sufficient international backing.

References

Aleklett, Kjell, Höök, Mikael, Jakobsson, Kristofer, Lardelli, Michael, Snowden, Simon, and Söderbergh, Bengt (2010), "The Peak of the Oil Age: Analyzing the world oil production Reference Scenario in World Energy Outlook 2008". *Energy Policy* (ISSN 0301-4215). Volume 38, February.

Aleklett, Kjell (2012). "WEO 2011 and IPCC Emission Scenarios", Power Point presentation, Bergen Energy forum. (http://org.uib.no/energiforumef/Foredrag/13022012-Aleklett.pdf).

Amazon Defense Coalition (2012). *"Summary of Overwhelming Evidence Against Chevron in Ecuador Trial"*. http://www.texacotoxico.org/eng/.

Banco Central del Ecuador (July 2012 and previous issues). Información estadística mensual. Banco Central. (www.be.fin.ec).

Bass, Margot, Finer, Matt, Jenkins, Clinton, et al. (2010), Global Conservation Significance of Ecuador's Yasuní National Park PloS ONE, Volume 5, Issue 1, January 2010.

Beicip Franlab (2004) Update on the ITT study. Upstream economic evaluation. Final report. Quito.

Berry, Albert. (2008). "Growth, Employment and Distribution Impacts of Minerals Dependency: Four Case Studies" in *South African Journal of Economics*, vol. 76: S2, August.

Bonaccorso, Elisa, Koch Ingrid and Peterson, Townsend V. 2006. "Pleistocene fragmentation of Amazon species' ranges". *Diversity and Distributions, (Diversity Distrib.)* 12, 157–164.

British Petroleum. (2008). *Statistical Review of World Energy*. London.

Energy Information Administration (EIA). (2011). Ecuador: Country Analysis Briefs. (www.eia.doe.gov, visited December 2011),

CEPALSTAT. (2011). (http://websie.eclac.cl, December 29, 2011).

Chivian, Eric and Bernstein, Aaron. (2010). *How our Health Depends on Biodiversity*. Boston: Harvard Medical School.

Correa, Rafael. (2007). Speech given to the United Nations Presidents' Forum on Climate Change, New York, September 22, 2007 (http://www3.presidencia.gov.ec , visited July 2009).

Deffeyes, Kenneth (2001). *Hubbert's Peak. The Impending World Oil Shortage.* Princeton: Princeton University Press.

ECLAC. (2009). *Statistical Yearbook for Latin America and the Caribbean*, Santiago: ECLAC.

—. (2010). *Statistical Yearbook for Latin America and the Caribbean*, Santiago: ECLAC.

El Comercio, July 16, 2012.

FAO. (2010). Evaluación de los recursos forestales mundiales. 2010. (www.fao.org , visited October 2011).

Gelb, Alan H et al. (1988). *Oil Windfalls: Blessing or Curse?* Oxford University Press: New York, Oxford.

Haffer, Jurgen.1969. "Speciation in Amazonian Forest Birds". *Science.* Vol. 165:131-137.

Haffer, J. and Prance, J. T. 2002. "Climatic Forcing of Evolution in the Amazonia during the Cenozoic: On the Refuge Theory of Biotic Differentiation". *Revista Estudos Avançados,* no 46, Sept-Dec., 2002.

Herbert, Bob. "Disaster in the Amazon". New York Times, June 4, 2010.

Hoorn, Carina. (2006). "The Birth of the Mighty Amazon" in Scientific American, May, p. 40-45.

INEC. Ecuador en Cifras. www.ecuadorencifras.com. Visited in July 2011.

INPE (Instituto Nacional de Pesquisas Espaciais, Sao Paulo, http://www.inpe.br. August 2012.

Larrea, Carlos. (2008). Yasuni-ITT: An Initiative to Change History. Quito: Gobierno Nacional de la República del Ecuador.

—. (2010). La estructura social ecuatoriana: 1982-2009. (http://repositorio.uasb.edu.ec) Visited December 29, 2011.

Larrea, Carlos and Warnars, Lavinia. (2009). "Ecuador's Yasuni-ITT Initiative: Avoiding emissions by keeping petroleum underground".

Energy for Sustainable Development 13 (2009) 219–223. Sachs, Jeffrey. (1995). "Natural Resource Abundance and Economic Growth". National Bureau of Economic Research, Working Paper 5398.

Meinhausen, Malte, Meinhausen, Nicolai, Hare, William, Raper, Katja, Knutti, Reto, Frame, David and Allen, Myles (2009). "Greenhouse-gas emission targets for limiting global warming to 2 ° C". Nature Vol 458/30 April 2009 doi:10.1038/nature 08017.

Sachs, Jeffrey (1995). "Natural Resource Abundance and Economic Growth". National Bureau of Economic Research, Working Paper 5398.

—. "Globalization in the Era of Environmental Crisis" (2011). In: The Road to Rio + 20: For a Development/led green Economy. Genève: UNCTAD.

SENPLADES. (2008). Plan Nacional para el Buen Vivir 2009-2013. (www.senplades.gob.ec, visited January 2012).

Solomon, Scott E. Bacci, Mauricio Jr., Martins, Joaquim Jr., Gonçalves Vinha, Giovanna and Mueller, Ulrich G. 2008, "Paleodistributions and Comparative Molecular Phylogeography of Leafcutter Ants (Atta spp.) Provide New Insight into the Origins of Amazonian Diversity". *PLoS ONE* 3(7): e2738. doi:10.1371/journal.pone.0002738.

Thorp, Rosemary (2009). "Dilemmas and Conflicts in the Mining Sector: What History Teaches". Keynote Address, Rethinking Extractive Industry Conference. Toronto, York University, March 5-7.

UNDP. (2011). Human Development Report, 2011. New York: Palgrave Macmillan.

UNEP. Ecosystems and Human Well-being: Synthesis, Millennium Ecosystem Assessment, Washington, D.C. Island Press, 2005.

Vallejo, María Cristina, Falconí, Fander, Larrea, Carlos and Burbano, Rafael (2011). *La Iniciativa Yasuní-ITT desde una perspectiva multicriterial.* Quito: PNUD-UASB-FLACSO.

World Bank (2003). Biodiversity and Forests at a Glance. (http://siteresources.worldbank.org/ visited July 2008).

WWF. Living Planet Report, Gland: WWW International, 2010. (wwf.panda.org/about_our_earth/all_publications/living_planet_report/).

ECOLOGICAL JUSTICE, CLIMATE CHANGE AND ANIMALS

ROBIN ATTFIELD

Cardiff University, Colum Drive, Cardiff CF10 3EU, United Kingdom
Postal address: ENCAP, Cardiff University, Humanities Building, Colum Drive, Cardiff CF10 3EU, United Kingdom
E-mail: attfieldr@cardiff.ac.uk

Abstract

Many hold that justice applies to human beings only, even if we have obligations to other species. This stance prioritises human interests, not least when responses to climate change are in question. But those who reject anthropocentrism and adopt either sentientism or biocentrism must accept the possibility of some human interests being superseded by some animal interests.

This possibility can be endorsed by virtue ethicists, for some of the virtues can be construed as upholding kindness and mercy towards nonhumans. It can be upheld by consequentialists, who relate our obligations to promoting states of value, including valuable states of animals. It could also be accepted by those deontologists who hold that it is a duty not to harm existing creatures. Rawlsian contractarians, however, must say otherwise; but their theory is implausible in any case.

It could also be rejected by some adherents of the image of God in man doctrine. But other strands in the New Testament, such as belief in the whole of nature being involved in God's plan of salvation (Romans 8), suggest that such a rejection is not required of Christians. Further, Psalm 104 and other Old Testament passages suggest that those who appeal to the Old Testament should not oppose it either.

This understanding of ecological justice does not require interventions to prevent predation in the wild. But it could involve action to prevent suffering among animals in farms and laboratories, and also interventions to rescue species endangered by anthropogenic climate change.

Ecological justice, climate change and animals

Most people agree that it is possible to be cruel towards nonhuman animals, and to neglect them, but is it possible to be unfair to them, either as individuals or as species? Could it be unfair (say) to a goat to sacrifice it in order to film a Comoro dragon having a meal, and could it even be unfair to (say) flightless birds to allow them to go extinct by introducing predators such as cats to their last remaining habitats?

It is widely held that ecological justice relates to inter-human dealings and relationships but not to transactions between species. The theory of justice of the late John Rawls supplies one example (Rawls, 1971), and many living writers, such as Tim Scanlon and Darrel Moellendorf, say the same. Some of them say that only humans have rights, and can thus be owed justice, while others make justice depend on being able to enter into a contract, however hypothetical, something that non-human animals cannot do. And even philosophers who are sympathetic to Peter Singer's claim that equal interests should be given equal consideration, whether they belong to humans or non-humans, and to its implication that some interests of nonhuman animals should be prioritised over lesser human interests, sometimes still hold a theory of justice that relegates nonhumans to a lesser realm of morality than justice, such as that of compassion, taking the view that justice still relates strictly to inter-human relationships only. Thus their theory of obligation, neutral between species as it is, can end up being trumped by an unquestioned traditionalist theory of justice.

But if we take the view that the basic interests of nonhuman animals sometimes count for more than the peripheral or trivial interests of humans, then our understandings of obligation and of justice need to be brought into line, so that what we say about obligation is not countered by our assumptions about the invariable priority of humans in matters of justice. Those who believe in animal rights will usually have no difficulties about this, as long as these rights are regarded as of comparable strength to at least some human rights. But many ethicists hesitate to affirm animal rights, in some cases because rights can be claimed and nonhuman animals are almost entirely unable to make claims, in some cases because they reserve the language of rights for the relations of parties who can enter into agreements with each other, something that, once again, animals cannot do, and in some cases because they are reluctant to base everything in morality on rights in the first place. This, however, makes them vulnerable to those who claim that justice applies to human beings only.

Here I want to explore what can be said about justice towards nonhuman animals by people who reject anthropocentrism, even if they are not adherents of animal rights. The relevant kind of anthropocentrism here is not metaphysical or teleological anthropocentrism, the claim that everything exists for the sake of humanity, but the normative or ethical kind, which claims that human interests are the only interests that matter in morality, and that human beings alone have moral standing. Most ethicists these days reject this kind of anthropocentrism, partly because they can do so without adopting any metaphsycial stance at all, but quite often for a range of good reasons besides.

I should acknowledge at the outset that rejecting normative anthropocentrism need not involve acceptance of obligations to animals, as it is compatible with holding that nonhman animals warrant moral consideration but have relatively slight moral significance. However, the usual grounds for rejecting normative anthropocentrism take us much further. The usual grounds include the wrongness of inflicting pain and suffering on non-human animals without a good justification, and these grounds imply that animal suffering is morally significant enough to be weighed against human interests, and to be found more serious than trivial ones, but perhaps less serious than vital ones. But this in turn suggests that, besides nonhuman animals having moral standing, their well-being has quite considerable moral significance, and enough to support obligations in their regard.

Certainly there are other grounds that can be advanced for rejecting anthropocentrism, and so the reasoning just rehearsed does not apply of itself to all those who share in this rejection. Yet for many of the grounds for rejecting it, there will be arguments that are counterparts of the one just advanced capable of sustaining the same conclusion. For example, biocentrists standardly hold not only that all living creatures have moral standing, but also that their good has intrinsic value. This view is compatible with the relevant degree of intrinsic value being slight. But when living creatures in their millions are at risk, as when a forest is threatened, this very approach implies that a very high degree of intrinsic value is in question, and therefore that agents who can prevent this have an obligation to prevent it happening and to preserve the forest.

Besides, the argument just presented has an interesting spin-off, for many would agree that intrinsic value is what we have obligations to promote. Let us adopt this assumption for present purposes. On this basis, both sentientists (who recognise negative intrinsic value in animal suffering) and biocentrists (who recognise positive intrinsic value in the well-being of nonhuman living creatures in general) have to accept that whenever the well-being of the creatures whose moral standing they recognise is at stake, obligations are incurred, other things being equal. Moral agents, in other words, have obligations with regard to (at least) sentient animals, and possibly with regard to nonhuman creatures in general, where there are no other moral considerations, or where other considerations exist but cancel each other out. But this conclusion has a bearing on our theory of justice; for a theory of justice which purports to set aside recognised obligations as insignificant cannot be accepted. If so, and even if we still hesitate to affirm animal rights, our theory of justice must sometimes prioritise nonhuman interests over some human interests. (See further Attfield, 2012.)

So far, I have argued that theories of justice are likely to need to be modified by acceptance of sentientism or of biocentrism. But what of other normative stances? The view that animals can be the focus of obligations can be supported by virtue ethicists, for some of the virtues can be construed as upholding kindness, mercy and loyalty towards nonhumans, despite costs to the human agent concerned, and thus as supporting not only acts of supererogation but also sometimes obligatory ones. There can undoubtedly be obligations of loyalty, and plausibly there can sometimes be obligations to be kind or to be merciful too. Virtue ethicists are not obliged to accept this, but their virtue-ethical stance enables them to accept it. Further, once obligations are recognised, then once again our understanding of justice needs to be modified accordingly. Virtue ethics, then, is compatible with fairness being owed to animals.

This can all likewise be accepted by non-anthropocentric consequentialists, who relate our obligations to promoting states of value, and include among these states valuable states of nonhuman animals; for this was implicit in my earlier reasoning. (Many sentientists, after all, are also consequentialists, as are some biocentrists.) It could also be accepted by those deontologists who hold that one of our duties is not to harm existing creatures. (They could, admittedly, have problems about what we owe to future animals, which do not yet exist. But this is not the occasion to tackle problems like this one.) While neither consequentialists nor deontologists are required as such to accept that fairness is owed to animals, their generic normative stances allow of this conclusion (just as virtue ethics does).

Rawlsian and other contractarians, however, can hardly accept this view about animals and fairness, since nonhuman animals cannot enter into contracts. But the theory that obligations are restricted to ones focussing on parties able to enter into contracts is implausible in any case. And this general conclusion also has a bearing on theories of justice, which, if based on contracts, whether real or hypothetical, will foreseeably be unduly restrictive and narrow.

Since this is an ecumenical conference, it is worth considering whether Christian ethics can go along with these conclusions. It is possible for those who emphasise the doctrine of the image of God in humanity in an exclusivist manner to resist or even reject such conclusions. But other strands in the New Testament, such as belief in the whole of nature being involved in God's plan of salvation, suggest that such opposition is at least not required of Christians. If the whole creation is groaning and travailing in expectation of salvation, and is to be set free from its bondage (Romans 8: 19-23), then the good of living creatures, and particularly of sentient ones, can hardly be disregarded, and should arguably be given some degree of priority over at least trivial human concerns. The Biblical text does not mandate this interpretation, but to say the least is does not sit easily either with normative anthropocentrism or with a humans-only understanding of fairness and thus of justice.

Indeed the same passage has also been interpreted as supporting belief in the eventual restoration of all creatures, as in 'The Book', a poem of Henry Vaughan's, who wrote:

> O knowing, glorious spirit! when
> Thou shalt restore beasts, trees and men,
> When thou shalt make all new again,
> Destroying only death and pain,
> Give him amongst thy works a place,
> Who in them loved and sought thy face! (Vaughan, 1976)

Believers in such a general restoration (or *apokatastasis*) are likely to resist an instrumentalist attitude to nonhuman creatures, and to be sympathetic to human agents having obligations with respect to all such creatures, and with respect to sentient ones in particular.

Further, passages such as Psalm 104 and other Old Testament passages suggest that those who appeal to that part of the scriptures would be unlikely to oppose any of this either. If God is concerned for the beasts as well as for humanity, and has made comparable provision for them, as this Psalm implies, then it hardly becomes human agents to relegate their interests below even trivial human interests. Other Old Testament passages support the same stance, such as Proverbs 12:10: 'A righteous man has regard for the life of his beast.' Since having regard involves according the object of this regard some degree of priority, this passage suggests that fairness to beasts, or at least to domestic animals, is in place. Indeed Judaic ethics is like Christian ethics, in having resources that provide for including animals within the sphere of justice.

Ecological justice, then, has a place both in secular and at least some kinds of religious ethics. (I am using 'ecological justice' simply to mean 'inter-species justice'; but will shortly include what are sometimes considered to be ecological implications.) It is time, accordingly, to inquire what are the implications of this branch of justice.

Some people hold that human beings should intervene to prevent predation, meaning not only that farmers should attempt to protect farm animals from wild predators, but also that we should intervene to prevent lions, eagles and crocodiles catching and eating their prey. But there is also a strong case against such interventions. For the very existence of predators depends on the process of predation continuing, and if there is value in the lives and the flourishing of predators as well as of prey, then it should be allowed to continue. Ecologists claim that the well-being of prey-species depends on this, as well as that of predators. But in any case the continuing operation of natural systems should itself be allowed, and widely depends itself on predation. There will be occasions on which injured animals, such as whales stranded on beaches, can and should be rescued, but no comprehensive policy of intervention is remotely justifiable.

But the position is very different where animal suffering results from humanly generated processes. Factory-farming, for example, involves vast amounts of avoidable suffering, as does the long-distance transport of live animals for slaughter. Another area for concern is that of animal experimentation; for while some of this is indispensable for discovering and testing medicines intended for human use, much of it is conducted rather to test cosmetics, and could be discarded without significant loss to human interests.

A further implicaton is that we should make more strenuous efforts than would otherwise be required to mitigate carbon emissions and stabilise levels of greenhouse gases in the atmosphere. Strong efforts are in any case required to do all this for the sake of human interests. But humanity may well be outlived on this planet by many non-human species, possibly for billions of years, as long as we curtail greenhouse gases in the present. Even if we do not, some nonhuman species would probably survive, but a great many would not, since these emissions are already driving many to extinction. If taking seriously the interests of nonhuman species is a matter of justice, then facilitating the survival of the species that are in danger of being driven to extinction, both for the duration of the survival of humanity and for the period beyond, is itself a matter of justice. This further practical implication of ecological justice should be taken as seriously by ethicists as the others.

Ethicists, I suggest, should not be silent about these matters. If we hold that animal ethics is just a matter of compassion, then it is understandable that it is often side-lined and subordinated to issues such as the relief of human poverty and the upholding of human rights. But if instead the vital interests of animals are more significant than the trivial or peripheral interests of human beings, or, in other words, the treatment of animals is a matter of fairness, then there are obligations to prevent avoidable and unnecessary animal suffering, and good grounds to regard this as a matter not just of compassion but also one of justice and of fairness. In arguing for this view, I have importantly not appealed to any considerations of animal rights; these obligations, I am suggesting, hold good even for those who reject such considerations, and the same holds for this particular field of justice.

Appendix

When questions about my presentation were invited, someone asked whether pests were included among the animals warranting ecological justice, and if so whether my theory forbids efforts to exterminate them. This question led to some related questions, some of which the reader may want to ask, and so I will include my answers here.

What my theory requires is that the interests of every affected party (animals included) be taken into consideration. It does not require, Jain-like, that none of them be killed, because justice to the others may sometimes be impossible otherwise. However, it calls for caution in identifying creatures as pests, for this term is sometimes used to imply that the creatures that it refers to have no moral standing and no intrinsic value, and this view clashes with the theory of justice for which I was arguing. It also calls for caution about plans and policies to exterminate animals when there are alternatives. For example, in parts of Britain there is a policy of killing badgers to prevent them spreading bovine tuberculosis, and yet there is a feasible alternative approach (in use in other areas), that of vaccinating the badgers. If the badgers can be saved from slaughter by such measures, then they should not be treated as 'pests' and exterminated accordingly.

This discussion led to someone asking whether human beings were not pests themselves, in view of their tendencies to cause the extinction of many other species and to pollute numerous ecosystems. (By contrast, another questioner asserted that there was no problem about according justice to nature, since nature would survive in any case, and that the real problem was human poverty. This questioner clearly did not envisage human beings as pests, but appeared to assume that nature as it has evolved, with its millions of species intact, and nature as depleted by human interventions were one and the same. He was also assuming the kind of anthropocentric stance that conflicts with humanitarian ethics and with the Old and New Testaments.) However, the question about human beings as pests required me to clarify why, despite the ecological damage caused by humanity, they should not be seen in this pestilential light.

My answer began with a reminder that justice for nonhuman creatures rests on their moral standing and the intrinsic value of their flourishing. If this is the status of nonhuman animals, it is, by parity, a status shared with human beings. To pretend that human beings lack this status would also be like saying that human beings are agents of justice, but do not deserve to be treated with justice themselves. Other ways of conveying this status concern recognition of human dignity, and recognition of the image of God in human beings, claims that were not in dispute when those topics arose in other sessions of the conference. It is true that human beings have caused extinctions on an unprecendented scale, but that does not make them a blight or a cancer on the face of nature. Indeed through their creativity they frequently enhance the natural creation, as well as changing it for the worse. Representing human beings as pests suggests that they are beyond redemption, and is one step away from willingness to accord them the kind of treatment that extreme nationalists and xenophobes accord to peoples of alien ethnic origin. Those of us who reject the inhuman treatment of human beings are well advised not to adopt language which represents humans as vermin or as pests, for the latter is prone to lead to the former.

A final questioner asked whether justice extends to ecosystems. Where the living creatures of an ecosystem merit just treatment, so it does, for relevant ecosystems are indispensable for survival, and morally significant creatures will not survive without them. But I cannot see the case for assigning rights or even moral standing to ecosystems themselves independently of that of the present and future creatures that comprise and depend on them. Ecological justice can be vigorously endorsed without adoption of an ecocentric value-theory.

References

Attfield, R. (2012), *Ethics: An Overview*, London and New York: Continuum

Rawls, J. (1971), *A Theory of Justice*, Cambridge, MA: Harvard University Press

Vaughan, H, (1976), *Henry Vaughan: The Complete Poems*, ed. Rudrum, A., Harmondsworth, UK: Penguin

SUSTAINABILITY, RELIGION AND ETHICS

COLLECTING, FRAMING, NEGOTIATING, AND CONNECTING IN TIMES OF A GLOCAL[1] "PROBLEMATIQUE"[2]: A HOLISTIC APPROACH TO COPE WITH GLOBAL CHALLENGES[3]

[1] GLocal: a neologism describing the linkages of effects across scales. One source was calling the dimension magnitudes: micro-meso-macro cosmos. The Scientific Director of the Global Change exhibition, Dr. Manfred Lange called this "depth" of a "possibility" space "GLocal" in early 1990. See Wikipedia http://en.wikipedia.org/wiki/Glocalization#History_of_the_concept and Pointers to possibilities: Ekistics, Dymaxion World and Eco-Cube, http://kairos.laetusinpraesens.org/lifedisx_1_h_8

[2] The „Global Problematique" and „Predicament of Mankind" or the central titles of the original reports leading to the formation of the Club of Rome in the late 60s. This article refers to this early work originated by Erich Jantsch, Hasan Oezbekhan, Alexander Christakis. We recommend the interviews with Christakis by the author in Europe's World: Learnings and Vistas based on revisiting 40 years "Global Problematique" - summing up 40 years "Predicament of Mankind" (Club of Rome early report), look-outs, and new forms of structured dialog and deliberation / decision cultures in an Interview with Alexander Christakis, and The Predicament of the Individual, Communities, and Humankind in the 21st Century - Deliberations about Structured Dialogic Design, Systems Thinking, Policy-Making, Multi-track Diplomacy, Empowerment and the Wisdom of the People.

[3] The authors have been aware of the Millennium Project since the mid 90s and have independently considered the "15 Global Challenges to Humanity" http://millennium-project.org/millennium/challenges.html as an ideal collection to which to apply the Structured Dialogic Design Process (SDDP), presented here and developed further by the 21stCentury Agora Institute. In APPENDIX A we provide more details about the Millennium Project, the Global Futures Intelligence System - GFIS the Union of International Associations - UIA and their Yearbooks and Encyclopedia, the MDGs and more. Have a look.

HEINER BENKING

Independent Journalist and Scholar, Council of Global Issues, Tagore-Einstein-Council, Germany

REYNALDO TREVIÑO CISNEROS

Institute of 21st Century Agoras
Strategic Planner, Instituto Nacional de Estadística y Geografía in Aguascalientes, Mexico

AND BETHANIA ARANGO HISIJARA

National Institute for Statistics and Geography, Communication Strategies Ibero-American Forum (FISEC), Mexico

Abstract: *In this paper a group of futurists discuss methods for understanding and resolving global issues, and related problems.*

1. Introduction

We dare to tackle big questions in this article. What are we "Collecting, Framing, Negotiating, and Connecting"? How do we frame and negotiate? If we can overlay and connect issues, how can we identify root causes and actions, avoid "data islands or silos" and parallel, uncoordinated and often contradicting efforts?

We live in times of "Big Data", but also "Big Noise and Confusion" and so the central issues around positions, perspectives, values, frames, scales, and orientations are neglected. Let us sum up, keeping Margeaux in mind, that not only data, but even values, are meaningless when out of context.

Please take note: This contribution is based on taking on a broad perspective, something we called a "Cognitive Panorama", a viewpoint across cultures and mindsets, times, cultures and spaces. This is published in depth in the International Encylopedia of Systems and Cybernetics[4] and elsewhere, and

[4] See Cognitive Panorama, International Encyclopedia of Systems and Cybernetics http://benking.de/systems/encyclopedia/newterms/#_Toc87362164. At this point we only want to bring to attention the dimensions, scales, width and broadness of

has been presented at the **Ecological Theology and Environmental Ethics** – ECOTHEE conference at the Orthodox Academy of Crete in June, 2008 under the title: "Missing Context and Orientations in Modern Times - Outlining the Problematique of the Human Predicament and sharing Commons in a global Embodied Covenant".

In the subtitle for the session we have underlined that we not only collect, frame, negotiate, and connect words, but meanings in their context. So we have to deal not only with words or "dots", but any kind of Signs, Models, Repositories, Spaces, and Minds[5],[6] - this is the original subtitle of the SAPREJ-12 session: Explorations into individual, socio-cultural collaborative and co-creative, and collective approaches towards tackling Grand Challenges.

So we are addressing a daunting task and challenge.

2. "Multi-Track, Multi-Level, Multi-Sector Deliberations, Diplomacy and Peace-Making"

This paper is a record of a collaborative process which received a mayor impulse at the International SDDP Facilitators Training School[7] at the Future

---

the approach we are taking here and refer to the recent workshop by the International Society for Knowledge Organisation (ISKO) and the paper on Formal Ontologies (forthcomming: Herre, H., Benking, H. 2013)

[5] This paper includes a constructive/critical review of Reynaldo Trevino Cisneros new publication introduced below and "Imagining Future Urban Challenges - Imagining Canada's Future 2030" by Peter Jones, OCAD. See the slides presented at the OAC and for example check later the multi-cultural applications as presented by Paul Hayes about the AIO/AMO exercises since 2004 and the work of the other SDP practitioners presented below. Including a featured special highlight of this synopsis was presented as a "HOLISTIC APPROACH TO COPE WITH THE MILLENNIUM PROJECT 15 GLOBAL CHALLENGES" outlining the "STRATEGIC ARTICULATION OF ACTIONS TO COPE WITH THE HUGE CHALLENGES OF OUR WORLD TODAY". See A SOCIAL SYSTEM APPROACH TO GLOBAL PROBLEMS by the Institute of 21stCentury Agoras, Monograph Series No 1, https://www.createspace.com/3977896

[6] For documentation and to include some visuals as unfortunately a video documentary is not available I urge you to check out powerpoint slides as presented at the OAC in Crete at the SAPREJ-2012: http://de.slideshare.net/benking/connecting-dotsspaces as pictures can tell more than 10.000 words – see infographics roots: http://benking.de/infographics/

[7] 3rd International Facilitators Training School for the (SDD) Process took place in Nicosia, Cyprus between the 25th and 27th July 2011.

Worlds Center (FWC) in Nicosia in 2011, and can be considered a preparatory session for the next symposium of The Hellenic Society for Systemic Studies (HSSS). Alexander Christakis urged us "to connect the dots". Yiannis Laouris added that this is not enough; he asked for a connection of the minds to better understand individual and collective reasoning and capacity building. This provided an extra "impulse" as I have long wondered about the positionality and perspectivity of "minds" or people and how they relate, and have been concerned with bridging the "incompatible" and "intangible" or the "odds" "along and across scales", wondering about the frames of references, the broader context and how it is communicated and considered in our thinking and reasoning.

In July 2012 the 8th National & International HSSS Conference 2012[8] took place in Thessaloniki with my virtual co-presenters at this SAPREJ-12 Conference at the OAC: Alexander Christakis, Yiannis Laouris, and with special mention and many thanks Maria Kakoulaki. I am very grateful because she agreed at the last minute to jump in as moderator and introduced the topic in the lecture room as I could attend only as a virtual conference participant.

The session provided an overview of the work from colleagues in Canada, Mexico, USA, Japan, ... Impossible to mention details; we have instead made the slideshow[9] of the video conference available online. I want only to highlight in this limited space the contributions of Alexander Christakis: "A MODEL DISCOVERED IS SUPERIOR TO A MODEL DELIVERED", and of Peter Jones, Yiannis Laouris, Jacqueline Wasilewski and Paul Hayes with their outstanding and avant-garde implementations of the structured dialogic design method in international projects over many years.

This paper is not another Structured Dialogic Design Process description. There are abundant books and conferences which have introduced the

---

http://futureworlds.eu/wiki/3rd_International_SDDP_Facilitators_Training_School

[8] 8th. National & International HSSS Conference 2012, 5-7 July, 2012, see: Democracy in [R]evolution: Why & How We the People Ought to Connect our Minds - Stakeholders' Strategic Perspective with Systemic Tools by Dr. Yiannis Laouris, Cyprus Neuroscience and Technology Institute, and "Why & How We the People Ought to Connect the Dots ..." http://www.2012.hsss.eu/1006_2/Keynote-Addresses

[9] Please see a collection of 100+ powerpoint slides for further details, maps, images and links for further study: http://benking.de/futures/connecting-dots&spaces.pdf

method over years. So we propose to visit the website of The Institute of 21st Century Agoras and check some links, get some books or see some videos.

Let us revisit the above mentioned papers at the HSSS with the same main title:: ***Democracy in [R]evolution:*** and Alexander Christakis and Maria Kakoulaki's subtitle: ***"Why & How We the People Ought to Connect the Dots …" - Strategic Thinking and Interactive Systemic Practicing*** explaining: by *connecting of the dots…" we mean the consensual "connection" of people's mental models, i.e., opinions, knowledge, understandings, wisdom and vision for the future,…,*."[10] and Yiannis Laouris' subtitle: ***"Why & How We the People Ought to Connect our Minds, Stakeholders' Strategic Perspective with Systemic Tools"*** *explaining how to develop "new concepts, scientific systemic tools and powerful social-media-based software capable of tackling contemporary multi-dimensional strategic complex problems."* and further down: *"Connecting our minds… will require new tools, which will enable the definition of shared problématique, as well as solution spaces through exploration and processing of diverse contributions and opinions regarding importance or influence relationships between statements. The new tools aim to support stakeholders to exploit their collective wisdom and reach consensus even when they are in large numbers".*

We can also resort here to an incredible source of practice and experience assembled in this Global Agoras group or around the GLOBAL AGORA sessions and SIS groups in Asilomar, or the IFSR Global Agora sessions in Fuschl. We are recording this on the 21st CenturyAgora website[11] and can build on it for our purposes here, just highlighting the Cyprus, Crete, Asilomar and Fuschl sessions over the years. As an "appetizer" this collaborative scripting of the boundary spanning narrative[12] might be be a point of departure for further study.

---

[10] Contribution "Why & How We the People Ought to Connect the Dots …", *Strategic Thinking and Interactive Systemic Practicing*, by Alexander N. Christakis & Maria Kakoulaki, http://www.2012.hsss.eu/1006_2/Keynote-Addresses

[11] 21stCenturyAgora.org – see for example the Asilomar or Fuschl proceedings: http://ifsr.ocg.at/world/files/fu2004_proceedings_v1.0.pdf

[12] http://redesignresearch.com/docs/Scripting_Collaborative_Narrative--Flanagan-2008.pdf

In this paper we can unfortunately only highlight a very small section of the work being done during the last 40 years and have to refer to the websites and long version of this paper in the internet.

See for example the multi-cultural applications already done since 2004 in the Pacific region doing “Northeast Asia Boundary-spanning Dialog Approaches” (BDA) under the umbrella of the International Christian University based in Japan[13] This “Boundary-Spanning Dialogue” made possible by an “Indigenous Leaders Interactive System” (ILIS) uses the same technology as the SDP mentioned above, and was developed by teams of Native Americans and later Maori tribes over nearly 30 years. So we are talking here about some expertise in multi-lingual, multi-cultural applications which included local and virtual meeting elements. Such pilot projects are milestones to tackle ultra sensitive historic issues and conflicts and help to explore different value systems, meaning and concepts. To give an example: the meaning of contrasting conceptions of Harmony in China and Japan. The procedure, outcomes and follow-up is published by the AIO in “The Ambassador” Newsletter.[14]

To outline the stage we have to include the above projects and build further on them to address the modern challenges like the use of heterogeneous sources, stakeholders being on and off-line in synchronous and asynchronous settings, and address issues along and across[15] scales, sectors, resorts and cultures!! Not a trivial task; as you see, the international community has been struggling for more than 100 years to find organizational and structural approaches to the “Problematique”. See the time-line laid out below.

So lets get on with the tasks and challenges ahead of us. We need to go back now again to ongoing projects to structure issues, to do Root-Cause-Mappings and try to tackle issues where you can not get all the stakeholders at any one time into one room. Such “Problématiques” more and more

---

13 Boundary-spanning dialogue for the 21st-century agoras, Alexander N. Christakis, Sabrina Brahms, http://onlinelibrary.wiley.com/doi/10.1002/sres.508/abstract

14 AIO-AMBASSADOR Newsletter http://open-forum.de/dialog/interview-Christakis-LaDonna-Harris-p4-5-Summer%202005.pdf

15 Access and Assimilation: Pivotal Environmental Information Challenges - Linking, Archiving, and Exploiting Multi-Lingual and Multi-Scale Environmental Information Repositories, GeoJournal March 1992
http://benking.de/ceptualinstitute/access-assimilation.html
http://link.springer.com/article/10.1007%2FBF02629811

surface and so we need to look into pragmatic steps widely scattered but maybe useful on the way.

Another arena of advanced international exercises in the direction we are targeting in this paper is the broad range of applications by the Future World Center (FWC) run by Yiannis Laouris (mentioned and introduced above) and the work of Peter Jones in Toronto with his "Imagining Canada's Future 2030"[16], as covered to some extend in the slide presentation at this event, highlighting the project in Mexico by Reynaldo Treviño Cisneros and Bethania Arango Hisijara: ***"STRATEGIC ARTICULATION OF ACTIONS TO COPE WITH THE HUGE CHALLENGES OF OUR WORLD TODAY".***[17]

You will find in **APPENDIX C** more about this report of Treviño Cisneros and Hisijara.

3. Summing Up

What we endeavour here in the texts and appendixes is to look at the "meta-level", onto the infrastructure and superstructure[18] of the "Problematique". The approach used by Christakis and his colleagues is designed to check for the deeper meaning, the deep drivers and hidden agendas. I fully resonate with the SDDP approach because only by looking into influences might we be able to go deeper and not only establish "root causes" but also come into a mindset of questioning and even changing one's own reasoning and knowing. One key contribution can be seen in what John Warfield, a colleague of Christakis, called "groupthink" and "clanthink"[19]. The work of

---

[16] Peter Jones, Shaping Tomorrow and Imaging Canada's Future 2030; see slide presentation: http://benking.de/futures/Presentations-Benking-OAC-GYC-etc.pdf

[17] See footnote 5 and Appendix C

[18] See: International Council for Science, The Committee on Data for Science and Technology (CODATA) 2005, http://www.icsu.org/what-we-do/interdisciplinary-bodies/codata Maps and Models, SuperSigns and SuperStructures ICSU CODATA MIST 2005 http://benking.de/systems/codata/CODATA-MIST2005.htm , and maybe also 'Granularity, Topicality, and Generalization of Information, Reflections about maps and models, orienting generalizations and their possible pragmatic and ethical implications and challenges'.

[19] WARFIELD, John N.: "Underconceptualization". Systemica, Amsterdam, 1 (6/8) 1989a

WARFIELD, John N.: "Groupthink, clanthink, spreadthink and linkthink". George Mason Univ., Fairfax, VA, 1993 this group around Information Management from the 70s and 90s is immense. I invite to check undercomplex approaches, overspecialisation, and oversimplistic approaches[20]. All understandable as human limitation and evolutionary survival factors, but definitely not appropriate for modern, non-point, and vague cultural and technological problem areas. See recent research on "unrealistic optimism"[21]. We are confronted here with a "quick-fix" attitude which ignores processes and effects and in particular the long view, any big picture approach, and the effects that bad decisions have for groups, nations, and all living things[22].

It is obvious that Humanity is at a crossroads in danger of following what is technologically feasible but not appropriate. We collect and follow non-transparent/opaque "indicators" or "signals" which are "interpreted" and selected on a statistical basis and might look awkward in living contexts. It seems mandatory to have algorithms and screen volumes of data, but how to understand and interpret outcomes and put them into perspective? It is fashionable to consider "swarm intelligence" that cheers numbers but eliminates the single impulse and decision, but is this wise? Isn't that only a "lemmings intelligence"? Not as earlier understood, that Lemmings all die on purpose and have no alternative, all hopping down the cliff like suiciders. But they are robust and many survive, and find solutions the hard way definitely including lots of suffering and hardship. Is this the only solution?

Another blind spot is seen in that our dualistic, nominalistic reasoning is only considering one position, one perspective, the eye of the ego, even when we know that we are "image and model making animals" (adapted from UNESCO) and can live "excentric positionalities" (Hellmuth Plessner)

---

[20] Please consult the Encyclopedia of Systems and Cybernetics, mentioned before and for the educational aspects early aspects to come to a transdisciplinary research agenda with reference to Nathan Keyfitz in the OAC - ECOTHEE publication mentioned above.

[21] http://bps-research-digest.blogspot.de/2011/12/brain-basis-of-unrealistic-optimism .html

[22] Are Humans not cultured enough to look into inflictions of suffering? There is for example a science called Panetics (http://en.wikipedia.org/wiki/Ralph_Siu), which looks into amount of suffering short- and long-term, severy and light suffering, so that alternative causes can be compared, and the invisible long range and qualitative factors can be considered as well. Nowadays after states looking into Human Happiness indicators, these old Eastern concerns of the dukkha and sukha – Suffering and Happiness become relevant again. LINK http://www.amazon.de/Panetics-Dukkha-Integrated-Infliction-Suffering/dp/1884437028

and communicate and keep in mind "Me, You, They and Others and their Position, Perspectives, and Models without need to agree on any single "right" view[23].

One aspect of the described SDDP deliberation process seems to me most appealing. People can not only change their minds individually but also in group sessions, making use of a very central effect or phenomenon of *Erroneous Priorities Effect* (EPE)[24]. Simply put, when you vote for priorities or you vote for influences you have diametrically different outcomes, as scientifically demonstrated as a spreadthink group pathology by Kevin Dye and as a central element of "evolutionary learning".

This key author was personally very concerned when policy advisers in 1988/90 for the Global Change Research agenda and exhibition were searching for the "Top 5" Problems, or the Top 5 Goals or Solutions. He always cites: "When you think you know but don't you get into trouble" and maintains that we can create more harm than ever before, faster and lastingly.

Problems are connected, vicious, hidden and such "quick-fixes" to the GRAND CHALLENGES might be misleading, as the root causes remain hidden. This can also be seen in the efforts to identify syndromes[25]. A clear indication of unscientific "searching in the dark", trend-extrapolation, and without any questioning of scales, proportions, consequences, dynamics and feedbacks. It is indicative that we have after 30 years a discussion about tipping points, but the term "Leverage Points" by Donella Meadows has not been taken up, that we discuss footprints, but ignore our handprints and mindprints[26].

---

[23] Positions and Identities in Global Contexts: Awareness of Self and Others, with me, you, we, they and "others" models. Club of Budapest Members Meeting at UNESCO Headquarters, Paris 1997, http://benking.de/cob-paris.html

[24] Dye, K. (1999). Dye's law of requisite evolution of observations, in Christakis, A.N. and Bausch, K. (Eds.): *How People Harness their Collective Wisdom and Power*, Information Age Publishing, Greenwich, CT, pp.166–169.

[25] Syndromes of Global Change: The First Panoramic View, Science for Policy Making, Matthias K.B. Lüdeke*, Gerhard Petschel-Held, and Hans-Joachim Schellnhuber, AIA - Ecological Perspectives for Science and Society, Volume 13, Number 1, 1 January 2004, pp. 42-49(8), oekom verlag

[26] www.mindprint.tv

We can at this point only maintain that human expressions and presentations and artefacts go far beyond words, and we need to consider all signs and senses and learn also to apply maps and models, schemas and performative expressions, as certain issues are between and beyond "words and lines". A short reference to George Lakoff "Whose Freedom?" and the "Spectrum of Meaning" by Herbert Stachowiak will help readers to follow that tangent.

We are presently working on "Collaborative Positive Futures Creation"[27] an Open Workspace where we look back 40 years, when MANKIND 2000[28] was founded by James Wellesley-Wesley, and also people mentioned in this article, people like Oezbekhan, Jungk, Galtung, … and celebrate 30 years of the World Futures Studies Federation, and start a collection of old and new efforts needed for Futures Creation.

4. Appendix A

The authors have been aware of the Millennium Project since the mid 90s and have independently considered the "15 Global Challenges to Humanity" http://millennium-project.org/millennium/challenges.html as an ideal collection to which to apply the Structured Dialogic Design Process (SDDP), presented here and developed further by the 21stCentury Agora Institute. As the report by Treviño Cisneros and Arango Hisijara was only available in Spanish, this session introduced the forthcoming expanded translation into English and included the then about to be announced "Global Futures Intelligence System" (GFIS). http:// millennium-project.org/millennium/GFIS.html and see: http://www.wilson center.org/event/launch-the-global-futures-intelligence-system-and-the-2012-state-the-future-report

The Millennium Project collected 10,000 pages of futures research and 1,300 pages of methods. The Union of International Organisations (UIA) later produced two Volumes of an International Encyclopedia of World Problems and Human Potential, and a recent 3rd Volume on Actions, Options, Strategies (see **APPENDIX B:** The Problem with Problems), and

---

[27] Here you find a growing space for collaborative approaches in Futures Creation. Many contributors of the session at SAPREJ-12 have also been involved in this exercise. See http://benking.de/futures/WorkshopSpace1-Outline-Background material-Rules-PDF.pdf and
http://www.newciv.org/nl/newslog.php/_v396/__show_article/_a000396-000237.htm

[28] http://www.m2000.org

conducted pilot projects over years to relate alternative measures and actions. Now in times of "Big Data" projects like the FutureICT http://www.futurict.eu/ are about to get launched to collect even more data (being even called BIG DATA – BIG SCIENCE in the announcements). As the International Agenda is a central concern here, we should also look into the Millennium Development Goals (MDGs) and now for the next decade, the Sustainable Development Goals (SDGs). It is obvious that such Goals should be addressed in a comprehensive and concerted way. Addressing health, education, security, justice, production, technology... policies in isolation is obviously doomed to failure. We recommend here the work of Anthony Judge on Intersectoral Strategic Dilemmas[29], tabled with the United Nations University before Rio 1992 and the work of the FWC in Nicosia, exploring options to address the Teach MDG's[30] in workshops with young Eastern European participants, or most outstanding in my view is the CARDIAC[31] project on research strategies for the European Commission.

What we are presenting here is one pilot project to connect issues based on the original work of Oezbekhan and others, and feel that focus should be given to connecting, synthesising and transcending instead of only applying automated, quantitative processes because number-crunching capacities allow processing volumes of data.

5. Appendix B

Tackling 100 years of the "problem with the problems" - The Problematique today and an excursion into the "PROBLEM WITH PROBLEMS" over the years.

We are confronted with a tough evolutionary problem. When confronted with overwhelming dynamics and complexity humans have developed a survival strategy of being "unrealistic optimists"[32] - they simply ignore anything beyond direct experience and touch, anything on another time-

---

[29] Inter-sectoral Strategic Dilemmas of Sustainable Development, prepared for the Rio de Janeiro Inter-Sectoral Dialogue 1992 and presented at various PrepComs in the year before http://kairos.laetusinpraesens.org/a11aa_et

[30] Teach MDG's: http://www.teachmdgs.net

[31] Coordination Action in R&D in Accessible and Assistive ICT FP7-Coordination Action – CARDIAC- http://www.cardiac-eu.org http://www.academia.edu/2888022/Technology_Transfer_Road-map_for_WP1

[32] See footnote 21

scale, anything on different levels, granularities, abstraction planes, anything they consider endangering. But as Doxiadis has shown, we can create maps and models. He could develop an ekistic grid to show how issues cross scale platforms, connecting patterns and issues in time and space. And this is how we came to our present title: Connecting Dots and Spaces. The philosopher Hans Jonas requested in his *Principle of Responsibility* "Ethics to be considered in Future with Space and Time Horizons".

"While the difficulties and dangers of problems tend to increase at a geometric rate, the knowledge and manpower qualified to deal with those problems tend to increase at an arithmetical rate."[33]

"Social institutions face growing difficulties as a result of an ever increasing complexity which arises directly and indirectly from the development and assimilation of technology. Many of the most serious conflicts facing mankind result from the interaction of social, economic, technological, political and psychological forces and can no longer be solved by fractional approaches from individual disciplines." (Bellagio Declaration on Planning. In: Erich Jantsch (Ed) *Perspectives on Planning*. Paris, OECD, 1969)[34].

Institutional difficulties in identifying problems

The "Bertrand Report", a recent major internal review of the difficulties afflicting the United Nations system (Maurice Bertrand. Some Reflections on Reform of the United Nations. Geneva, UN Joint Inspection Unit, 1985, JIU/REP/85/9) notes:

> *"In short, it is the sectoralized, decentralized and fragmented structures of the System that are the reason for its failure to adapt to the solution of development problems."* (para 104) *"The countries concerned need a World Organization capable of facilitating syntheses, organizing co-ordination, helping to find long-term financial arrangements, and granting many-sided aid to solve the most urgent problems. What the United*

---

[33] Yehezkel Dror. Prolegomenon to policy sciences, AAAS symposium, Boston, 1969, See also the Club of Rome Report CAPACITY TO GOVERN and 6 recommendation presented at the UN Climate Summit 1995. Co-authors Anthony Judge and Heiner Benking.
http://www.benking.de/Global-Change/governance.html

[34] See Significance: Acknowledgement of the universe of problems - World Problems Project http://www.un-intelligible.org/projects/problems/11sign1.php

Nations System offers them is a series of divergent and contradictory recommendations, some 30 bodies whose action has to be coordinated with that of some 20 sources of bilateral aid, but it does not help them to solve their medium and long-term financial problems." (para 106)

"In other words, since the Organization here is confronting the essential mission it should fulfil, we have to ask ourselves whether it is properly equipped to do so; whether the results obtained so far are satisfactory or negligible; and whether the Organization really does possess the organs capable of reflecting upon and identifying the problems and the framework of negotiations which the modern world needs. The replies to these questions are inevitably negative; the machinery of negotiation is not easily identifiable and separable from the rest of the activities under the various sectoral programmes and does not constitute a coherent system. The results achieved relate only to a few limited fields and do not represent solid progress in the direction of changing world consensus. This situation has its political reasons, which are well known, but they do not explain everything. Actually, it is the structure of negotiations offered by the World Organization that is ill-adapted to solving the problems of the modern world." (paras 107-8)

"They call for considerable preliminary efforts to identify the problems which are susceptible to negotiation before any negotiations can begin. This work of identification is complex, and it comes up against difficulties of a cultural, technical, ideological and semantic kind; it can often only be concluded when a preliminary agreement is beginning to take shape on a given concept; so that it is no longer surprising that it implies attempt after attempt at formulation, often clumsily done, and that it is a source of endless talk. Negotiation among 160 parties presents specific technical difficulties other than those of the size of the meeting chamber or the organization of simultaneous interpretation. It involves the definition of interest groups whose composition and dimensions vary according to the subjects dealt with, and the method of representation of these groups." (para 109)

9. Absence of consensus concerning problem priorities

http://www.un-intelligible.org/projects/problems/11sign1.php

6. Appendix C

STRATEGIC ARTICULATION OF ACTIONS TO COPE WITH THE HUGE CHALLENGES OF OUR WORLD TODAY[35].

See also the powerpoint slides and the recent English translation of the expanded original study.

Beside the presentation slides available in the internet from this SAPREJ-2012 session pls. See also below the contribution: REFLECTIONS AROUND A PLAUSIBLE GLOBAL STRATEGY TO HARMONIZE OUR PLANET EVOLUTIONARY PROCESS by Reynaldo Treviño Cisneros, Aguascalientes, Ags., January 31, 2013

Since the creation of the Club of Rome around 1970, the need for a holistic conceptualization of the world problematique and of strategies to overcome in a collective yet systematic way the challenges this problematique presents to humankind took form in the minds, first of scientists and planners, and later of many people who thought that our planet Earth could be considered a web of intrinsically interrelated facts all along its pattern of evolution.

Were those facts totally separate from each other? Were they connected in such a way that anything occurring in one corner of the planet exerted a measurable influence over the rest of facts involved in ulterior states of its evolution? Could there be a fact or group of facts influencing the Earth to enter into irreversible action, and bring it to an unstable and degrading form of existence? Are we humans responsible for installing on Earth facts contributing to harmony or to wreckage? Thinking of today global conditions, which of the consequences would be the most likely to occur? How can we become responsible for Earth's sustainability as a whole? These questions hit from time to time the minds and hearts of social leaders and society members, especially if they have connected themselves to cooperate with many others to change whatever states of affairs have been perceived, understood or judged as wrong, bad, or unfair to actual and future generations.

---

[35] See footnote 5 and this Institute of 21stCentury Agoras Monograph Series No 1, https://www.createspace.com/3977896

MAIN METHODOLOGICAL IMPUTS

- The Millennium Project 15 Global Challenges as identified in 2010, by the big group of international collaborators guided by Jerome Glenn, Theodore Gordon and Elizabeth Florescu, after a large Delphi process that began in 1996.

- The 49 Continuous Critical Problems presented in 1970, in the paper by Hasan Ozbekhan "The Predicament of Humankind", to the Board of Members of the Club of Rome, which in our opinion still are valid in our present global context.

- The Structured Dialogic Design methodology and software, both created by Alexander Christakis, based on the methodology of Interactive Management by John Warfield and Alexander Christakis himself, to cope with highly complex problems in a democratic decisions environment.

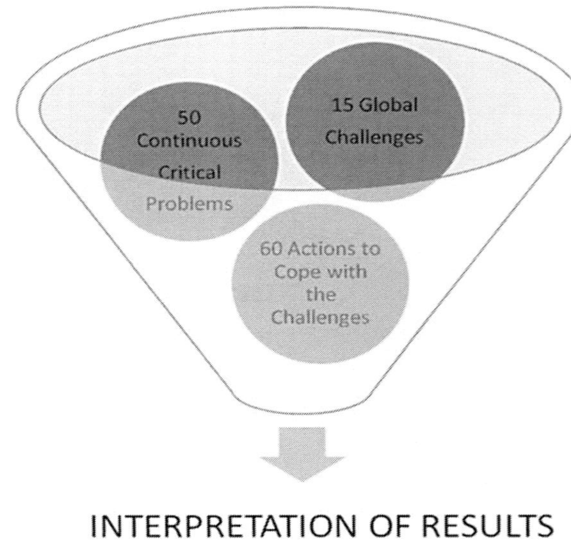

A trial to collect the best answers to those questions for the benefit of Earth as a whole emerges as a newly-born imperative. We are now demanding impartial, honest and applicable answers generated by an ever increasing number of human communities that can finally make a difference in re-directing Earth's evolution to its most harmonious path of development.

In Strategic Articulation of Actions to Cope with the Huge Challenges of Our World Today, Bethania Arango Hisijara and Reynaldo Treviño Cisneros invite people from different walks of life in the world to reflect as follows: The analogy of the Web (Earth) as an intricate tangle of threads of different diameters and longitudinal sizes preventing the use of its components as a highly functional configuration (with fluid interactions) leads us to think that we need to dismantle or loosen certain knots, to untie the threads, classify them, order them, and arrange them in order to make possible to weave them in a new configuration usable by the whole of the Web in its evolutionary process. The demand of a process of harmonization between the subsystems and between the Web and its environment becomes predominant. Its main effect should be global ecological balance and, consequently, the survival of the Web.

A list of the subsystems here implied could be the following: ecological, political, socio-demographic, economic, physical-chemical, biological, psychic, ethical and juridical-legal. They can be influential in areas of justice management, religion, communication, techno-science, and culture. Subsystems are permanently interdependent and interacting through flows of matter, energy, information, knowledge, emotions, behaviours and values. Besides, all the subsystems through their own interactions at different paces might incorporate some factors into the Web, whose random combination could produce determinate effects in certain stages of its evolution. No factor alone might be the cause of a global challenge, i.e., none might become a necessary and sufficient cause to produce complex effects of the nature and intensity of the global challenges. The random combination of sets of factors is responsible for raising continuous critical problems to the rank of global challenges. Sometimes, the continuous critical problems are bonded together and produce bottlenecks and vicious cycles that are constituents of the global challenges. There is the need to try to produce, through strategy, diverse combinations of factors that might dissolve bottlenecks and investigate factors that are capable of eliminating vicious cycles in the subsystems and in the Web.

It is a huge endeavour. The methodology employed by Reynaldo and Bethania is reproducible in its fundamental procedures. However, it might be enriched through a structured dialogic process, which fosters boundary-spanning processes and encourages collective intelligence, intra, multi and trans-discipline perspectives, and the emergence of progressive and more embracing insights.

The design of a Global Strategy cannot remain inside the established frontiers given in this research. New design efforts should go beyond these results and give free play to uncountable questions such as

- the "motivational why",
- the "transformational what for", and most importantly to
- the "how, where, for whom, with whom, in spite of what obstacles, with which resources,
- what are the departing points, what resources might be added in intermediate stages, what indicators and detection systems might be employed to evaluate and ponder progress and possible achievements,
- what sub-products or collateral effects emerge,
- what values are shared all along the implementation process,
- how often the monitoring is required and, also,
- what emergent properties might become usable to improve, accelerate or decelerate according to new necessities, or possibly, to change the route after the learning has been obtained.

One cannot conclude that a Strategy of the dimensions here articulated will be deployed without the contributions of additional systems for detecting errors and deviations. Many actual events (Fukushima and other nuclear plants in the world, the civil war in Libya, Haiti and the scourging poverty of its inhabitants) call us to continue reflecting over the Web itself and our interventions in it. The price of not doing this is the increased burdens we levy on future generations, and possibly the irreversible decline of the Web.

Nevertheless, the possibility of a suitable Global Strategy raises the hopes of mankind for new ways to reach the harmonization of subsystems with the Web, and the harmonization of the Web with its environment and in its evolutionary process.

MYSTICAL ECOLOGY: FOOD AND DRINK AS SPIRITUAL NOURISHMENT IN THE MYSTICAL ORTHODOX TRADITION

CONSTANTINOS ATHANASOPOULOS

Research Affiliate, Department of Philosophy
Open University, UK
c.athanasopoulos@open.ac.uk

Abstract

After I discuss what I mean by Orthodox Mysticism in Philosophy and Theology, I sketch the traditional contemporary philosophical and theological views about Ecology and the use of food and drink for bodily nourishment. Following this, I present the views of three Orthodox mystics on food and drink as spiritual nourishment and how they differ from traditional contemporary philosophical and theological views. Finally, I discuss why we should be based on the Orthodox mystical views on the use of food and drink for a more coherent and ethically and theologically justified ecological perspective.

1. The Context of Orthodox Mysticism in Philosophy and Theology

If one searches in a dictionary, one can find a common idea behind possible definitions of mysticism: there is a direct and not externally verifiable and observable communication with the divine, a communication, I want to claim, that does not follow the laws of traditional logic (chiefly among them the law of non-contradiction), nor is the result of deductive reasoning. If seen in this way, it is not surprising that mysticism has met with considerable suspicion by contemporary philosophers and theologians and either has been totally disregarded or has been discussed via the methods of traditional science, i.e., as an aberrant phenomenon of sociological or psychological religious behaviour.

It is no accident, for example, that Jerome Gellman in the *Stanford Encyclopedia of Philosophy*, trying to discuss how philosophers use mysticism, writes:

> 'Philosophers have focused on such topics as the classification of mystical experiences, their nature in different religions and mystical traditions, to what extent mystical experiences are conditioned by a mystic's language and culture, and whether mystical experiences furnish evidence for the truth of their contents. Some philosophers have begun to question the emphasis on experience in favor of examining the entire mystical complex (See Jantzen, 1994 and 1995, and section 9 below). Since this article pertains to mysticism and philosophy, it will concentrate chiefly on topics philosophers have discussed concerning mystical experience.'

It is important to note here that Gellman focuses on the philosophical importance of the mystical *experience* (which for him in its philosophical, narrow, sense is a unificatory experience; see Gellman, 2004); it is also important that he decides to discuss the (false for my perspective) dilemma of mystical experience vs. the 'entire mystical complex' as rejected by feminist authors (Jantzen), who think that mysticism is permeated with *androcentric* bias. Gellman, in another work of his (Gellman, 2004), explains that even though the *perennial* (i.e., the identification of common mystical experiences across cultures) and the *constructivist* (i.e., the reduction of the mystical experience into a social construct) approaches to mysticism are not very successful, his favoured *naturalistic* interpretation of mystical experience, i.e., that the mystical experience cannot be explained by supernatural means, is more attractive for a contemporary philosopher. However, I have to point out that his reducing mysticism into an *experience*, i.e., as something that can be analysed, described and discussed by the rationalistic and quantifiable methods of traditional science is both an evidence of this philosophical and theological 'suspicion' towards mysticism and a reason for Gellman's dilemma: if x is something that has been discussed primarily by *androcentric* science, then of course it can be classified in 'either-or' terms, when compared to feministic approaches to religion.

Other contemporary philosophers try to establish the rules and norms of the *language use*, when it attempts to describe mystical experience, only to find that:

> 'This leaves us with a discomforting worry. If logic is inapplicable to the mystic's discourse, does that not come very close to saying that discriminations cannot be made in this field between sense and nonsense, the sound and the unsound? The literal approach must be, for a philosopher, a desperate measure, a last resort only. To treat it as anything else would be methodologically perverse. Apart from the difficulties of discrimination, where logic is inoperative, the approach demands an unshakable prior conviction that the mystic's paradoxes are to be taken at their face value as reports of veridical insights. Here there is much that can be challenged.' (Hepburn, 2006, p.456).

So, one can safely draw the conclusion here that contemporary philosophical approaches to mysticism attempt to reduce the phenomenon into quantifiable and rational (i.e., following the laws of traditional logic) forms of describable behaviour. But is this valid? I think not. Primarily because the mystics and more particularly the Orthodox mystics think that this is impossible (as we will soon see). But are the contemporary theological discussions about mysticism in a better state?

Some may say they are, if one takes into consideration the views of Ralph Norman, who concludes that:

> 'Mysticism, then, is a fundamental aspect of the Christian tradition. I have shown how it is intimately linked with a number of historic Christian doctrines, and how a re-appreciation of mysticism can lead to fruitful and creative insights into central Christian doctrines like the Trinity, Incarnation and Ascension that pose a challenge to much modern theologizing. Further, I have shown how the Christian mystical tradition creates a useful locus for theological exploration of the postmodern. If Christian doctrine is to interact with contemporary thought, then this is certainly an area that needs further investigation. Perhaps most importantly, my description of the transformation of the definition of mysticism in recent years reminds modern theology of the usefulness of historical theology. All too often contemporary theology looks back on earlier periods through the lenses of received opinions that can be shown to be faulty. And yet, when the tradition is revisited, one discovers there a basis which informs the most up-to-date thought. Perhaps, after all, the premodern and postmodern are peculiarly alike, and if theology is to find a way through the thought of a Derrida or Deleuze, its best bet is to re-adopt the insights of the Christian mystical tradition.' (Norman, 2004, p. 464).

So it can be claimed, on the basis of the views of the theologian referred to, that the theological view on mysticism is more close to the Orthodox one, since: a) he thinks that mysticism is a valuable and irreplaceable element in Christian theology, and b) it cannot thus be reducible to the *experience* of the mystical. But is the majority opinion of contemporary theologians in agreement with this? A much too hasty embrace of the postmodern theological discourse, which is evident in the above passage, is an indication that its author is in a minority position, trying to appease the spirit of theological postmodernists or at least postmodern sympathizers. Taking into consideration that postmodernists have severely criticized both Christianity and the Christian ecclesiastical tradition, emphasizing and highlighting the significant differences in interpretation and relativizing theological truth, the conclusion mentioned above seems a bit self-contradictory and more of a compromise at all costs. But, the obvious question seems to be here: *why we should compromise truth*?

As we have seen from our brief excursion into the philosophical and theological literature discussing mysticism, if we try to find mysticism's true value and importance through much of contemporary philosophy and theology, we have a difficult (if not impossible) task on our hands, since, for the majority of contemporary philosophical and theological discussions, either mysticism is reduced into an experience that can be described by the methods of natural science, or at best it has to be accommodated in the conceptual and ideological contexts of contemporary highly debatable viewpoints (such as postmodernity) which create more problems for the appreciation of the value and importance of mysticism than actually help us to understand it in its own right. Having thus arrived at an impasse, we can proceed now to how we can think in positive terms about the Orthodox approach to mysticism.

2. The Areopagitic Texts

Some of the first important texts to consult for an Orthodox understanding of Mysticism are those attributed to the famous mystical father of the Church, St. Dionysius the Areopagite, the famous student of St. Paul the Apostle. I shall not enlarge on the issue of whether these texts belong to St. Dionysius the Areopagite, or to another much later author anonymous to us. I shall only mention that according to standard commentaries, the texts have some anachronisms and refer to historical events much later than the life of St. Dionysius; suffice it to say however, that they are in spirit very close to St. Paul's teaching about Christian dogma (and this argument has been advanced by commentators like John of Skythopolis and St. Maximus the Confessor, who were more closer in time to the time of the first circulation of the texts) and thus, I shall, on the basis of this evidence, refer to them as St. Dionysius the Areopagite's texts and thus "areopagitic" (see also on this Florovsky 1987). Now, a few things need to be stressed in relation to these texts.

Firstly, they discuss a peculiar form of epistemology, which does not obey traditional laws of reasoning and logic. It is important to highlight that one of Dionysius' great contributions to Mystical Theology, as we know it today, is his theory about *Γνόφος Αγνωσίας*: *Γνόφος* or *gnofos* has been translated as 'the divine gloom' [or 'cloud' or 'mist']; *Αγνωσία* or *agnosia* can be best translated as 'non-knowledge'. We should emphasize that this *Agnosia* is not ignorance (as some have translated the term). It is a peculiar form of knowing that is ineffable and unknowable to the non-Mystic; it is however communicable from one mystic to the other, through their common point of reference: the triune God.

This indicates that we have here a peculiar form of epistemology that is quite alien to traditional philosophical approaches to knowledge and reasoning.

A *first key feature* in this epistemology is its peculiar *doxastic proportionality*. We can see this idea outlined in the very first chapter of the text *On Divine Names*.

In the following passage, we see Dionysius' claim that his Theology is not based on the persuasiveness of man's wisdom, but that it is *proportionally* limited by the theologian's superior union of reasoning and intuition:

> (*On Divine Names*, Chapter 1, Section 1):
> '...we shall establish the truth of the things spoken concerning God, not in the persuasive words of man's wisdom, but in demonstration of the Spirit-moved power of the Theologians, by aid of which we are brought into contact with things unutterable and unknown, in a manner unutterable and unknown, *in proportion* to the superior *union of the reasoning and intuitive faculty and operation within us*. [...]' (transl. J. Parker)

A *second key feature* is the role of Agnosia in this new epistemology. In Areopagitic terms:

> (On *Divine Names*, Chapter 1- Section 1)
> '...For Agnosia, [non-knowledge] of its superessentiality is above reason and mind and essence—to it must we attribute the superessential science, so far aspiring to the Highest, as the ray of the supremely Divine imparts itself, whilst we restrain ourselves in our approach to the higher glories by prudence and piety as regards things Divine [...].

> For, if we must place any confidence in the All Wise and most trustworthy Theology, things Divine are revealed and contemplated in proportion to the capacity of each of the minds, since the supremely Divine Goodness distributes Divinely its immeasurableness (as that which cannot be contained) with a justice which preserves those whose capacity is limited [...].
> [...] For, as things intelligible cannot be comprehended and contemplated by things of sense, and things uncompounded and unformed by things compounded and formed; and the intangible and unshaped formlessness of things without body, by those formed according to the shapes of bodies; in accordance with the self-same analogy of the truth, the superessential Illimitability is placed above things essential, and the Unity above mind above the Minds; and the One above conception is inconceivable to all conceptions; and the Good above word is unutterable by word...' (transl. J. Parker)

A *third key feature* of this new epistemology is that it is embedded within the new metaphysics, ontology, ethics and language that the areopagetic texts outline; the use of specific divine names (such as 'Light' or 'Good') is symbolic of the humanly unknowable realm of the triune God: man can know *only what is revealed to him via God's grace and only while in ecstatic love with Him.* We see this feature outlined in the passages below:

> (*On Divine Names*, Chapter 4, Section 5)
> 'Let us now then celebrate the spiritual Name of Light, under Which we contemplate the Good, and declare that He, the Good, is called spiritual [33] Light, on the ground that He fills every supercelestial mind with spiritual light, and expels all ignorance and error from all souls in which they may be, and imparts to them all sacred light, and cleanses their mental vision from the mist which envelops them, from ignorance, and stirs up and unfolds those enclosed by the great weight of darkness, and imparts, at first, a [39] measured radiance; then, whilst they taste, as it were, the light, and desire it more, more fully gives Itself, and more abundantly enlightens them, because "they have loved much," and ever elevates them to things in advance, as befits the analogy of each for aspiration.'

> (*On Divine Names*, Chapter 4, Section 13)
> But Divine Love is ecstatic, not permitting (any) to be lovers of themselves, but of those beloved. They shew this too, the superior by becoming mindful of the inferior; and the equals by their mutual coherence; and the inferior, by a more divine respect towards things superior. Wherefore also, Paul the Great, when possessed by the Divine Love, and participating in its ecstatic power, says with inspired lips, "I live no longer, but Christ lives in me." As a true lover, and beside himself, as he says, to Almighty God, and not living the life of himself, but the life of

> the Beloved, as a life excessively esteemed. One might make bold to say even this, on behalf of truth, that the very Author of all things, by the beautiful and good love of everything, through an overflow of His loving [49] goodness, becomes out of Himself, by His providences for all existing things, and is, as it were, caused by goodness and affection and love, and is led down from the Eminence above all, and surpassing all, to being in all, as befits an ecstatic superessential power centered in Himself. [...] And, in short, the lovable is of the Beautiful and Good, and Love preexisted both in the Beautiful and Good, and on account of the Beautiful and Good, is and takes Being (transl. J. Parker).

A *fourth key feature* is its unifying character. Trying to circumvent the specific problems in Hellenistic philosophy in terms of human will and agency, and more specifically how knowing and willing good can be one and the same while in ecstatic love, Dionysius emphasizes the unifying force of the mystical illumination. We can see this in the following text:

> (*On Divine Names*, Chapter 4, Section 6).
> The Good then above every light is called spiritual Light, as fontal ray, and stream of light welling over, shining upon every mind, above, around [34], and in the world, from its fullness, and renewing their whole mental powers, and embracing them all by its over-shadowing; and being above all by its exaltation; and in one word, by embracing and having previously and pre-eminently the whole sovereignty of the light-dispensing faculty, as being source of light and above all light, and by comprehending in itself all things intellectual, and all things rational, and making them one altogether. For as ignorance puts asunder those who have gone astray, so the presence of the spiritual light is collective and unifying of those being enlightened, both perfecting and further turning them towards the true Being, by turning them from the many notions and collecting the various views, or, to speak more correctly, fancies, into one true, pure and uniform knowledge, and by filling them with light, one and unifying.

3. St. Symeon the New Theologian

The mystical tradition that was created in the East, as a direct influence of the areopagitic texts, found its most profound Eastern Orthodox defender in St. Symeon the New Theologian (949-1022), a mystical theologian of such an influence on the Orthodox Church that he is proclaimed as 'the New Theologian', indicating that his work is of comparable validity to the work of St. John the Apostle and Evangelist and St. Gregory Nazienzen. St. Symeon not only consolidated and further elucidated the areopagitic corpus and school of thought and theology, but he enriched it with poems and treatises of great value and importance especially for the development of the Hesychastic Movement in the 13th and 14th centuries (within the context of which Nicholaos Cabasilas, the third key Orthodox mystical author that we will discuss here, matured and wrote his treatises). Key ideas expressed through the texts of St. Symeon are the following:

1. Humans (even though created) can see and unite with the triune God during this life (even though this union is short-lived while in the body).
2. The mystical experience is not only of the soul but encompasses the whole person (both mind and body). More specifically, for St. Symeon, through the mystical union of man and God, man gains via divine grace something from the characteristic unity of the triune God, becoming, in essence, a threefold hypostasis, fulfilling thus the '*κατ' εικόνα*' or 'according to Thy image' (*τρισυπόστατος*: *νους* or *nous*, *ψυχή* or soul, *λόγος* or *logos*, reason): man becomes one with God (θέσει) with his body, his soul, and the Holy Spirit of Whom he partakes (see *Cat.* 15, 72-77; *Hymn* 44 (324), and Krivocheine 1986, p.289)
3. The heart (and not the intellect nor the mind) is elevated as the organ that we have to be primarily concerned with and it is through it and with it that the mystical experience can be achieved in this life (indicating a firm belief in the psychosomatic mystical experience and transformation).
4. Finally, absolute freedom for both God and man is confirmed: man can only attract the mystical vision through the purification and strengthening of the heart and the senses and he can only achieve it if and when God wills it. It is not only dependent on the continuous struggle of the man but also on the will of the triune God. The following two texts show these ideas in detail. In the first, the union with the triune God and its mode of attainment is outlined in a mystical poem:

St. Symeon's mystical poem: 'What is this awesome mystery'

What is this awesome mystery
that is taking place within me?
I can find no words to express it;
my poor hand is unable to capture it
in describing the praise and glory that belong
to the One who is above all praise,
and who transcends every word...
My intellect sees what has happened,
but it cannot explain it.
It can see, and wishes to explain,

but can find no word that will suffice;
for what it sees is invisible and entirely formless,
simple, completely uncompounded,
unbounded in its awesome greatness.
What I have seen is the totality recapitulated as one,
received not in essence but by participation.
Just as if you lit a flame from a flame,
it is the whole flame you receive.
(transl. by John Anthony McGuckin)

In the following text, the importance on the use of the heart in prayer is further highlighted:

St. Symeon the New Theologian: The Three Ways of Attention and Prayer
[The third way of attention and prayer.]
The third way of attention and prayer is then this: the mind should guard the heart in the time of prayer and always stay inside it. From there, from the depths of the heart, it should then lift up the prayers to God. For once it tries inside the heart and tastes and is soothed--as the Lord is good!--then the mind will never want to leave the place of the heart. It will there repeat the words of Peter the apostle: “It is wonderful for us to be here!” [Mt 17:4, Mk 9:5, Lk 9:33] Then it will always wish to look inside the heart, remaining there and pushing aside and expelling all the concepts which are planted by the Devil. To those who have not realised this work of salvation and remain unaware of it, this will most of the times seem very hard and unpleasant. But those who have tasted its sweetness and enjoyed the pleasure inside the depths of their hearts, they all cry together with Paul: “What could ever come between us and the love of God?” [Rm 8:38-39]
[...]
[QUESTION. Why is it not possible to achieve all this through the first and the second way?]
As we have repeatedly said, the first and the second way do not bring any spiritual advancement. When we want to build a house, we do not make the roof first and then lay the foundation--for this is impossible!--but we firstly lay the foundation, then build the house and then add the roof. We should do the same in spiritual matters: first lay the foundation, which is to guard the heart and cast out its weaknesses; then build the spiritual house, which is to cast out the evil spirits fighting us through our senses; finally, having overcome the war as soon as possible, add the roof, which is to depart from all things earthly, and give ourselves completely to God. Thus we complete our spiritual house in Christ our God, to whom all glory is due, unto the ages of ages. Amen.’ (transl. by Demetrios S. Skagias)

From the above discussion of Orthodox Mysticism it is evident that there are five main features in Orthodox Mysticism which make it a *distinct mystical approach:*

a) It cannot be described in terms of subjectivism vs. objectivism (both in terms of epistemology and ethics).
b) The mystical experience is *veridical* (i.e., we can say when we have it), satisfying in this way partly the demand made by some contemporary philosophers (e.g., see Hepburn above), but it does not obey the laws of traditional logic and reasoning (there is a new epistemology involved).
c) Body and soul are not divided, but they are unified in the mystical union of God and man.
d) There is *no intellectualism nor a conceptualism* of any kind, because all reasoning, imagination and memory of sensory images are to be avoided; the organ of chief importance for the mystic is the heart and not the mind. It is important to note in relation to this feature, that Orthodox mysticism is not some kind of sensualism or emotionalism, since emotions and senses of the body are also to be avoided in the attempt to unite with the triune God.
e) The mystical union can be achieved while alive and with the body; later hesychastic mystics, such as St. Gregory Palamas (1296-1359) and Nicholaos Cabasilas (1319/23-c.1391) emphasize that this union is achieved via the divine energies and not the divine essence. The Ninth Ecumenical Synod (1341-1351) made this new mystical idea part of the Orthodox dogma and made prevalent the mystical tendencies in the Orthodox Church. The Orthodox Church celebrates this every year during the fasting period before Easter: on the Sunday after the Sunday of Orthodoxy (during which we celebrate the canonical re-institution of the ecclesiastical use of the icons) it is part of the *Typikon* (the ecclesiastical order of the daily services) that the Decisions of the Synod are read in the Church, so that the Orthodox believers remember that this is part of Orthodox Christian dogma.

4. The Philosophy of Food and Drink

In contemporary philosophical discussions about the ethics of consumption of food and drink, there is a growing tendency to provide arguments in terms of the *rights* and *duties* of the people and beings involved, either recognizing in them *intrinsic qualities*, which can justify the existence of our duties towards them and their rights for respect by us, or by linking them to the rights and duties of others, who have some interest in the consumed animals' fate or well-being. So, for example, in the case of some philosophical arguments put forward by vegetarians and/or vegans, animals should not be consumed and in general all suffering of theirs should be restricted, if not avoided at all costs, on the basis of their inherent moral qualities and their wellbeing. This places ethical restrictions on what we can eat and drink and in some way places a *fast* upon us on moral considerations. Thus, in general, the *deep ecology* and *vegan* argument for an abstention from all meat and dairy products is that we should not eat meat nor drink milk, because this action makes animals feel pain or discomfort. For anthropocentric ecologists the argument becomes that eating meat/ milk/ dairy products reduces the well-being of other people in the developing world (since the resources used to feed animals can be used more efficiently to feed starving people through agricultural products that are not derived from animals). In both cases there is some sort of a contract involved (between us and the animals, or between us and the other people in less developed countries) and through the consumption of specific foods and drinks we break the moral rule that binds us to obey the rules of this contract.

Some philosophers (for example, Peter Singer and Tom Regan) actually go to the point of making parallelisms between the 'speciesism' that existed between people of different colours and races in the recent past and the *biological speciesism* that exists between humans and the rest of the animal kingdom (cf. Singer 1975; Regan and Singer 1976). They make the claim that the same moral rules that bind the different human races and form a contract that is sanctioned by international legal documents and agreements (such as the United Nations Declaration of Human Rights), bind humans as well to the rest of the animal kingdom. This approach has influenced a large part of the relevant current bibliography on the ethics of food and the 70s texts discussing the 'speciesism' approach are now standard textbooks in Environmental Ethics and any set bibliographies on the Philosophy of Food and Drink.

It has been pointed out however, by critics from the early 80s onwards, that the Singer-Regan approach cannot offer convincing arguments in its support. It is not only that the contract theory that it alludes to is fallaciously entertained (since there is a need for a further proof that animals – or other people concerned – can take part in such a contract and that they can both represent and promote their interests through this contract); it is also that for many adherents of vegetarianism or veganism, the contract theory (or any theory based on duties and rights) is not a good way to put forward their viewpoint and arguments for the intrinsic value and sacredness of animals, plants and the other beings that exist in the world, but are not created nor produced by humans. Such critics make the point that even if a human being was the last person on this earth with the possibility of destroying earth and all the beings on it after his/her death, it would be morally wrong to do so, not through any contracts (which would not hold after death in any case), but on the grounds that nature and every non-human living creature is intrinsically valuable with a value that is independent of any human consideration, making them *sacred*. They also point out that resorting to anthropocentric values and moral qualities (for example moral virtues or elaborate rule and/or contract theories) would not satisfy their requirements, since they claim that non-human creatures cannot be examined in human terms and human categories of thought.

It is my considered opinion that the contemporary moral-philosophical discourse, as we know it today, cannot cope with the pressure of such arguments and it needs to have access to a religious perspective, which will allow for the *re-sanctification* of nature that is necessary to get philosophers out of the philosophical impasse.

But not just any religious perspective will meet the demand for a significantly different approach. The religious perspective that will bring the philosophers out of this theoretical difficulty must be a perspective that allows for a *significantly non-human centered/directed perspective* to come into the investigation and help in the formulation of a *less anthropocentric* approach (both in terms of epistemology and in terms of ethics) to the issue of respect towards natural beings in this world.

This leads us back to the Orthodox mystical way of seeing our role in the universe and our goal in life. As we indicated above, this mysticism provides a new epistemology and ethics, which is significantly different from the contemporary philosophical and religious alternatives, both in terms of the less anthropocentric aims and the more *soteriological* perspective it gives for creation and the role of creation in the achievement of the mystical union between God and man. Let us see this in more detail.

5. The Orthodox Mystics on Food and Drink as Spiritual Nourishment: St Nicholaos Cabasilas on the Mysteries of the Church

From the above discussion of the main characteristics of Orthodox mysticism, it is obvious that Orthodox mysticism cannot be anthropocentric nor follow the established and traditional rules of human reasoning, if it is to achieve the desired goal of union with the triune God. It is also evident that humans can achieve a new awareness of their role in creation and of how they should behave towards other creatures, during the ***ecstatic rapture*** that the mystic way entails. This is clear from both the texts of St. Dionysius the Areopagite and those of St. Symeon the New Theologian.

Actually, St. Symeon the New Theologian emphasises that any kind of animosity or improper attitude towards creation is the result of Adam's fall from heaven and comes as a result from it. Mystical union with the triune God will not only return man to the mode of existence he had before the Fall, but it will return in him the wonder and full respect with which he needs to regard God's Creation (see *Hymn* 32, and Kesselopoulos 2001, p.81). He further on claims that the use of material goods and creation for man's survival is not evil but it should be carried out with no 'desire of the flesh', no 'desire of the eyes' and no 'vain thoughts' (*Catechetical Orations*, 5, and Kesselopoulos 2001, p.121). This further supports St. Symeon's conviction that, if a man is to be united with God mystically, he must do so with a complete transformation of his desires, arrived at through ecstatic love towards the Creator and the consequent abstention from any thought and behaviour that is not directed by this ecstatic love towards the Creator. Man thus becomes not only a guardian of Creation through asceticism but considers it as an invaluable means for his salvation; this forces upon him a transformation of his desires and the way he looks at Creation. This mystical perspective has many advantages over the traditional contemporary approaches to the problem of feasting in a world that requires fasting: Firstly, it is not dividing the world into the realm of the subjective and the realm of the objective, or (put in other terms) between the body and the mind, the material and the immaterial. As we saw, both matter (body) and spirit are united in the mystical union in a new form of hypostatic existence. Secondly, it allows for a new non-anthropocentric epistemology and ethics: man does not exist for himself, but *for God*, and Creation is the place which God created for man's salvation. In his ecstatic love towards God, man cannot regard the world as existing for his own means and interests but through his ecstatic love towards God.

Even though these advantages are a real improvement on the philosophical and theological impasse in contemporary debates, how is the demand for sanctification of nature to be met with this perspective in mind?

Deep Ecology vegetarians and vegans here may protest that this mystical standpoint may be interpreted as indicating that man has the role of a simple attendant or house keeper and, thus, Nature may be regarded as part of the household items such a house keeper attends to, with no further responsibility or value to him.

Nicholaos Cabasilas claims that it is here that man's role in the Divine Eucharist as returning Creation to its rightful place can be most clearly envisaged. Man, in the process of the Divine Leitourgy, and while performing the sacrament of Eucharist, offers to God what is made by God (the priest announces in the Eucharist while offering the Holy Gifts that he offers them as of God's to God: '*τα σα εκ των σων σοι προσφέρομεν*'). Cabasilas stresses that the part of Creation that is offered on the altar is not man's but God's. By offering to God what is God's, man enters into a mystical union with the triune God and becomes united with Him through grace, placing Creation in the proper *soteriological* perspective. By re-consecrating nature and creation within the process of the Divine Eucharist, man finds his true role in Creation, becomes light himself and mystically unites with God, radiating light back into creation. Man does not exist to use nature and creatures but to consecrate them and refer them back to their rightful Owner and Creator. And man, according to Cabasilas, can only do this once he lives through the mystic life of the consecrating Church and its mysteries. It is only through the mystical realization of man's failure to be a true son and creature of God that man can realize and achieve the mystical union with God (Cabasilas, 1960, pp. 116-8).

Through the mysteries of the Church and most importantly the Divine Eucharist, where bread and wine are offered and become the Blood and Body of Christ, man becomes holy again and, with the return to his holiness, he can further sanctify nature and offer it to God. Cabasilas here is in full agreement with the Orthodox Mystics before him, in claiming that the 'naming' of Creation by man during Creation (in the book of

Genesis) is part of this re-sanctification of Nature. Man was created to sanctify Nature and not destroy it, according to the Orthodox Mystics.

It is important to emphasise that what is offered in the Eucharist is not owned by man. Man is utterly naked, defenseless, with no justification and without pretence, when facing his Creator. The mystery of the all powerful grace and mercy, thus, becomes the true *soteriological* and *sanctifying* force.

Man performs a *reasonable* sacrifice of his will over things and himself to gain the object of his ecstatic love, without losing his personal and hypostatic characteristics, so that he can find what he ecstatically loves and desires. Trying to explain the words 'reasonable sacrifice' in the Divine Leitourgy, Cabasilas claims that the sacrifice is *reasonable* because it is *mystical*: the priest, by pronouncing the appropriate words, does all that is necessary to make the offered part of Creation holy again and suitable for sacrifice to God, bringing man back into his true role in Creation, in accordance with the book of *Genesis* in Old Testament (Cabasilas, 1960, p.116). According to Cabasilas, 'the *sacrifice* is truly *an act* and *a reality*' and even though the priest only says specific names and words, these become significant and mean far more than perceived by human ears, through the mystery of divine grace, in the same way that in *Genesis*, man by naming nature is not just saying names and words empty of mystical and theological significance. It is important to emphasise that this *use of names* and *words* lies within the mystical tradition of the Areopagetic texts. It is through the names of God that we can gain a mystical awareness of His *apophatic* and *cataphatic* realization and, thus, be led into a mystical union with Him. In the same way, it is through the *naming* process and, via this process, the re-sanctification of Nature, that we can achieve the mystical union of God with creation. Man thus becomes *synergos* (i.e., collaborator) in salvation with God, gaining himself a new *soteriological* perspective.

Cabasilas, in a text where he explains the difference between the new and the old Adam (man after and before the Fall), put forward the following ideas: the true bread (*ο αληθώς άρτος*) is Jesus Christ, who, in the same way that the usual bread strengthens man's heart and gives him physical strength, came down to us and brought us a new form of life and strength. It is Him (Jesus) that we should strive to incorporate within us constantly, to be preserved in the constant and continuous time of hunger (λιμόν) that we live during our current lives (*Patrologia Graeca*, vol. 150, pp.680-684).

This comes in striking similarity with Dionysius' *Epistle to Titus* and what is claimed there about the symbolic and mystical meaning of food, solid and liquid:

> *Epistle to Titus, Hierarch* (Section IV)
> But what is the solid food and what the liquid? For the Good Wisdom is celebrated as at once bestowing and providing these. I suppose, then, that the solid food is suggestive of the intellectual and abiding perfection and sameness, within which things Divine are participated as a stable, and strong, and unifying, and indivisible knowledge, by those contemplating organs of sense, by which the most Divine Paul, after partaking of wisdom, imparts his really solid nourishment; but that the liquid is suggestive of the stream, at once flowing through and to all; eager to advance, and further conducting those who are properly nourished as to goodness, through things variegated and many and divided, to the simple and invariable knowledge of God. Wherefore the divine and spiritually perceived Oracles are likened to dew, and water, and to milk, and wine, and honey; on account of their life-producing power, as in water; and growth-giving, as in milk; and reviving, as in wine; and both purifying and preserving, as in honey. For these things the Divine Wisdom gives to those approaching it, and furnishes and fills to overflowing a stream of ungrudging and unfailing good cheer. This, then, is the veritable good cheer; and, on this account, it is celebrated, as at once life-giving and nourishing and perfecting.

6. Conclusions

From our brief excursion into Orthodox Mysticism, as found primarily in the mystical theology and philosophy of the Areopagetic texts, St. Symeon the New Theologian, and Nicholaos Cabasilas, we can see that there is another way of dealing with the issues of feasting and fasting that avoids the difficult impasse traditional contemporary philosophical and theological discourses are faced with, when having to deal with an apparent lack of respect towards creatures and nature in general.

Feasting and fasting, as primarily concerned with the way humans grow and sustain themselves, should not be treated as unimportant nor as not directly concerned with the way we can know and understand ourselves. Mystical Metaphysics and Ontology here are interconnected with Mystical Anthropology and Ethics. Within Orthodox Mysticism, Asceticism exists not for the mortification of the flesh, but for the re-orientation of our intellectual and volitional powers. Food and drink thus become means of spiritual nourishment: the priest in the mystery of the Divine Eucharist offers parts of creation to God in the form of a 'reasonable' sacrifice, but neither the gifts offered nor the words used belong to the priest, who is with no pretence and no justification.

Creation becomes autonomous and has a separate existence from God, only when it lacks the ecstatic love, the true and chief criterion of the mystical way. Creation and man, in this world, void of ecstatic love, thus, become both alienated and alienating. Both feasting and fasting should become one and have as their goal the mystical transformation of the human will, so that it can unite and reside in God's will.

Now, can the result of our investigation into Orthodox Mysticism be compatible with the philosophical and theological views on mysticism, with which we started our investigation? Well, partly yes. Firstly, we have established that in Orthodox Mysticism we can be veridical about our mystical union and this is one of its key features. This can be used to avoid difficulties stemming from the vague and relativistic attitude of many contemporary philosophical and theological schools of thought. Secondly, we saw that another key feature of Orthodox mysticism is its unificatory aspect and its simplicity. Within the mystical union the mystic does not lose his personhood; however his will and intellect do unite with God's. In this way both simplicity and unity are maintained and achieved.

All in all, however, we need to emphasise that in the new ontology, epistemology and ethics that is put forward in Orthodox Mysticism, traditional philosophical and theological categories need to be revised and transformed in order to accurately describe what is achieved in and through the Orthodox mystical union.

References

Dionysius the Areopagite, *Works* (1897), transl. by John Parker, James Parker and Co. Publ., London (can be also found in: Christian Classics Ethereal Library, http://www.ccel.org/ccel/dionysius/works.html)

Florovsky, George (1987), *The Byzantine Ascetic and Spiritual Fathers*, Belmont, MA.

Gellman, Jerome (2005), 'Mysticism', *Stanford Encyclopedia of Philosophy*, http://plato.stanford.edu/entries/mysticism/#1 (accessed 15/11/08).

Gellman, Jerome (2004), 'Mysticism and religious experience', in William J. Wainwright, ed., *The Oxford Handbook of Philosophy of Religion*, 2004, pp.138-167.

Hepburn, Ronald W. (2006), 'Mysticism: Nature and Assessment of', in Donald M. Borchert, ed., *Thomson-Gale Encyclopedia of Philosophy*, 2nd ed., Thomson Gale Publ., London , p. 453-461.

Kesselopoulos, A. (2001), *Man and the Environment*, transl. by E. Theokritoff, St. Vladimir's Seminar Press, N.Y.

Krivocheine, Basil (1986), *In the light of Christ*, transl. by A. P. Gythiel, St. Vladimir's Seminar Press, N.Y.

McGuckin, John Anthony (2003), *The Book of Mystical Chapters: Meditations on the Soul's Ascent, from the Desert Fathers and Other Early Christian Contemplatives*, Shambhala.

Nicholaos Cabasilas (1960), *A Commentary on the Divine Leitourgy*, transl. by J. M. Hussey and P. A. McNulty, SPCK, London.

Norman, Ralph (2004), 'Rediscovery of Mysticism', in Gareth Jones, ed., *The Blackwell Companion to Modern Theology*, Blackwell, Oxford, pp.449-465.

Regan, Tom, and Singer, Peter, eds. (1976), *Animal Rights and Human Obligations*, Prentice-Hall.

Singer, Peter (1975), *Animal Liberation*, New York Review, N.Y.

Symeon the New Theologian, The *Three Ways of Attention and Prayer*, transl. by Demetrios S. Skagias, *Myriobiblos*, http://www.myriobiblos.gr/texts/english/symeon_threeways.html (accessed 10/11/08)

QUANTIFICATION OF ENVIRONMENTAL SUSTAINABILITY FOR TIMBER AND STEEL CONSTRUCTIONS USING HIERARCHICAL MODELS AND FUZZY INFERENCE TOOLS

D.N. KAZIOLAS,[1] G.E. STAVROULAKIS,[2] I. ZYGOMALAS,[3] D. EMMANOULOUDIS[4] AND C.C. BANIOTOPOULOS[5]

[1] Technological Educational Institute of Kavala, Greece dnkazio@teikav.edu.gr
[2] Technical University of Crete, Chania, Greece gestavr@dpem.tuc.gr
[3] Aristotle University of Thessaloniki, Greece izygomal@civil.auth.gr, ccb@civil.auth.gr
[4] Technological Educational Institute of Kavala, Greece demmano@teikav.edu.gr
[5] Aristotle University of Thessaloniki, Greece and University of Birmingham, England ccb@civil.auth.gr

Abstract

Sustainable development aims at the optimization of the whole of human activity in terms of environmental, economic and social impact. Under this holistic approach the techniques used for the construction, usage and recycling of timber and steel structures can be optimized in order to enhance their overall impact. The quantification of the various factors involved, such as raw materials, construction methods and waste management, is difficult, since the parameters are of different natures. A combination of mathematical tools and user's experience is required at this point. Fuzzy inference and hierarchical decision-making models provide us a suitable framework to add the contribution of each factor and, finally, compare various alternatives. A short introduction of this method is presented in the paper.

1. Introduction

The term 'sustainable development' is nowadays used about various instances, sometimes with a varying capacity and context. Construction is one of the sectors where construction product manufacturers often present products that are 'sustainable' without always taking into account any specific measures or processes. Although these cases should be regarded as the exception to the rule, it is nevertheless necessary to examine the specific content of environmental sustainability and its requirements and implications for the construction sector. In such a way, the quality and sustainability of construction products and services will be higher, thus rendering construction a more efficient and sustainable sector.

The problem of holistic design in timber and steel structures is considered in this paper, account being taken of classical as well as environmental aspects. First suitable design criteria and goals are listed. Then the optimal design problem is categorized within modern methods of management decision-making using hierarchy approach techniques and fuzzy inference. This way a number of suitable techniques that allow us objective choice of materials, methods and management of a steel or timber structure for the duration of its life are outlined. Technical details for possible implementations can be followed through the references of this paper, and will be presented elsewhere.

2. Sustainable Development and Environmental Sustainability

In recent decades sustainable development has been studied in various contexts. It seems that environmental sustainability constitutes one of the most important aspects of sustainability. The historical development of this field in connection with civil engineering projects and, in particular, timber and steel construction, has been presented in Kaziolas et al. (2012). In the present paper computational tools that are available to assist the solution of practical decision-making problems are outlined. The model application concerns the decisions that must be taken during the design of timber and steel constructions.

3. Timber and Steel Construction

Timber and steel construction constitute two sectors of construction that are characterized by increased potential in terms of environmental sustainability. This potential can be attributed to the differences that separate the two techniques from traditional ones which are primarily based on concrete. Since concrete is a material that cannot be efficiently recycled or reused, its use leads to the generation of large amounts of waste. By contrast, wood and steel can be retrieved and recycled or reused after the decision for demolition has been made.

3.1. Raw Materials

One of the major issues is raw materials. Although the protection of the environment was originally introduced at a very general level, it has nowadays become closely associated with the urgency to minimize the use of natural resources, and thus decrease the amounts of raw materials extracted. Since both timber and steel construction projects utilise significantly large amounts of materials, it is necessary to find ways to optimize their use, along with the use of energy or other related environmental inputs.

The capacity of both wood and steel for recycling and reuse enables the minimization of newly extracted raw materials required for a construction project. To fully exploit the recycling and reuse options, establishing a well-organized and efficient network of collection and sorting points would prove most beneficial. Wood, steel and iron that can be used for the production of new construction materials are a key ingredient of the sustainable nature of the two sectors, as this eventually leads to the minimization of newly extracted quantities of raw materials.

The influence of the appropriate raw materials in terms of environmental impact was highlighted by Zygomalas and Baniotopoulos (2011), who calculated the carbon emissions into the environment that are generated for the acquisition of 1 kg of hot-rolled structural steel members, intended for use in steel construction projects in Greece. Three acquisition routes were examined which cover the local manufacturing of the steel members, their import as ready-to-use products from foreign suppliers, and the inland processing of imported semi-finished steel products such as steel billets. The results show that the imported products (also including transport to Greece) are responsible for significantly higher carbon emissions, while the steel members manufactured locally cause the lowest emissions across all three categories.

3.2. Construction Time

Another basic requirement concerning environmental sustainability and timber and steel structures is the minimization of the use of resources during the construction stage. The more time and resources required for the completion of a project, the greater the environmental impact of the whole project. It is therefore necessary to shorten project delivery times. This can be achieved and further improved for timber and steel construction, since both techniques allow for the formation of off-site elements that are only assembled on site. This advantage has to be exploited to the highest degree since it directly decreases the total amount of time required for the completion of a timber or steel structure.

3.3. Waste Management

Environmental sustainability has shown that project designers should not concentrate only on the initial phases of a construction project but also on the last. In an age of rapid economic transformations it is very common for the service life of a construction project to be finished well before its scheduled ending. New owners, new functions and new uses are all reasons for interrupting the service life of a structure and proceeding with interventions or even reconstruction activities. In view of environmental sustainability such options increase the environmental impact caused, since large amounts of waste are generated and new materials are required. It is therefore crucial to be able to recycle or reuse the retrieved construction materials from the existing building during its demolition. Wood and steel are indeed characterized by this potential and can subsequently minimize the amount of waste which cannot be utilized and has to be sent to landfills. This can be observed in a study conducted by Zygomalas et al. (2009) in which the environmental impacts caused during the life-cycle of a simple steel shed were calculated. According to their findings, one third of the environmental burden caused by the construction of the steel shed was deducted from the total life-cycle impact due to the environmental benefit from the recycling of retrieved materials.

3.4. Quantitative evaluation of sustainability

Sustainability involves a lot of uncertain information and vague declarations. Therefore suitable modelling tools must be used. Fuzzy inference seems a suitable environment for the evaluation and ranking of various alternatives and finally for the choice of the best ones. Fuzzy sets and their calculus, fuzzy logic and fuzzy inference have been based on the seminal works of L. Zadeh. These techniques allow for the usage of imprecise information and linguistic reasoning, both including vagueness and uncertainty, in classification, optimization and control. Applications of fuzzy evaluation on the environmental sustainability assessment of various activities have been published recently [Phillis et al. (2001), Andriantiatsaholiniaina et al. (2004), Amindoust et al. (2007), Pislaru et al. (2011)]. The evaluation of the environmental sustainability of structures taking into account their life-cycle has not been considered till now, to the best of our knowledge. Research work in this direction is planned for the near future from our group.

4. Design of Structures: Hierarchical and Fuzzy Approaches

4.1. Formulation of Model Problem and Solution Methods

Let us consider the design of an industrial structure, where timber or steel structural and nonstructural members are coming into consideration. The materials used for various parts of the structure, the construction methods as well as the complete life-cycle policy are factors that must be optimally designed. Possible solution procedures include:

Experience

Engineering experience and knowledge is the most common method used worldwide. Unfortunatelly experienced engineers are not available in all places. Furthermore the decision-making process is influenced by several factors (objective evaluation of alternatives). Last but not least corruption may play a role in some cases or at least the decision is subjected to criticism.

Classical optimization formulation

Within the structural optimization framework one may include all desired factors and criteria that are necessary. Cost, strength, environmental aspects, life-cycle evaluations etc. The arising optimization problem is most probably difficult to solve, since usual mathematical assumptions like continuity and differentiability of all involved functions cannot be guaranteed. Nevertheless, in principle, usage of some applicable global optimization tools, like genetic or evolutionary algorithms, could solve such a problem.

Fuzzy decision-making (sorting of alternatives)

Decision-making and comparison of alternatives can be performed by using classical fuzzy inference. The various factors that affect the solution (e.g. cost, environmental impact, recyclability etc) are transformed into fuzzy variables with suitable classes of characteristics. For example cost may be characterized as very low, low, medium, high etc and recyclability may be characterized as impossible, very low, low, acceptable, or excellent. The decision-making process is based on a number of linguistic declarations that are based on the previous categorization of every involved variable. For example one looks for a solution of low cost AND high recyclability. Every available alternative (technical solution) is evaluated within this fuzzy scheme and at the end the alternatives are sorted with respect to the expressed comparison rules.

This approach requires one or more experts in order to perform the characterization of each variable (fuzzification), and in order to write down the fuzzy inference rules. For several variables this step is not always a trivial one. In other words we are trying to solve a very complicated problem in one step, while in the hierarchical approaches outlined in the sequel we split it in several smaller steps.

Hierarchical approach (AHP, ANP)

The various factors are grouped into several levels within a logical hierarchy related to the decision-making. The required factors (importance of each factor etc) are again gathered from experts, based on pairwise comparisons and systematic extraction of their knowledge through e.g. questionnaires. Then the problem is solved in steps by following one of the available techniques. Recent applications of this technique in civil engineering can be found in Ugwu et al. (2006), Reza et al. (2011).

Fuzzy hierarchical approach (fuzzy AHP, fuzzy ANP)

The hierarchical approaches are enhanced by using fuzzy variables in all of certain selected variables. These variables are most suitable for storing the experts' knowledge. As a consequence all operations involved in the decision-making process follow the corresponding hierarchical model and fuzzy arithmetic operations. Concrete applications can be found, among others, in Maleckly et al. (2010), and Pan (2008).

4.2. Fuzzy Inference

Fuzzy inference is based on the theory of fuzzy sets, after Zadeh, (1965). Instead of using sharp definitions like YES/NO, BLACK/WHITE, it uses the degree of membership notion that allows a gradual change from one class to another. Fuzzy variables are suitable for the description of quantifiers that are imprecise or statements which include some uncertainty (for example, 'George is tall' or 'This room is warm'). Furthermore linguistic expressions and rules are written within the fuzzy framework and provide us with a technique to describe quantitatively decision-making processes (fuzzy inference).

Fuzzy inference has been applied in a variety of industrial applications (controllers of cameras, washing machines, financial modelling). Furthermore a number of software packages, like the Fuzzy Inference Toolbox of MATLAB, are available for numerical applications. For applications in real-time environments or processing of massive data, hardware realizations are available.

Advantages

Robustness against imprecise knowledge
Easy integration of linguistic expressions of knowledge (e.g. warm, tall, expensive)
Near to user's experience, since it is based on natural language expressions

Disadvantages

No theoretical support for the design and optimization of tools (e.g. controllers)
An expert (or several) must provide information (rules, weights etc).

Steps for practical realization

1. Fuzzification
every classical variable is tranformed into a fuzzy variable (with classes of descriptors)

2. Rule expression
an expert writes the rules that combine each class of one variable to every class of other input or output variables

3. Rule evaluation
3.1 evaluation, by calculating the membership function for every class in every variable
3.2 Output evaluation

4. Aggregation
Combination of active rules' contributions to form the fuzzy result

5. Defuzzification
From the fuzzy result a classical value of the output variable(s) is estimated.

4.3. Hierarchy Decision-Making

Within this approach we must define the goal of design, as well as various criteria and subcriteria. The weights of each element can be defined from the evaluation of suitable questionnaires, filled in by experts in the field. Finally every alternative is evaluated within the hierarchical decision-making scheme and the result provides a ranking of the studied alternatives with respect to the adopted criteria and subcriteria. The hierarchical structure of these techniques allows us to split a complicated problem into smaller subproblems, for which comparison is more easily understandable by an expert in the field. From the theoretical point of view the complexity of solving the optimal design problem is reduced by solving it within the hierarchical scheme.

Let us consider a model problem that concerns the design of an industrial building (goal).

A set of possible criteria and subcriteria can be proposed: see Figure 1.

Using this framework we can compare three possible alternatives concerning the design of the structure, as described in Figure 1. Each one of the alternatives is evaluated and finally the alternatives are sorted with respect to the given criteria.

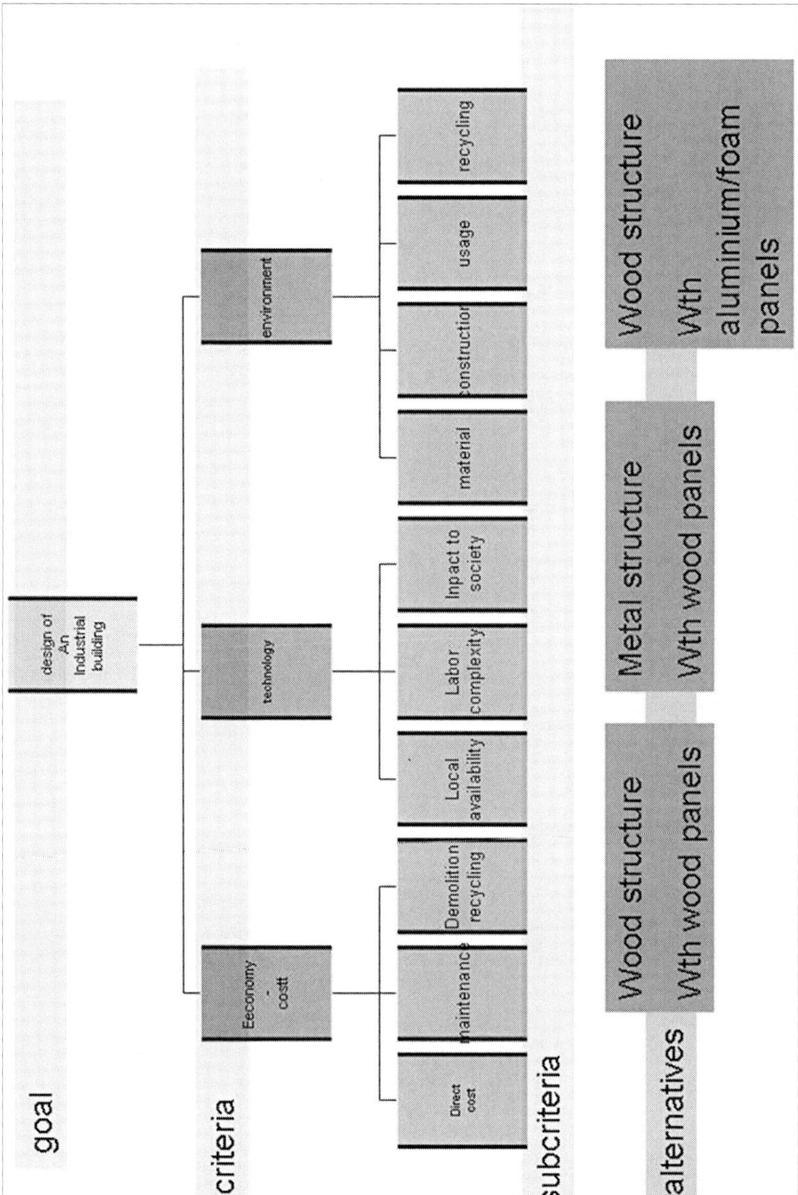

Figure 1. Analytical hierarchy process for the design of an industrial building

Analytical hierarchy process (AHP)

In the AHP one or more experts are asked to compare the criteria between each level in the decision tree (see Figure 2). The method calculates from this information the weights required for the numerical application.

Analytical network process (ANP)

In the ANP one or more experts are asked to compare pairwise all criteria between the same as well as different levels in the decision tree (see Figure 3). The method calculates from this information the weights required for the numerical application. It is clear that the quantity of information required for the usage of ANP method is higher than that required for AHP.

Fuzzy AHP and ANP

Both previously outlined hierarchical models can be extended by making each involved variable fuzzy. This way the robustness and flexibility of fuzzy logic are introduced into the system. At the same time, since the involved calculations must be performed using fuzzy calculus, the calculation complexity increases.

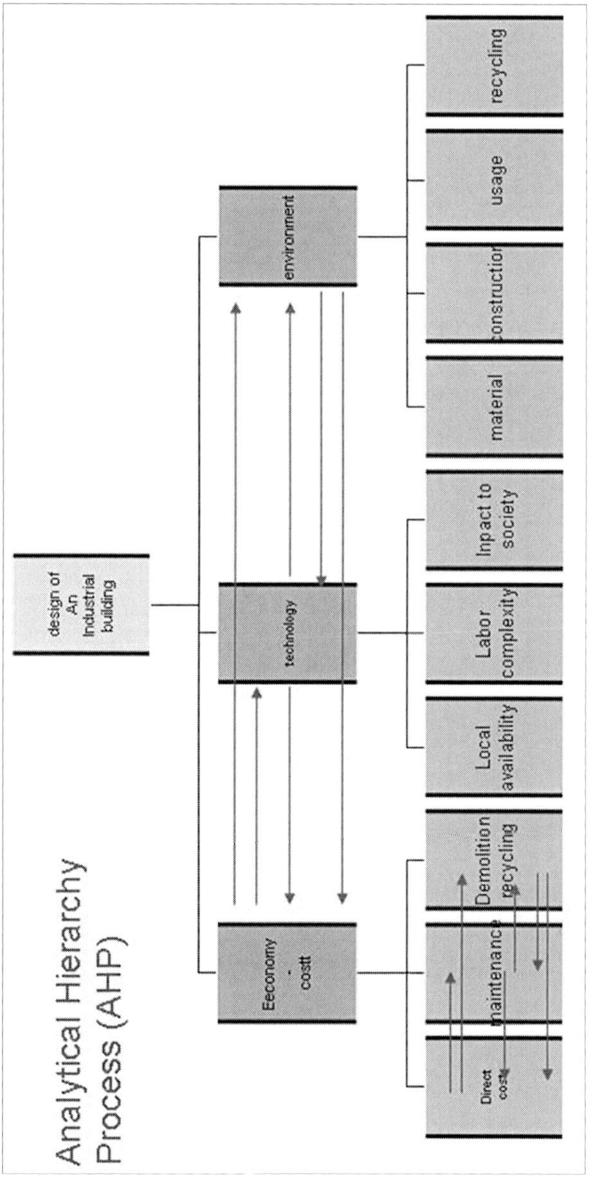

Figure 2. Analytic hierarchy process requires the quantification of comparisons between elements of the same level within the hierarchy.

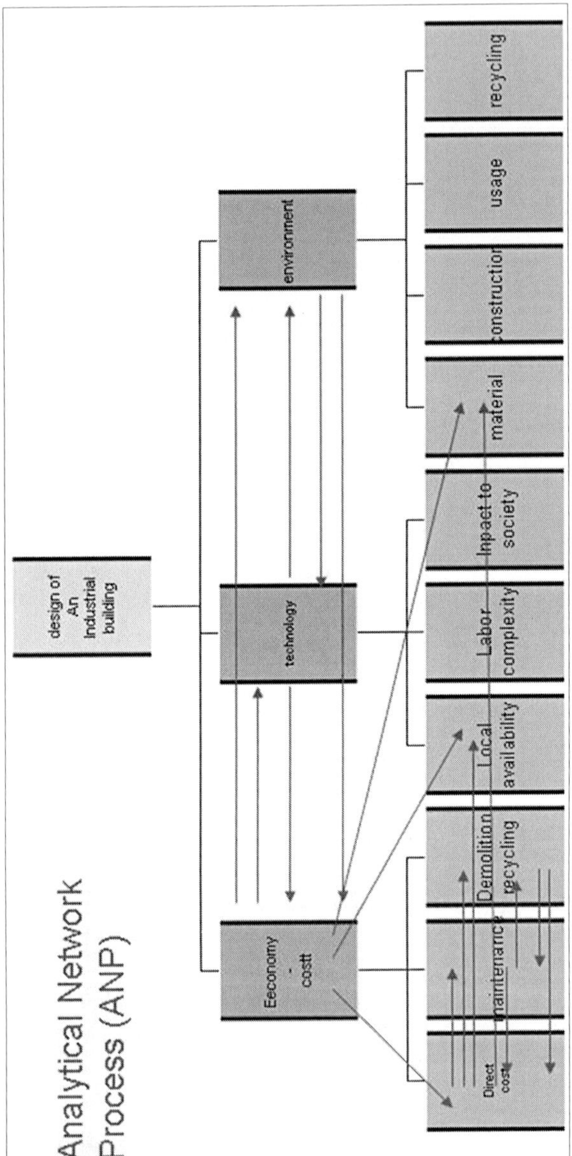

Figure 3. Analytical network process differs from the AHP since it allows for interaction between elements belonging to different levels within the hierarchy.

5. Conclusions

The concept of environmental sustainability was first introduced as an urgent requirement for the protection of the environment against the consequences of human activity in view of finite resources. In the following years, environmental issues gained increasing attention and were being prioritized in international agendas for future policies. Ultimately, the overarching goal of sustainable development was set to ensure that the future of the next generations would not be compromised. Environmental sustainability was defined as one of the three dimensions of sustainable development, the remaining two being economic and social sustainability. A number of key events and publications contributed to the development of the concept and led to its current establishment.

The construction sector can play a critical role in ensuring sustainable development, since it is responsible for the consumption of about 50% of raw materials within the EU, and generates the largest waste streams as well. It was found that most of the materials in these waste streams are coming from onsite construction based on traditional materials such as concrete. In view of this, steel and timber construction were found to present increased environmental sustainability potential, due to the properties of the materials and their capacity for recycling and reuse.

The main implications and requirements for timber and steel construction that were identified concern raw materials, the construction stage of a project and the end of its life-cycle. The urgency to minimize the use of natural resources and thus decrease the amounts of extracted raw materials can be balanced with a well-organized and efficient network of collection and sorting points which ultimately enables the minimization of newly extracted raw materials required for a construction project. Environmental impact during the construction stage can also be minimized through the shortening of project delivery times, as timber and steel construction both allow for the formation of off-site elements that are only assembled onsite. At the end of a construction project life-cycle and after the decision for its demolition has been made, timber and steel also provide increased potential in terms of environmental sustainability. The recycling and reuse of materials retrieved can minimize the amount of waste which cannot be utilized and has to be sent to landfills.

In this respect, it can be argued that timber and steel construction do not actually produce 'waste' with the traditional meaning of the term but rather collected quantities of raw materials which can be used for the manufacturing of new wood and steel construction products.

Taking into account all these facts, the application of modern decision-making tools based on AHP and ANP, as well as their fuzzy extensions, seems reasonable. This paper has outlined a possible application with the concrete example that concerns the design of an industrial building using timber and steel elements. A thorough investigation and examples based on this technique will be presented in the future.

Acknowledgements

This research work is co-funded by the European Union (European Social Fund) and National (Greek) Resources – **ARCHIMEDES III**.

References

Amindoust A., Ahmed S., Saghafinia A., Bahreininejad A. (2012), *Sustainable supplied selection: A ranking model based on fuzzy inference system.* Applied Soft Computing 12:1668-1677.

Andriantiatsaholiniaina L.A., Kouikoglou V.S., Phillis Y.A. (2004), *Evaluating strategies for sustainable development: fuzzy logic reasoning and sensitivity analysis.* Ecological Economics, 48(2), 149-172.

Kaziolas, D.N., Zygomalas I., Stavroulakis, G.E., Emmanouloudis D., Baniotopoulos C.C. (2012) *Evolution of environmental sustainability for timber and steel construction.* Proceedings of the *International Conference on Sustainability in Energy and Buildings 2012,* Stockholm, Sweden, 3-5 September, 2012.

Malekly H., Meysam Mousavi S., Hashemi H. (2010), *A fuzzy integrated methodology for evaluating conceptual bridge design.* Expert Systems with Applications, 37, 4910-4920.

Pan Nang-Fei. (2008), *Fuzzy AHP approach for selecting the suitable bridge construction method.* Automation in Construction, 17, 958-965.

Phillis Y.A., Andriantiatsaholiniaina L.A. (2001), *Sustainability: an ill-defined concept and its assessment using fuzzy logic.* Ecological Economics 37:435–456

Pislaru M., Trandabat A., Avasilcai S. (2011), *Environmental assessment for sustainability determination based on fuzzy logic model.* 2nd International Conference on Environmental Science and Technology IPCBEE, vol. 6, IACSIT Press, Singarpore.

Reza B., Sadiq R., Hewage K. (2011), *Sustainability assessment of flooring systems in the city of Tehran: An AHP-based life cycle analysis.* Construction and Building Materials, 25, 2053-2066.

Ugwu O.O., Kumaraswamy M.M., Wong A., Ng S.T. (2006), Sustainability appraisal in infrastructure projects (SUSAIP). Part 1. Development of indicators and computational methods. Automation in Construction, 15, 239-251.

Zadeh L.A. (1965), *Fuzzy Sets,* Information and Control, 8, 338-353.

Zygomalas I., Efthymiou E., Baniotopoulos C.C. (2009), *On the Development of a Sustainable Design Framework for Steel Structures,* Transactions of FAMENA, issue 2, volume 33, Zagreb 2009, pp. 23-34.

Zygomalas I., Baniotopoulos C.C. (2011), *Extending an existing LCI database to include the complete range of steel members acquisition.* Proceedings of the 2nd International Exergy, Life Cycle Assessment and Sustainability Symposium, Nisyros island, Greece, 19-21 June 2011, pp. 641-648.

HUMAN ACTIVITY AND THE NATURAL ENVIRONMENT: A PHILOSOPHICAL APPROACH

K. KALACHANIS,* I. MATIATOS, C. KOTAMPASI, P.STAMATIS AND P.NICOLOPOULOU-STAMATI

University of Athens, Medical School,
M.Sc. "Environment and Health, capacity building for decision making"
*To whom correspondence should be sent at kalahanis@hotmail.com

Abstract

Human activity is directly related to the physical environment, while it also shows the attitude of man towards nature. In Greek antiquity man considered himself as an integral part of nature, a perception that originates from both philosophical thought and religious tradition. By contrast, modern civilization treats nature just as a source to be exploited. As a result many health problems have occurred, while many countries have been condemned to poverty.

1. Introduction

The relationship that human beings develop with the natural environment is not simply related to survival through the use of natural resources, but a relationship aiming at economic development and improved quality of life. It is manifest that man has not been content just to survive in a hostile environment, but also has attempted to improve his life conditions by using natural resources. This objective has determined his activity and its impacts on the natural environment. In this paper we are trying to examine the way in which the natural environment has been affected by human ways of thinking and acting. Specifically, we are going to study human activity from antiquity up to the industrial revolution and the modern era.

2. Man and Environment in Antiquity

During ancient times, people thought of themselves as an integral part of the natural environment, having observed that their lives depended on the plant and animal kingdoms. Greeks in antiquity understood this relationship between humans and Mother Earth, who was indeed worshipped as a supreme Goddess, being honoured with festivals, praising the rebirth of nature in spring. One such festival was Anthesteria, which used to take place at the end of month Anthesterion (from late February to the first few days of March). The name of the festival comes from the ancient Greek word 'anthos', which means 'flower'. It is apparent that the name of the festival was intended to praise the rebirth of nature, something which created a very cheerful and pleasant atmosphere for the participants.

This perception about the connection between human beings and nature was accepted by ancient Greek philosophers and especially those of the Presocratic period (6th to 5th centuries B.C.). The main characteristic of Presocratic thought was the attempt to find naturalistic explanations of the world, by attributing the origin of the Universe and of all beings to an initial first principle. Philosophers like Thales of Miletus, Anaximenes, and Empedocles, having observed the life-cycle of all beings, came to the conclusion that not only the nature we understand, but the whole universe is one single entity. Those philosophers from the city of Miletus (in Asia Minor) used elements from their everyday life in order to symbolize this entity. Specifically Thales considered *hydor* (water) as the first principle of all beings, while Anaximenes thought that the first cause was *aer* (wind). Also, Anaximander attempted to show the transcendental nature of the first cause, by naming it 'apeiron' (without borders) [Dioneges Laertius (1964)]. So, scholars from Miletus introduced a philosophical conception known as monism, according to which the fundamental character of the universe is unity.

Apart from Thales, Anaximander and Anaximenes, another scholar from Ephesus (a Greek city in Asia Minor), Heraclitus disagreed with the philosophers from Miletus, since he claimed that the first element of the Universe was fire [Diogenes Laertius (1964)]. However, the basic idea of his thought was the origin of the Universe from an initial first cause.

As we may infer, the ancient Greeks established a holistic approach which was extended later though Christianity. St. Gregory, Bishop of Nyssa, a great Byzantine philosopher and theologian of the 4th century AD, supported the view that humanity is a miniature cosmos, being rightly praised with this name [St. Gregory of Nyssa (1972)].

It is apparent that this view of the Universe had a very serious impact on human activity, since their buildings were in total harmony with the environment. In particular, temples of the god of medicine, Asclepius, were usually built in places where the location, the lush vegetation, the climate, the availability of natural sources of clean water and theplace's natural beauty made it suitable not only for treatment, but also for worship [Stavropoulos (2000)].

Hippocrates, the famous doctor from the island of Kos, had already understood that since humans interact with the environment, then there must be a direct impact on their health, if this relation is disturbed. That is the reason. This is why he considered that human activity, especially when a city is being built, should take into account the impact suffered by human health. Therefore he considers it essential for the maintenance of human health to study the climatic factors that affect it, namely the quality of wind and water, which is important to take into account when a city is being built. So, the Hippocratic vision of man as integrated in the cosmic system and the environment should be a guideline in how human activity should take place, namely with respect for nature and health [Kalachanis (2011)].

However, is this unity of man and the Universe a fact? All chemical components of the human body (carbon, nitrogen, oxygen, etc.) are supernovae remnants, formatted in star nuclei. It is know from the field of astrophysics that the explosion of a star of great mass expels much of its material at a velocity of up to 30,000 km/s (approximately 10% of the speed of light), driving a shock wave into the surrounding interstellar medium. This shock wave sweeps up an expanding shell of gas and dust, called a supernova remnant. This gas consists of hydrogen, carbon, oxygen and helium. When these gases collapse through gravity, new stars are created and also new solar systems and of course life. It is obvious that every single molecule of the human body consists of chemical bonds created in deep space. The stellar origin of the matter composing the human body shows our connection with the universe [Danezis, Theodosiou 1999].

Summarizing, it is clear that according to the ancient Greek thought every human activity should take into account the fact that human beings are parts of a whole, which is the environment. Therefore, any harmful intervention in this system may also have harmful impacts on human health.

3. Human Activity and the Environment in the Modern Era

The concept of the harmonious coexistence of man with his physical environment was gradually changed due to significant changes in technology, which in turn triggered changes in economic structures. While the economy from ancient times was based on craft, agricultural production and trade (marine and land), the rapid development of technology generated significant changes.

A significant landmark of this process was the Industrial Revolution (1750-1850), which initially took place in Britain. During this period the development of the means of production necessitated the use of natural resources. Massive changes in agriculture, manufacturing, transportation, and technology had a profound effect on the social, economic and cultural conditions of the times. Since then, the production of goods has become enormously greater, and nowadays appeals to wider groups of people; so it becomes imperative for industry to find source materials at its behest.

The transition to this new era of industry undoubtedly contributed to improving people's quality of life, but also has contributed to the alteration of the human attitude towards the environment. Nature is not any more the "home" of man, but rather seen as useful for exploitation. Man, instead of considering himself as part of nature and embedded within this harmony, placed himself at the summit of the environmental system. Beyond the undeniable deterioration of the physical environment, human activity has also resulted in the occurrence of health problems. Impacts of environmental contamination have been observed in many systems of the human body, such as the reproductive, respiratory and endocrine systems, and a significant increase in cancer rates has also been observed, especially in the developed industrial world [Nikolopoulou-Stamati (2007)].

Besides health problems, environmental degradation due to uncontrolled human activity has resulted in economic disaster in many countries of the Third World. Millions of people are condemned to poverty for the sake of economic profit. Therefore humanity needs to examine the priorities of its civilization. Man should not only consume, but care for our home – Earth – because our whole activity has serious impacts on our planet. Also, an economic growth is needed not only aiming at profit, but also at the protection of the natural environment. Such a strategy, if adopted by decision makers, will have as its basic parameter the assumption that man is an integral part of the environment, both requiring mutual respect. This attitude will combine economic growth with protection of the natural environment, and will eliminate poverty. In this respect the primary objective of harmonious living in a natural environment will be fulfilled.

4. Summarizing

It is manifest that from ancient times to the modern era the human attitude towards the environment has changed dramatically. While the natural environment was first treated as the "home" of man, the development of industry and the overconsumption of natural resources not only has caused environmental deterioration, but has also condemned millions of people to poverty. Therefore a sustainable development is required which will combine both the protection of the natural environment and the quality of life.

References

Danezis, E. and Theodosiou, E. (1999): The Universe That I Loved- Introduction into Astrophysics, Diavlos Publications, Athens [in Greek].

Diogenes Laertius, (1964): *Vitae Philosophorum*, ed. H.S. Long, Clarendon Press, Oxford, books 1-2 and 9.

Kalachanis, K. (2011): Environmental Impact on Health in the Work of Hippocrates, Final Thesis, Post-Graduate Program "Environment and Health, Capacity Building for Decision Making", University of Athens, Medical School [in Greek], pp. 22-30.

Nicolopoulou-Stamati, P. (2007): "Introduction: Concepts in the Relationship of Congenital Diseases with the Environment" in: Congenital Diseases and the Environment, Springer pp. 1-14.

St. Gregory of Nyssa. (1972): De creatione Hominis sermo alter: ed. H. Horner, Leiden, Brill, 65, 13-14.

Stavropoulos, S. (2000): Asclepeia of Peloponnesus: Aeolos editions, Athens, pp. 20-21 [in Greek].

Greed Index Measurement: An Alternative Tool to Promote Sustainability and Eco-Justice

Lucas Andrianos[1]

Institute of Theology and Ecology
Orthodox Academy of Crete, Chania, Greece
E-mail: laoac@otenet.gr
http://greedline.webs.com

Abstract

Greed is the greatest of all plagues against justice, peace and sustainability. The society of 21th century relies on unrestricted structural greed and promotes it through unlimited growth, overconsumption and individualistic competitive behaviour. This paper aims at analyzing greed by offering an empirical tool to measure, monitor and control the root causes and effects of greed on a global, national, institutional as well as on an individual level. This holistic approach of greed is referred to as structural greed.

The findings give answers to critical questions such as "what is greed?"and "how can we measure and control it?" We have developed a new model called GLIMS, which stands for Greed Lines and Indexes Measurement System. It uses fuzzy logic reasoning and inputs from statistical indicators of natural resources consumptions, financial realities, economic performances, social welfare and ethical and political facts. The outputs are concrete measures of three primary indexes of ecological, economic and socio-political greed (ENV-GI, MON-GI, POW-GI) and one overall multidimensional structural greed index (MSGI). The results are greed index scores that are expressed in a scale of zero to one hundred. A greed index score equal to 100 corresponds to the maximum level of greed for the subject of analysis. In contrast to the poverty line, the

---

[1] World Council of Churches, sustainability consultant for the Greed line group study

GLIMS model allows the formulation of discrete greed lines and the computation of an overall multidimensional structural greed index (MSGI). Greed measurement is necessary to control the effects of greed in systemic and structural aspects. It is indispensable for policy making and for raising awareness on structural as well as on individual standard.

Starting as a pilot project of the World Council of Churches in linking poverty, wealth and ecology, it is the first time that crisp measurements of greed are proposed, using a set of eleven greed indicators for selected economies and individuals. The vision is to control structural greed by offering training and practical tools for policy-making through time-series and sensitivity-analysis of greed indexes for all economies, institutions, churches and individuals.

1. Introduction

As a follow-up to the Alternative Globalization Addressing People and Earth (AGAPE) process, which concluded with the AGAPE Call presented at the 9th General Assembly of the World Council of Churches[2] (WCC) in Porto Alegre in 2006, the WCC initiated a program focused on eradicating poverty, challenging wealth accumulation, and safeguarding ecological integrity based on the understanding that Poverty, Wealth, and Ecology (PWE) are integrally related. In this work, we analyze greed and its measurement, using fuzzy logic evaluation. The methodology is inspired from the SAFE[3] model which was initially created by the same authors (Andriantiatsaholiniaina, 2001, Andriantiatsaholiniaina et al., 2004). The root causes of greed and its consequences for global crisis are examined from the perception of Christianity belief and the theological conception of the Trinitarian nature of creation as a whole. The conceptual description of structural greed is presented in correlation to the findings of the WCC greed line group study on poverty, wealth and ecology.

In this paper, the first part will give an overview of the GLIMS model and describe the concept of structural greed. The following section will present the methodology for shaping greed lines using fuzzy logic reasoning and mathematical standardization of greed indicators in the GLIMS model. After the explanation of the set of greed indicators and the greed line

---

[2] http://www.oikoumene.org/en/

[3] Sustainability Assessement by Fuzzy Evaluation (Andriantiatsaholiniaina, 2004); www.sustainability.tuc.gr

functions from empirical data, the next section will show sampling results from the GLIMS model using MATLAB on national and individual levels. This last section will answer critical questions such as "how greedy is a country or a person?", and raise discussions and recommendations. The conclusion is the possibility to monitor and measure multidimensional greed indexes on global, societal, institutional and individual levels.

2. Brief Insight on Greed Assessment

Jesus Christ warned us to *"take care to guard against all greed, for though one may be rich, one's life does not consist of possessions"* (Luke 12:15), and ***Einstein*** once said "*Three forces move the world: Stupidity, fear and greed*". We will look at greed, initially through historical insights and then through wise sayings on greed measurements and temptations in the past. According to Greek mythology, King Midas was trapped by the greed for gold, and lost his "golden" daughter. Also Croesus, not hearing Solon and the virtue of moderation, but following the advice of the greedy priests of Apollo to attack Cyrus and accumulate more property, was duly punished.

"The earth produces enough to satisfy human needs but not greed," was once said, very well, by ***Gandhi***. So what is this trait (congenital or acquired), so particularly evident in the human species, which emerges pervasively but is most intense at the beginning of civilization? Ziggurats, pyramids, empires, all indicate some form of greed in the promoters. Because the healthy mind (logic) regulates the measure, so with a sound spirit (wisdom) it is possible to get the optimal measure (index). What is the "optimal measure"? It is the mean between two opposite, avoiding extremes[4].

In Ancient Greek philosophical terms, there are many lessons which refer to greed and the necessity of its measurement and moderation, to preserve human happiness and to avoid mass destruction. ***Plato***, (427-347 B.C.), urged the need for determining wealth limis and said: "*The form of law which I propose would be as follows: In a state which is desirous of being saved from the greatest of all plagues (GREED)—not faction, but rather distraction—there should exist among the citizens neither extreme poverty nor, again, excessive wealth, for both are productive of great evil . . . Now the legislator should determine what is to be the limit of poverty or of*

[4] Dictionary Tegopoulos Smith

wealth." The ***"cup of justice"*** which was invented centuries ago by the wise Pythagoras, is also called *greedy cup*. The message it conveys is as crystal clear as is the water that we can drink from it: "You may drink little or more, even a little more, if you like. You may share with others and satisfy your needs. But do not wish to fill the cup to the rim (greed line), in order to drink more than the others. Because then you will lose it all! [5] .

Respect for the "greed line" should be the rule in all dimensions of systemic decision-making on all levels; otherwise the consequence is catastrophic for human beings and for the Earth.

3. The Concept of Structural Greed

a. Descriptions of greed

There are many descriptions of greed according to subject: individual, institutional, national, corporal and global. One of the hallmarks of human behaviour is greed. In Greek the word "greedy" is *'a-plistos'*. "*Aplistos*" is derived from the privative "*a*" and "*plistos*" which means complete or full. Therefore "greedy" is having more, insatiable, through unfulfilled desire. The opposite of greed is the "plistos" which means "full-integrated" or theoretically defined as a standard value, because supposedly doing well with his or her situation. The wholeness of human being consists in the fulfillment of a balanced threefold need: material, mental and ethical or spiritual. Jesus Christ said, "It is written: '*Man shall not live on bread alone, but on every word that comes from the mouth of God*' (Mathew 4:4; Deuteronomy 8:3). According to Ancient Greek philosophers, happiness can be reached if all needs are satisfied in moderation, avoiding extremes[6]; one of the seven sage of ancient Hellas stated "*everything is best in moderation*" [*ΜΕΤΡΟΝ ΑΡΙΣΤΟΝ*]).

b. What is greed?

(Raiser, 2011) Greed could be defined as the *desire to have more than one's legitimate share of material goods and power* (mental and psychological). In contrast to poverty which deals with needs that can be objectively defined and even quantified, greed is about desires which are "*difficult to contain*" and involve an "*emotional energy that seeks to*

[5] http://greedline.webs.com/ (February 2013]

[6] Cleobulus of Rhodes (6th cent. BC)

transgress or disregard limitations" and which are consequently difficult to circumscribe and measure.

c. The concept of structural greed and its measurement

"*In today's complex economy where people often fail to recognize the structural connections between their desire to improve their living standards (status) and the poverty suffered by others* (Raiser, 2011), Christian churches and ecumenical organizations have the task of making visible – and lifting up the voices of – those people who are in the socio-economic margins." The systematic approach of greed focusing on the holistic interconnections between its potential causes and its effects upon global society or its manifestation in the Trinitarian nature of human being is attributed to structural greed (Andrianos, 2011).

The development of *multidimensional greed indicators (MGI)*, as a counterpart to the multidimensional poverty indicators first developed by Oxford University, was proposed by Michael Taylor (2011). The indicators could focus on categories of health, education, empowerment, relationships, environment and security and, in each case, would refer to the potential greed (status) of an individual and its consequences for others (desire/trends). The MGI basically would address the questions: *Am I greedy? How am I greedy?* In summarising approaches to developing a greed line, a distinction between static and *dynamic approaches* has been made, with the latter showing changes – growing enrichment versus growing impoverishment – *over a period of time* (Goudzwaard, 2011). Moreover, choosing a particular approach would depend on "*whom we like to address*" *and the availability of data*. Ideally, racial, gender and other forms of discrimination should be captured. Also it is recommended to develop a social ethical consumption function that factors in inequitable socio-economic conditions and ecological limits, and in deriving the greed line from said function (Larrea, 2012). Finally, it is suggested that defining a greed line using a fuzzy logic approach by computation using "linguistic values" and developing a multidimensional structural greed index by "fuzzy combination" (simulation with uncertainty) of the three pillars of society (ecological sustainability, economic-financial performance and socio-political justice) can give a solution for a practical monitoring of structural greed (Andrianos, 2012).

Bible study has pointed out the Pauline teachings revealing that *human greed is a sin that has adverse consequences not only on our neighbors (natural ecosystems and humankind), but also on Creation as a whole*[7]. With regard to individuals, structural greed ought to take into account the effects of greed on the balance of threefold human need: *material, mental and spiritual (Andrianos, 2012).* And at a national level, it must include the effects of greed on the three basic pillars of society: *ecological sustainability (material overconsumption), monetary accumulation (economic-financial) and power inequality (socio-political).* The following figure (Fig.1) summarizes the methodology for greed index measurement using fuzzy evaluation.

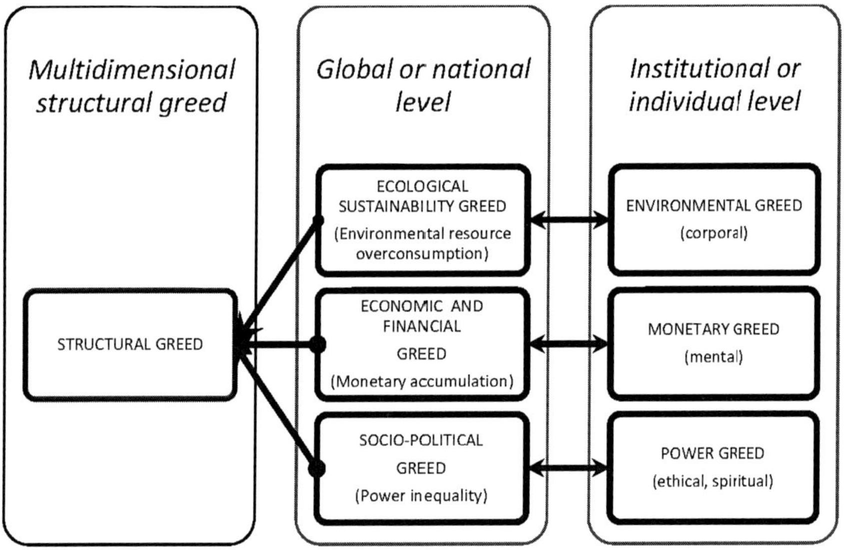

Figure 1: Concept and dimensions of structural greed

d. Greed lines concept

If greed is "having too much" money, resources and power (in contrast to describing poverty as "having too little"), when does one "have too much"? (Peralta, 2011) It was proposed that the point or level when individuals or societies "*have too much*" is approached or describes a situation *(status)*, first of all, *when other individuals and societies have too few resources to live by and, second of all, and when the accumulation of wealth and power undermines the common good or threatens (in*

[7] [Rom. 8:20].

tendency) the global commons. While the poverty line is drawn at the point of personal consumption allowing for the satisfaction of basic needs, the *greed line could be drawn at "the highest point of personal consumption which can be obtained without negatively affecting the welfare of society and that of future generations* (Larrea, 2011)." Greed lines are the levels of resource consumption, money accumulation or power seizure over which societal or individual behaviours may harm human well-being and the integrity of Creation. These negative effects of behaviours transgressing greed lines could be expressed in term of relative poverty, or socio-economic injustice, or offences to human feelings, and/or environmental destruction.

e. Towards a multidimensional structural greed index (MSGI)

The WCC group began to explore the possibility of identifying *multi-dimensional indicators of greed at the structural level which could be further developed into a structural greed index*. The indicators could have as its basis people's economic, social and cultural rights enshrined in the United Nations human rights conventions, which essentially define the protective limits for maintaining human life and promoting human development[8]. As with the MGI proposed by Michael Taylor (2011), the indicators ought to be simple and manageable enough *(amounting perhaps to not more than 15)*, so as to be able to effectively communicate a message to a targeted audience of churches, policymakers, business establishments and citizens. Aside from ***raising awareness*** among the general public, the indicators are envisioned to eventually lead to the development and implementation of policies and measures ***(decision-making)*** to avert structural greed.

4. Measuring a Multidimensional Structural Greed Index (MSGI) using Fuzzy Logic Evaluation

a. Why a fuzzy logic approach for greed measurements?

The following two basic features justify the use of fuzzy logic reasoning. ***(a)*** Fuzzy logic has the ability to deal with ***complex and polymorphous concepts***, which are not amenable to a straightforward quantification and contain ambiguities. In addition, reasoning with such ambiguous concepts may not be clear and obvious, but rather fuzzy. ***(b)*** Fuzzy logic provides the ***mathematical tools to handle ambiguous concepts and reasoning, and finally gives concrete answers ("crisp" as they are called) to***

[8] UNDP, Human Development Report, 2011. http://hdr.undp.org/en/

problems fraught with subjectivity. Greed is, indeed, quite subjective. What appears greedier for an environmentalist may be less greedy for an economist, and the ingredients signifying greed may differ for these specialists. Another important aspect of fuzzy logic is that it uses linguistic variables, thus performing computation with words. If a traditional mathematical approach towards greed assessment were adopted, such as cost-benefit analysis or algebraic formulas, then certain factors which are impossible to quantify would be left out. There exist, however, aspects of greed which cannot be quantified, and yet are very important as, for example, moral values and opinions. In this area of human thought fuzzy logic performs successfully[9].

Fuzzy logic is a scientific tool that permits ***simulation of the dynamics of a system without a detailed mathematical description.*** Knowledge is represented by IF-THEN linguistic rules, which describe the logical evolution of the system according to the linguistic values of its principal characters that we call linguistic variables. Real values are transformed into linguistic values by an operation called "fuzzification", and then fuzzy reasoning is applied in the form of IF-THEN rules. A final crisp value is obtained by "defuzzification", which does the opposite to fuzzification. A simple example of IF-THEN fuzzy approximate reasoning is the assessment of human happiness based on popular feeling about the importance of health. Choosing money and health as the principal factors of happiness, the fuzzy rules might be:

- ***IF*** one has "***much***" money ***AND*** "***good***" health, ***THEN*** he is "***very***" happy,
- ***IF*** one has "***much***" money ***AND*** "***bad***" health, ***THEN*** he is "***insufficiently***" happy, and
- ***IF*** one has "***little***" money ***AND*** "good" health, ***THEN*** he is "***satisfactorily***" happy.

"Much" and "little" are linguistic values of the linguistic variable money; they correspond to the fuzzification of a fixed amount of money. (Good, bad), and (very good, satisfactorily, insufficiently) are, respectively, linguistic values of the state of health and happiness.

b. Linguistic variables and dimensions of structural greed

Briefly, a linguistic variable is defined by four items: (a) The name of the variable (e.g. income), (b) Its linguistic values (e.g. "*very little*", "*little*", "*sufficient*", "*much*" and *very much*"), (c) The membership functions of the linguistic values, and (d) The physical domain over which the variable

[9] (Zimmermann, 1991; Zadeh, 1994).

takes its quantitative values. The membership function of a linguistic value gives the degree to which any quantitative value belongs to the linguistic value. For example, the membership functions of "much" and "little" could be exponential functions of the amount of income per month in dollars, and the range of income is the physical domain of the variable. Our assessment of greed is based on Christian teaching, which sheds light on the Trinitarian nature of human being (Andrianos, 2011). Greed is the desire for the fulfillment of the threefold needs of a human being: material (ecology), mental (economy) and spiritual (socio-political) needs. Therefore, greed assessment should have three simultaneous targets:

- Ecological sustainability: Measure of natural resources consumption in supporting human needs and economical growth;
- Economic financial performance: Quantification of money accumulation which is an assessment of mental achievement of a human being to secure standards of living; and
- Socio-political justice: Evaluation of power inequality and ethical implications for the improvement of human happiness and survival now and for the generations to come.

The Multidimensional Structural Greed index (MSGI) of the system whose greed level we are asked to appraise has three major dimensions: environmental or ENV-GI (ecological sustainability), monetary or MON-GI (economic financial) and socio-political or POW-GI (power inequality) [Fig.2].

Analytically, the multidimensional structural greed index (MSGI) of a country/individual is a combination of three primary components of structural greed:

- **Environmental component (ENV-GI)**, referred as the ecological sustainability greed index,
- **Monetary component (MON-GI)**, measured as the economic financial greed index, and
- **Power greed component (POW-GI)** that is the socio-political greed index.

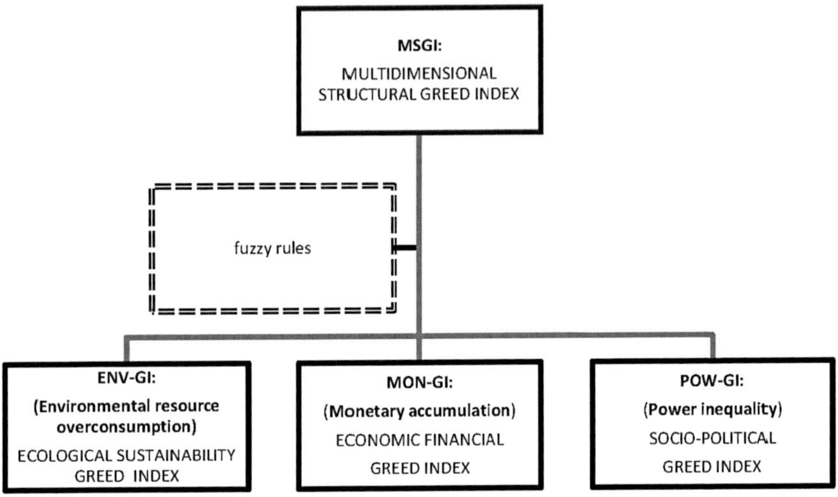

Figure 2: Linguistic variables for the multidimensional structural greed index measurement

The physical dimensions of the three primary greed indexes comprise five secondary structural greed indexes, which are:

- **Ecological sustainability greed index (ESUS-GI),**
- **Financial greed index (FINA-GI),**
- **Economic greed index (ECON-GI),**
- **Social greed index (SOCI-GI), and**
- **Political greed index (POLI-GI).**

Each secondary greed index is then assessed using the "Status-Desire/Trends" approach, which assumes that greed is computed by the assessment of the current achievement (status) of accumulation or consumption of goods, money or power and the "desire" to increase or to reinforce that situation (desire/trends). Therefore the five secondary greed indexes are the result of the combination of ***eleven tertiary variables*** of structural greed, called "***greed indicators***".

- **Ecological sustainability greed index (ESUS-GI)** comprises only one greed indicator:
- (1) The ***global ecological footprint (on national or individual level)***, because it is an evaluation of both status and desire aspects of resource use including "land use", "biodiversity use", "water use", "energy use" and CO2 emissions;

- **Financial greed index (FINA-GI)** comprises two tertiary components or financial greed indicators, which are:
  - (2) The ***bank assets ratio*** (national level of financial assets share) or ***personal financial assets*** (on institutional or individual level) as status indicator for money accumulation;
  - (3) The ***country real interest rate*** (national level) or ***financial interests rates*** (on institutional or individual level) as desire/trends indicator for money speculation greed;
- **Economic greed index (ECON-GI)** comprises also two tertiary components or economic greed indicators, which are:
  - (4) The ***PPP GNI*** (purchase per parity gross national income) or ***annual revenue*** (on institutional or individual level) as status indicator for wealth sustainability;
  - (5) The ***governmental debt as percentage of GDP*** (national level) or ***households debts*** (on institutional or individual level) as desire/trends indicator for wealth production vs. consumption;
- **Social greed index (SOCI-GI)** comprises four tertiary components or social greed indicators, which are:
  - (6) The ***poverty ratio*** (national headcount percentage) or ***living standard*** (on institutional or individual level) as status indicator for social greed with respect to human rights;
  - (7) The ***top 10% of national income*** (national level) or ***social class*** (on institutional or individual level) as desire/trends indicator for socio-economic inequality;
  - (8) The ***child mortality rate*** (national level) or life expectancy ***health standard*** (on institutional or individual level) as status indicator for healthcare and respect for human rights;
  - (9) The ***years of schooling*** indicator (national level) or ***education level*** (on institutional or individual level) as desire/trends indicator for social solidarity.
- **Political greed index (POLI-GI)** comprises two tertiary components or political greed indicators, which are:
  - (10) The ***corruption perception index*** (national level) or ***morality standard*** (on institutional or individual level) as status indicator for global ethic;
  - (11) The ***civil liberties indicator*** (national level) or ***personal freedoms*** (on institutional or individual level) as desire/trends indicator for power seizure and dignity inequality;

To build the fuzzy rules within the GLIMS model, membership functions, greed line functions and greed index function should also be attributed to all greed variables of primary, secondary and tertiary greed indexes.

c. Greed line functions and fuzzy rules

The greed line function could be a decreasing or increasing function of the values of indicators. It might be a decreasing function of environmental protection, incidence of democracy and morality and might be an increasing function of the ecological footprint, resource consumption, income per capita, poverty, inequalities and disrespects for human rights.

The knowledge ruling the computation of MSGI of any system is represented by fuzzy rules whose general form is: "***IF*** (PREMISE) ***THEN*** (CONCLUSION)".

The rules are expressions of the role of interdependencies among various dimensions of greed. They are combinations of IF-THEN rules operating on rule bases derived from expert knowledge of the system integrity. By their nature, such functions are highly non-linear. The term 'integrity' is defined as the degree to which each greed variable fulfills criteria of greed lines. Criteria of greed lines are recommended critical targets that each greed indicator should pass to reach a greedy status. These rules are the results of multidisciplinary analysis about greed and its polymorphous effects. Economists, ecologists, theologians and other experts agree that the three components of greed should be given identical weight in an overall measurement[10]. Knowledge acquisition methodologies, such as interviews or questionnaires, can also be used to build the rules[11].

5. Empirical Results on Individual Levels: How Greedy is One Person?

Illustrative explanation of the set of structural greed indicators at individual and institutional (churches and business) level are summarized in the following table 3.

---

[10] (IUCN / IDRC, 1995).

[11] (Zadeh, 1973; Ericsson and Simon, 1984).

Table 3: Explanations for the greed indicators at individual level.

Dimensions of greed	Greed indicators	Personal /institutional data values					
		Legitimate target	Greed line interval		Greed line unit	Indicator definition and link to greed	Data sources
ENVIRONMENTAL	1-PERSONAL ECOLOGICAL FOOTPRINT	Lower better; moderation between min=0.4 and max=10.7	0.9	1.1	1= one planet Earth; World average is 3.1 hectares per person (2011)	The ecological footprint is the amount of biologically productive land and sea area necessary to supply the resources person /institution consumes, and to assimilate associated waste. Red light when ratio is > 1 Planet Earth.	Personal ecological footprint can be calculated at http://www.footprintnetwork.org

FINANCIAL	2-FINANCIAL ASSETS	Lower better	80	100	Percentage of revenue	Financial assets are the amount of money that an individual or an institutional has in the financial market. It is valued in comparison to the total assets or to the total revenue (%). High level of shareholders corresponds to a big amount of money accumulation which is the result of higher greed for money. Red light when ratio is > 100%.	Questionnaire.

	3-FINANCIAL INTEREST RATE	Lower values mean less greedy institution or individual	0	5	Percentage	Depending on the desire for making more money in a short time, a higher interest rate is considered to be result of greedier person or an institution. A positive interest rate is a form of greed so the greed line could be set at 5% (world average interest rate) Red light when ratio is > 5%	Questionnaire

ECONOMIC	4-ANNUAL INCOME	Higher values express higher greediness for wealth sustainability (between min=530$ and max=37910)	30000	40000	Money in US .$ based on purchasing power parity (PPP)/ capita)	Personal revenue is income converted to international dollars using purchasing power parity rates. An international dollar has the same purchasing power over GNI as a U.S. dollar has in the United States. Total revenue is the value of all final goods and services produced by a person or an institution Red light when ratio is > 40000 US .$ /capita/year	Questionnaire

	5- HOUSEHOLDS DEBT	Lower values mean less greedy institution or individual	80	100	Percentage of revenue (installment vs. income ratio).	Debt includes domestic and foreign liabilities such as currency and money deposits, securities other than shares, and loans. Because of imbalanced behavior (production vs. consumption), the limit over which a person or an institution is considered as greedy (prone to crisis) can be set to 100% Red light when ratio is > 100%	Questionnaire

SOCIAL	6-LIVING STANDARD	Higher better but moderation (close to average) is optimum;	0,6	0,8	In a scale of zero to 1	Access to adequate living standard is a fundamental need for human rights but luxury living is offensive and becomes a form of greed when in excess compared to the average standard. Red light when ratio is >0,8	Questionnaire
	7-SOCIAL CLASS	Higher better but moderation (close to average) is optimum;	0,6	0,8	In a scale of zero to 1	Adequate socio-economic standard is a fundamental need for human rights but excessively high level of wealth is a form of greed with respect to inequality when compared to the average standard. Red light when ratio is >0,8	Questionnaire

	8-SHOOLING YEARS	Higher better but moderation (close to average) is optimum;	0,6	0,8	In a scale of zero to 1	Education a fundamental need for human rights but it becomes a form of greed (solidarity) when in excess compared to the average standard. Red light when ratio is >0,8	Questionnaire
	9-HEALTH STANDARD	Higher better but moderation (close to average) is optimum;	0,6	0,8	In a scale of zero to 1	Adequate health standard is a fundamental need for the fulfillment of human rights but it becomes a form of greed (dignity) when in excess compared to the average standard. Red light when ratio is >0,8	Questionnaire

POLITICAL	10-PERSONAL MORALITY	Lower better. Higher morality values are for lower level of greed	0	5	In a scale of -10 to +10	Morality index intends to provide a time-base measure of the moral behavior of individual. The red light is 0, under which the bad behavior is threatening human life and peace. Righteousness is toward the top and evil is toward the bottom. So upward trends are good, and downward trends are bad. Red light when ratio is <0	*http://www.moralityindex.com/* Questionnaire

	11-PERSONAL FREEDOM	Higher better but moderation (close to average) is optimum;	0,6	0,8	In a scale of zero to 1	Adequate health standard is a fundamental need for the fulfillment of human right but it becomes a form of greed (human rights) when in excess compared to the average standard. Red light when ratio is >0,8	Questionnaire

a. Table of greed indexes measurements for selected individuals (5 persons)

To evaluate individual greed indexes, we apply the GLIMS model to five individuals (two men, two women and one youngster), which could be representative of individual behaviour trends. The results are compiled in the following table (Table 4) and charts:

Table 4: Values of greed indexes measurements for selected individuals (2012 survey)

INDIVIDUAL GREED INDEXES	Person 1 N (Male)	Person 2 L (Male)	Person 3 A (Female)	Person 4 G (Female)	Person 5 S (Young)
ESUS-GI	40.26	41.28	55.29	50.41	44.28
ENV-GI	**40.37 (E)**	**41.53 (E)**	**54.90 (G)**	**50.36 (G)**	**44.77 (E)**
FINA-GI	20.37	22.80	56.91	20.37	20.37
ECON-GI	46.20	50.41	41.69	36.93	26.05
MON-GI	**38.31 (F)**	**40.53 (E)**	**49.79 (E)**	**35.48 (F)**	**32.19 (F)**
SOCI-GI	43.17	53.74	55.82	44.31	43..40
POLI-GI	37.29	32.08	59.21	37.28	28.63
POW-GI	**44.33 (E)**	**45.43 (E)**	**51.88 (G)**	**44.69 (E)**	**41.04 (E)**
Overall MSGI score	**45.50 (E)**	**46.23 (E)**	**50.37 (G)**	**46.73 (E)**	**44.94 (E)**

Linguistic values for greed indexes: L = Low; F = Fair; E = Enough; G = Greedy; VG = Very Greedy; EG = Excessively greedy

In a scale of 0 to 100, the discrete values corresponding to linguistic scores for "Greed indexes" are the same as for the national level, as follows:

- L = LOW for $0 < MSGI < 20$
- F = FAIR for $20 \leq MSGI < 40$
- E = ENOUGH for $40 \leq MSGI < 50$
- G = GREEDY for $50 \leq MSGI < 60$
- VG = VERY GREEDY for $60 \leq MSGI < 80$
- EG = EXCESSIVELY GREEDY for $80 \leq MSGI \leq 100$

One individual is considered as greedy with respect to a specific aspect of greed when its greed index score is greater than 50. The greed line for MSGI measurements is then [50-60] and the red light is 40 (enough level). Data for the whole set of greed indicators for each person are obtained via surveys (Annex 2) that are based on questions as follows (Box 1: Personal Greed Index Sample Questions).

Box 1. Personal Greed Index: 10 Sample Questions

1. How big is your ecological footprint (to calculate click on http://www.footprintnetwork.org/en/index.php/GFN/page/calculators/)?
2. Do you have stock shares or savings capital in the bank? If yes, how much?
3. How much is your annual income?
4. How much credit card debt do you have? How much is your monthly installment vs. your monthly income?
5. How much do you spend on luxury goods in a year? How many cars do you have? How big is your house?
6. Would you consider yourself to be part of the upper, middle or lower income class? Do you aspire to a higher social class? If so, which class?
7. What is the average life expectancy in your family? How much do you spend on private healthcare per year?
8. What level of education do you have (in years of schooling)? Did you go to a public or private school?
9. How much tolerance do you have for corruption?
10. How much do you value money and power over friendships, ecological harmony, and wisdom?

For the case of person 1, it is shown that the environmental, economic and social greed levels need corrections for they have passed the red light level of "enough" [40]. Individual 1 should give first priority to controlling economic greed, then to solving social and environmental greed as next priorities.

6. Conclusions

Church leaders, policy makers and any believers would need a scientific tool to clarify the effects of greed and establish policies for an economy of life with more justice and sustainability.

We present a new model for greed assessment, called "Greed Lines and Indexes Measurements System" or GLIMS, in an attempt to provide an explicit and comprehensive description of the concept of structural greed. Using linguistic variables and fuzzy linguistic rules, the model gives quantitative measures of several greed indexes, which are then combined into an overall multidimensional structural greed index or MSGI. The model allows the measurement of greed indexes at national, corporate and individual level. Therefore it is helpful for anti-greed policy-making and ethical behaviour reflections.

A sensitivity analysis of the model permits to determine the evolution of greed variables subject to perturbations in the values of greed indicators. Then, the problem of overcoming greed in policy-making becomes one of specifying priorities among critical greed indicators and designing appropriate policies that will guarantee more justice and development.

To overcome greed, recommendations differ from economy to economy and corrections from individual to individual. More developed countries need to focus mostly on the effect of their environmental greed whereas less developed countries should strive to correct both the monetary and socio-political system.

The greed index approach using fuzzy evaluation provides new insights of tackling greed at its roots. It may serve as a practical tool for decision-making and policy design at individual, communal or national levels. Such approaches are urgently needed nowadays if we want to attack the problems of justice, peace and sustainability systematically.

References

Andrianos, L. (2012), "Setting the greed line using biblical insights and sustainability ethics in order to avert ecological crisis". In *Ecological Theology and Environmental Ethics* Vol. 2 (ECOTHEE-11) OAC Publications, 2012. ISBN 978-960-86383-7-2.

—. (2011), Environmental ethics and sustainable development: a fuzzy approach. Latvian Christian Academy publication.

—. (2011), Structural Greed and Creation: A Theological Reflection. The Ecumenical Review, 63: 312–329.

Andriantiatsaholiniaina L. A., 2001. Sustainability Assessment using Fuzzy Evaluation. Technical University of Crete, Chania, Ph.D. Thesis.

Andriantiatsaholiniaina L. A., Kouikoglou V. S., and Phillis Y. A., (2004). "Evaluating strategies for sustainable development: Fuzzy logic reasoning and sensitivity analysis," Ecological Economics, 48(2), 149-172.

Global Footprint Network (http://www.footprintnetwork.org/en/index.php/GFN/page/footprint_for_nations)

Goudzwaard, Bob (2011, unpublished), "Problems and Possibilities of Indicating Poverty, Wealth and Greed," paper presented at the 2nd meeting of the Greed Line Study Group, March 2011, Crete

Larrea, Carlos (2011), "Inequality, Sustainability and the Greed Line: A Conceptual and Empirical Approach," Ecumenical Review Vol. 63, Issue 3.

Mshana, Rogate (2007), Poverty, Wealth and Ecology: The Impact of Economic Globalisation, WCC: Geneva, retrieved from http://www.oikoumene.org/fileadmin/files/wcc-main/documents/p3/poverty_24p.pdf

Mshana, Rogate, ed. (2009), Poverty, Wealth and Ecology: Ecumenical Perspectives from Latin America and Caribbean, WCC: Geneva.

Mshana, Rogate, ed. (2012), Linking Poverty, Wealth and Ecology in Africa, WCC: Geneva.

Peralta, Athena, ed. (2010), Poverty, Wealth and Ecology in Asia and the Pacific: Ecumenical Perspectives, WCC, CCA and PCC: Geneva, Chiang Mai and Suva, retrieved from http://www.cca.org.hk/resource/books/olbooks/poverty_wealth_and_ecology_in_ap.pdf.

Phillis, Y.A. and Andriantiatsaholiniaina, L.A., 2001. Sustainability: an ill-defined concept and its assessment using fuzzy logic. Ecol. Econ., 37: 435-456.

Raiser, Konrad (2011), "Theological and Ethical Considerations regarding Wealth and the Call for Establishing a Greed Line," Ecumenical Review Vol. 63, Issue 3.

Taylor, Michael (2011), "On Greed: Toward Concrete and Contemporary Guidance for Christians," Ecumenical Review Vol. 63, Issue 3

The MathWorks Inc., 2012b. http://www.mathworks.com.

UNDP, Human Development Report, 2011. http://hdr.undp.org/en/

World Council of Churches, http://www.oikoumene.org (visited February 2013)

http://www.footprintnetwork.org/en/ (visited February 2013)

Zadeh, L.A., 1994. Fuzzy logic, neural networks and soft computing. Commun. ACM, 37(3): 77- 84.

Zimmermann, H.J., 1991. Fuzzy Set Theory and Its Applications. Kluwer, Boston.

CLIMATE CHANGE, ECO-JUSTICE AND HEALTH

ECOJUSTICE: VISIONING THE NEXT "NEW NORMAL"

JOY KENNEDY

Former Poverty, Wealth & Ecological Justice Coordinator,
The United Church of Canada
E-Mail: kennedyjoy1@gmail.com

The following text is based on a powerpoint presentation with illustrations.

Abstract

All life on this small planet is interdependent and interconnected. Air, water, land and energy sources provide the necessary environment to sustain life. But the structures created by humans for living within the carrying capacity of Earth's ecosystems are stretched and fragile. A renewed theology of Earth as God's sanctuary and our Home helps to break the vicious grip of the prevailing paradigm of unfettered growth and economic globalization. We can learn to move into a new way of being Earth Community, living out God's mission, being the change we wish to see. We know that poverty, exclusion, wealth concentration, financial crises, climate change, pollution and ecological destruction require moral, ethical and spiritual solutions. What is "normal"? We live in a socially constructed world that interacts with the natural environment. Human systems and norms are constructed on underlying belief systems, or ideologies, with values at the core. Systems created by humans can be and constantly are re-created by humans. We envision Ecojustice as a higher principle and claim an ancient vision of the "new Earth." New choices must be made – choices to uphold sustainable communities and responsible growth that recognizes human responsibility for Creation and for future generations. The gospel is about the coming of the new age, "on Earth as in heaven". Resurrection is the hope and invitation to transform global society to live in unity, solidarity and sustainability; with equity, harmony, compassion, and love.

Then I saw a new heaven and a new Earth... Behold, I make all things new. (Rev. 21:1,5)
In and with God
We can direct our lives toward right relationship
with each other and with God.
We can discover our place as one strand
in the web of life.
(*A Song of Faith*, The United Church of Canada)

1. We Are a Part of the One Earth Community

How do we SEE the face of the Earth? of God? and of each other? Turn! Look to your neighbour and see their face! S/he is the image of God. We begin by acknowledging that God is present in our midst – we are not alone. We can also glimpse the renewed face of the Earth - in beautiful nature - in windmills, solar panels - and green pastures and still waters – in a restored Creation, the gift of a loving Creator.

1.1 The problem and the realization

Life on the planet is indisputably faced with an imminent danger. An impending threat is the real and rapid onset of climate change, with devastating consequences and suffering for the most vulnerable – and eventually all. The issue of climate change is not simply a political, social, economic or environmental issue. At the root it is the profound moral issue of who we think/believe we are, and how we act in relation to the planet – our home, the Creation, and to the Creator.

1.2 Reading the Signs of the Times

(My favourite theologians – Calvin and Hobbes)

1.3 More Signs

1.4 Seeking a New Narrative: Singing a Song of Faith on a Troubled Planet

As we truly believe, so we act. If we believe in our hearts that the planet Earth is just a resource bank, "natural capital", there for all-powerful humans to exploit, excavate, extract, and dump, we will treat it that way and make a heavy footprint.
If we believe in our hearts that we are a part of the sacred Creation and depend on its wondrous gifts for life, and that human activity requires respect and responsibility, we will act differently. We will love and protect, value and serve our home, the *oikos*.

1.5 What is "normal" and what is held as the highest value?

What do humans value most - and worship as god? the market, money? or community and relationships, and a sustainable home? or God the Creator of all?

Wealth and power get in the way of spiritual consciousness. People in affluent and technologically-dependent societies seem to believe that they are above Creation, can overcome the laws of Creation, and those powers have been allowed to become the norm.

Many prominent economists acknowledge that the dominant operative economic principle is greed and self-interest and the dominant behavioural norms reflect that.

But now for our survival as a human species the principles of equity, equality, fairness and justice need to be worked out among all peoples. We need a new ethic and a new understanding that we are beings made from dust, from Earth (whence we will return); we are water, we are air, we are fire/energy. The understanding of who we are as natural beings needs to be restored.

1.6 *All Creation is groaning, waiting for the revealing of the children of God.* (Romans 8:19)

In the season of Advent Christians wait with expectation for new birth. We long for the day of celebration, as we will enter into a new Season of Creation, when bells will ring out "Rejoice!"

The Church continues to participate in global processes and is present at all United Nations climate conferences, as a community of faith, with many other people of faith, in the midst of the assembly of the global community. With faith in something bigger than history, bigger than all of us, bigger than fear or despair, and faith in the future and our ability to change our behaviour, with a call to transform our global society, to live in unity, solidarity and sustainability; with equity, harmony, compassion, and love.

2. A Theology of Home

Earth is more than a ball of water and dirt
and air spinning through space.
Earth is more than a speck of stardust
that spawned life.
Earth is God's sanctuary.
Earth is a sacred site in the cosmos.
(Norman Habel, Australian theologian)

But we must acknowledge that the Earth Community is in Crisis!

The Economic World is in crisis; the Environment is in crisis; the Global Society is in crisis. Symptoms include: war / terrorism, insecurity, poverty, flood, drought, disease, lack of access to potable water, climate change, etc., add your own…

Ecological conflicts abound. Where is Climate Justice? Where will Pacific Islanders go when the sea level rises? Which country will take millions of climate refugees? What happens when the waters dry up forever? When the glaciers are no more?

2.1 How did we get here? Whose fault is this?

Empires of all kinds conquered for access to resources – water, land, minerals, labour.

Colonization – was about control of resource extraction.

Globalization – is also about control and manipulation of wealth through resource exploitation and financial speculation.

Modern empire is about controlling the wealth of the Earth by and for a few. *And* Industrialization required: combustible materials, i.e. fossil fuels; raw materials; abundant water sources; cheap labour; licence to pollute.

Along with its consumable products it produced emissions of greenhouse gases (GHGs) like CO2 which were released and accumulate in the atmosphere, causing climate change.

2.2 Prognosis?

"How can we speak of the urgency of global crises without inducing either passivity and hopelessness or unrealistic escapism?" [Reuther (1992)] While the protection of the *oikos* – ecosphere, atmosphere, biosphere, hydrosphere, troposphere and sociosphere – from irreversible deterioration should be the overriding mission of the planet's inhabitants, the political willpower to act with one heart and mind to ensure the Earth Community's very survival continues to elude the nations of the world.

The present world is ruled by the "Domination System" within which Christians are to work to prepare the conditions for God's Domination-free Order. (1 John 2:15-17). Those liberated from the tyranny of the old order receive a new, holy spirit – "not the spirit of the Domination System (*kosmos*), but the Spirit that is from God. (1 Cor. 2:12): "the Spirit of truth, which the Domination System is not able to receive, because it can neither recognize it nor comprehend it. You know it, because it is already in your midst, and will be inside your very beings. (John 14:17) This new Spirit resocializes believers into the New Reality." [Wink. (1992)]

2.3 EcoTheology, EcoSpirituality and Resolutionary Theology

We need to: re-interpret, re-articulate, enliven our living out of new resolutions; lift up those involved in this living out; re-energize ourselves, communities and churches and resist in the face of imperial "truths" such as "we can't do anything" and "there are no alternatives".

2.4 Is Another World Possible? There is no Planet B!

Do our values make us insatiable?

> "We are consumed by aliens, alien values and alien loyalties, and we eat our way into exile." (cf. Hosea 7:9 – also Leviticus 26:26, Hosea 4:10) [Brueggemann. (1977)]

2.5 The Economic Choices Made ™ ™ ™

™ The errant belief in the capacity of "the market" to solve the crisis of justice and sustainability has resulted in only deepening the problems of the majority of the world.

™ An ideological confrontation has emerged as intolerance; insecurity and fear are now being exploited by a new kind of imperialism, an anti-terrorism front which targets not only terrorists but any who oppose the enforcement of the "total market" economy.

™ The imposition of monolithic financial and trade regimes excludes millions from the world economy. Those victimized and those acting in solidarity are resisting injustice and inequality, for generations to come.

2.6 Systems Created by Humans can be Re-created by Humans

Possibilities - based on New Norms & New Worldviews:

- New Social Covenants
- Economic Cooperatives
- Fair Trade
- Socially Responsible Investment
- Millennium Development Goals
- Payment of Ecological & Climate Debt
- Green Energy
- Keep the Oil in the Soil
- Add your own…

2.7 *The Accra Confession: Covenanting for Justice in the Economy and the Earth* (2004)

States: (my emphasis)

> We believe that God is sovereign over all creation. "The earth is the Lord's and the fullness thereof." (Ps 24.1) Therefore, **we reject the current world economic order imposed by global neoliberal capitalism** and any other economic system, including planned economies, which **defy God's covenant by excluding the poor, the vulnerable and the whole of creation from the fullness of life**. We **reject any claim of economic, political and military empire** which subverts God's sovereignty over life and acts contrary to God's just rule.
>
> We believe that God has made a **covenant with all of creation** (Gen 9.8-12). God has brought into being an earth community based on the vision of justice and peace. The covenant is a gift of grace that is not for sale in the market place (Is 55.1). It is an **economy of grace for the household of all of creation**. Jesus shows that this is an inclusive covenant in which the

poor and marginalized are preferential partners and calls us to put justice for the "least of these" (Mt 25.40) at the centre of the community of life. All creation is blessed and included in this covenant (Hos 2.18ff).

Therefore we **reject the culture of rampant consumerism and the competitive greed and selfishness of the neoliberal global market system** or any other system which claims there is no alternative.

We believe that any **economy of the household of life** given to us by God's covenant to sustain life is accountable to God. We believe the **economy exists to serve** the dignity and wellbeing of people in community, within the bounds of the sustainability of creation. We believe that human beings are called to **choose God over Mammon** and that **confessing our faith is an act of obedience**. Therefore we **reject the unregulated accumulation of wealth and limitless growth** that has already cost the lives of millions and destroyed much of God's creation.
We believe that God is a God of justice. In a world of corruption, exploitation and greed, God is in a special way the God of the destitute, the poor, the exploited, the wronged and the abused (Ps 146.7-9). **God calls for just relationships with all creation.**

Therefore we **reject any ideology or economic regime that puts profits before people, does not care for all creation and privatizes those gifts of God meant for all**. We reject any teaching which justifies those who support, or fail to resist, such an ideology in the name of the gospel.

2.8 We Must Create and Proclaim Alternatives

The community of faith believes that we need not despair – that we can turn the world around, that there are always alternatives and paths to justice. People everywhere demonstrate that in their daily lives and speak out for a new way. Living faithfully in the midst of empire is essential to who we are as Church.

3. A Theology of Home

O God, our help in ages past, Our hope for years to come,
Our shelter from the stormy blast, And our Eternal Home.
(Traditional Hymn)

O beautiful Gaia; O Gaia calling us Home.
O beautiful Gaia, Calling us Home.
(United Church of Canada Hymnal)

Each culture, tradition, family and individual has a different interpretation and experience of "home". Think about what you believe or feel or imagine as "home". What does that indicate about your relationship to it and how you would name it? How might that contribute to how you live in it? How might that influence how you understand Creation and the Earth in particular?

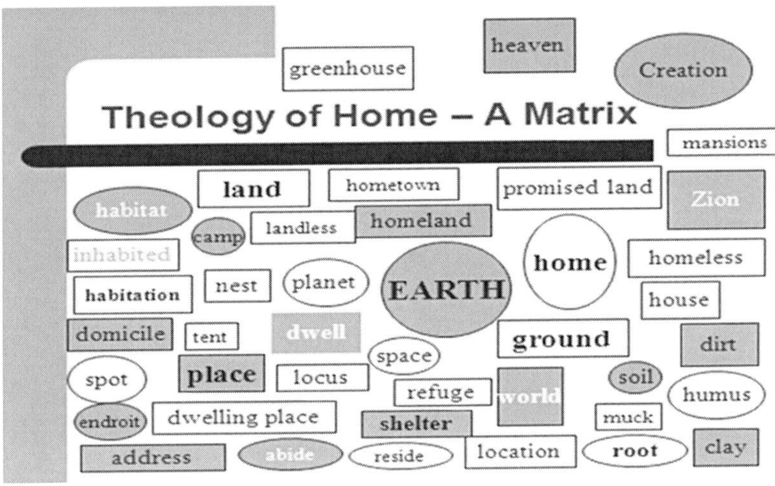

3.1 Are we "at home"?

Martin Buber projected "…epochs of *Behausung* in which man feels at home in the universe, and epochs of *Hauslosigkeit* in which man feels homeless, marooned in the universe, and inclined to regard himself as problematical."…Either condition is provisional and always on the way to the other. That constitutes the central issue of biblical faith and perhaps in our own societal experience. [Brueggemann]

If this is true, where are we as a global people today?

3.2 Theology of Place and of Land

"Land is never simply physical dirt but is always physical dirt freighted with social meanings derived from historical experience. The yearning for land is always a serious historical enterprise concerned with historical power and belonging….Our humanness is always about historical placement in the Earth, but that historical placement always includes excess meanings both rooted in and moving beyond literalism….Land is a central, if not the central theme of biblical faith. Biblical faith is a pursuit of historical belonging that includes a sense of destiny derived from such belonging." [Brueggemann]

3.3 Theology of Whole Earth Justice

I came that they may have LIFE in abundance - not wealth, or stuff or labour saving devices – but *LIFE* [Jesus)] which includes clean air to breathe, water to drink, food, and sustainable community.
Salvation = "to be a saving health to all the nations" = Proclaim God's Jubilee

- Good news to the poor
- Release to the captives
- Recovery of sight to the blind
- Freedom for the oppressed

3.4 Just Finance and the Economy of Life [World Council of Churches (2009)]

- The era of financial globalization and economic expansion is increasingly driven by GREED.
- We need a new paradigm of economic development and re-conceptualization of wealth based on relationships, care and compassion, solidarity,
- and love, aesthetics, and the ethics of life, participation and celebration, cultural diversity and community vitality,
- and including responsible growth that recognizes human responsibility for creation and for future generations – an economy glorifying life.

3.5 Eco-justice and Ecological Debt [World Council of Churches (2009)]

- Ecological debt is a spiritual issue
- It is accumulated damage caused over time to ecosystems, places and people, through production and consumption patterns; and the exploitation of ecosystems at the expense of equitable rights of other countries, communities or individuals.
- It is also the debt owed by economically and politically powerful national elites to marginalized citizens; the debt owed by current generations of humanity to future generations, and on a more cosmic

scale, the debt owed by humankind to other life forms and the planet. It includes social damages such as the disintegration of indigenous and other communities.

- Ecological debt, including climate debt, disproportionately owed by countries of the global North to the global South, and to future generations, due to historic and current resource plundering, environmental degradation, and appropriation of common ecological space to dump toxic emissions, is owed by all of us to the Earth itself.

3.6 Climate Debt / Climate Equity / Climate Justice

The primary responsibility for climate change lies with the industrialized North and the greater suffering caused to the South in the form of: displacement of peoples; loss of sources of livelihood; food insecurity; reduced access to water; forced migration; ecological degradation.

Churches have been part of ecological destruction and advancing climate change through their own consumption patterns, and through perpetuating a theology of human rule over nature. Now is the time for *metanoia.*

Churches need to deepen their climate justice campaigns to include climate debt and climate equity and advocate for their repayment by applying the ecological debt framework.

3.7 How will it be repaid / redeemed?

In the light of biblical teaching (cf. Matthew 6:12), we pray for repentance and forgiveness, but we also call for the recognition, repayment and restitution of ecological debt in various ways, including non-market ways of compensation and reparation, that go beyond the market's limited ability to measure and distribute.

WCC recognizes the need for a drastic transformation at all levels in life and society in order to end the ecological indebtedness and restore right relationships between peoples and between people and Earth. This warrants a re-ordering of the economic paradigms from consumerist, exploitative models to models that are respectful of localized economies, indigenous cultures and spiritualities, Earth's reproductive limits, as well as the right of other life forms to blossom beginning with the recognition of ecological debt.

4. Whole Earth Justice

Members of the body of Christ in this world have been trying to faithfully interpret the signs of the times and speak to the world about God's love, Christ's redemption, and the covenantal call to Ecological Justice and Whole Earth Justice. Our great work is now to create a new normal for EcoJustice to thrive.

4.1 Climate Change and the expression of the deepest longing of our hearts

The people are waking up! Huge demonstrations/manifestations around the world call for climate justice.

- Candlelight vigils prayerfully ask for a "real deal" – a transformation from business-as-usual to a global Fair Ambitious and Binding agreement.
- In 2009, bells rang out around the whole world, each rung 350 times calling for a return to 350 parts per million of carbon (CO2e) concentration in the atmosphere.
- On 10/10/10 – there were Global Work Parties – in over 140 countries!
- In 2011 and 2012 there were Power Shift mobilizations around the world.

You with eyes to see and ears to hear, the world's citizens are awake and taking action!

4.2 What we believe

In Christ all things are possible. To face climate change we need system change. To stop unstoppable climate change we must unleash an unstoppable movement of people for climate justice around the world that governments and corporations cannot constrain. We need to replace

- Competitive globalization *with* multilateral cooperation,
- Neo-liberal capitalist exploitation *with* sustainability, equality, equity, sufficiency and subsidiarity,
- Global commons commodification *with* conservation of the sacred gift Earth,
- Voluntary global agreements *with* legally binding global treaties,
- Economy based on greed *with* economy of enough.

4.3 Paradigm shift needed

Climate Change is an opportunity for a paradigm shift to sustainable development: transition or leap-frogging to a low carbon pathway. Indicators of emissions reductions will ultimately depend on community action. Advocacy is meaningless without real change at the grassroots.

4.4 Plea Bargaining – with God? – with Mother Earth?

Mother Nature doesn't compromise.

4.5 How to get out of the mess we are in?

The WCC statements address the ecological issues of economic relationships. Jesus warned "You cannot serve both God and wealth". (Luke 16:13)

- "Economics", like "ecology", and "ecumenism", comes from the Greek *oikos*, and is about global housekeeping.
- We need some global house*cleaning*.
- The global financial crisis is an opportunity which is in danger of being missed.
- Originating in the richest parts of the world, it points to the immorality of a system that glorifies money and has a dehumanizing effect by encouraging acquisitive individualism.
- The resulting greed-based culture impoverishes human life, erodes the moral and ecological fabric of human civilization, and intoxicates our psyche with materialism.
- There is growing awareness of our common vulnerability and the limits of our current way of life.
- Those most affected – who we should listen to as modern-day prophets - are women, who bear a disproportionate share of the burden; young

people and children, as doubts are raised and their sense of security for the future is eroded; and those living in poverty, whose suffering deepens.
- The crisis we face is at the same time systemic and moral.

5. The Way Forward

- Solutions to the Climate crisis must take into consideration the poverty crisis.
- Principle of a just and equitable framework is necessary for a lasting global solution.
- Re-order the 'global ecological space' ethically and fairly.
- Create conditions for the energy-deprived world to transition to a low carbon economy and urgent action on de-carbonization in the North and everywhere.
- Challenge consumerism culture: too much material "wealth" is counter-productive to sustainability.

5.1 What is "Normal"?

All human systems are constructed on underlying belief systems, or ideologies, that have values at the core. Beliefs and values change societies – but also change as societies evolve. Behavioural norms based on these beliefs and values also evolve – for good or ill.

5.2 How do we change our ways? A 12-Step program for the planet?

Many of us humans currently have some very bad habits

- dependence on unbridled growth and unfettered finance;
- belief in an ideological theology of "dominance", or human rule over Creation - and others (which we may call "empire");
- repressive and oppressive thinking and global systems that shackle many and need to be overcome;
- addiction to violence and resource extraction, fossil fuels and profligate energy use.

Therefore we need a new ethical imperative and commitment to moral choices on climate change issues that benefit all people and the whole Creation. We need a power analysis, ànd power sharing with a rights-based mandate for inclusiveness. We need mandatory *not* voluntary limits and global rules and new models of inclusive Earth Community.

5.3 Visioning EcoJustice as a higher principle

- envision the New Earth and claim an ancient Biblical vision
- proclaim a new EcoEthic and create a new EcoEthical reflection
- truly live into inclusion and diversity with Interfaith partners
- use thematic lenses:
    - global commons, climate change & energy, ecological debt, Indigenous justice, empire, resource extraction, sustainable economy
    - create sustainable communities
    - practice Whole World Ecumenism and honour Earth Community

5.4 A new global financial architecture for a sustainable planet

That is

- ethical, just, and democratic;
- grounded on a framework of common values: honesty, social justice, human dignity, mutual accountability, ecological sustainability;
- accounts for social and ecological risks in financial and economic calculation
- reconnects finance with the real economy;
- sets clear limits to, as well as penalizes, excessive and irresponsible actions based on greed.

We need a new ethos and culture that

- reflects the values of solidarity, common good and inclusion;
- replaces the anti-values of greed, individualism and exclusion;
- uses new indicators of well-being (other than GDP – Gross Domestic Product);
- e.g. Human Development Index; ecological footprint calculations; Gross National Happiness (GNH)

5.5 New economic systems based on different values

For example: A Gross National Happiness Index would hold as the highest priority, and reflect in practical terms, indicators such as: quality and pattern of life; good governance (true democracy); education; health; ecological resilience; cultural diversity; community vitality; balanced use of time; psychological and spiritual well-being.

5.6 Re-Drawing the Line(s)

There are current international agreed understandings of the *"Poverty Line"*. The WCC is working to create and propose a *"Greed Line"* with indicators to measure how much is enough, sufficient – or too much. A *"Green Line"* must include no more than 1.5°C mean average global temperature rise and 350 parts per million of CO2e concentration in the atmosphere.

5.7 Follow the Money – Some Solutions

- Admit that the current system is broken, with huge social and ecological costs.
- Revise and create new taxation systems to incentivize green technology and dis-incentivize the fossil fuel industry and keep petroleum in the ground.
- Redistribute wealth – the North pays up its climate debt – a proxy for ecological debt.
- Cancel the illegitimate international debt of poor countries.
- Create new systems of national accounts and indicators.
- End institutional complicity – including our own church complicity – in unjust and unsustainable structures.
- Emphasize participation of women, Indigenous Peoples, differently-abled, youth, poor, previously excluded.
- Promote new values for sustainable economy and Whole Earth Justice.

5.8 What must we do

To restore the sacred balance, repair the breach, and redeem the ecological debt that we humans have built up? The ecumenical community believes that **justice** and **equity** must be at the heart of any sustainable economic, social or environmental system supporting Earth Community. The ecumenical community affirms the cultural and spiritual **values** that must undergird **eco-justice** and **eco-sustainability**.

The ecumenical community says:

□ We believe that the global market economy, which has been dis-embedded from the life and control of nation states, must be re-embedded in sustainable communities, in order to serve the needs of all the people, not satiate the greed of the few, and in order to protect and preserve the assets of the Global Commons.

□ We believe global and national governance structures must be transformed, redrawn and recreated.

□ Our focus on sustainable communities leads us to critique the current trends toward neo-liberal economic globalization, including:

- concentration of power in the hands of a minority
- the widening gap between rich and poor
- devastating impacts of globalization on the most vulnerable, including Indigenous Peoples, women, the aged and children
- regional and global threats to the environment, and
- a weakening of political institutions and their legitimacy at the national and international level.

□ New choices must be made. For the world to be "made new" we should decide what we mean by the "world" - the human social construct of reality? or the Earth itself, our planetary home?

□ Eco-perspectives must be brought to bear on Climate Change.

□**There is *Always* an Alternative.**

5.9 We need Planetary Values

Signposts for Sustainable Communities and Preserving the Global Commons:

Common but Differentiated Responsibilities – realizing historical and capability realities

Diversity – respecting, embracing complementarity of different cultures, genders, species

Equity – sharing benefits and responsibility to preserve the Global Commons for future generations

Full and meaningful participation – acknowledging and making space for all stake-holders, especially those most affected, in decision-making

Reciprocity – correcting and equalizing unfair power imbalances through sharing, or put positively, allowing for power to be shared equitably

Mutual accountability – ensuring peer review, full disclosure, monitoring and verification

Precaution – do no harm

Solidarity – rebuilding relationships and standing with those in particular who have been disempowered, marginalized and made voiceless

Subsidiarity – appropriately assessing roles and responsibilities at the level closest to where they are required, from local to global
Sufficiency – meeting needs before wants and not allowing greed and abuse to outstrip available non-renewable resources

6. What Must Communities of Faith Do for Climate Justice?

- Advocate, educate and demonstrate commitment to belief in *Whole Earth Justice.*
- Help build moral courage to get where we need to go as a global community.
- Repudiate a destructive theology of dominance and embrace one of humility and being an integral part of the Creation.
- Confess, repent and make restitution.
- Green Sacred Spaces; be self-critical; measure contributions to emissions reductions.
- Ensure *All* voices are included in the solutions: participation of women, Indigenous Peoples, differently-abled, youth, poor, marginalized, previously excluded people.
- Be the change you want to see.
- Be the messenger to herald a new world that awaits.
- Model hope.

6.1 What does God require of us now in the *Oikos*?

It is up to us humans to decide. Do we know what we want? Eco Nightmare or Eco dream? What we are prepared to work and struggle and live and die for? Will we respond to the call to truly work together to renew the face of this troubled Earth, this Gaia, this polluted, degraded, denigrated, battered, hurt and abused, but beautiful Home of ours – the only one we have?

6.2 Resurrection

Resurrection is "the gift of power to the powerless and the invitation to the dispossessed to enter into new power, freedom, and life, that is, *"turf"*. The gospel is about the coming of the new age, the new kingdom, the new land. In that there is no anxiety, no sorrow (John 16:20). There is only trust in the promise of a land of rest and joy. But surely such a gift is a scandal!" [Brueggemann. (1977)]

"Our kinship with all earth creatures is global, linking us to the whole living Gaia today. It also spans the ages, linking our material substance with all the beings that have gone before us on earth and even to the dust of exploding stars. [Reuther. (1992)]

6.3 Together We will Make a Difference!

God, our Creator, Redeemer and Sustainer is calling us to a new way of being and living for Whole Earth Justice with EcoJustice as One Earth Community.

Let us mobilize and with the blessing and power of the Holy Spirit all join in the building of a movement for sustainable, loving, compassionate, resilient and beautiful Earth Community as we honour our Creator and all of Creation and we take responsibility for our lives on this fragile Planet Earth – our island home.

References

Presentation based on original PowerPoint presentation to UNFCCC, COP15 (2009) by Kennedy, Joy. (2009), *Renew the Face of the Earth: A Faith-based Approach to Climate Justice,* Downloadable PowerPoint presentation at: http://regserver.unfccc.int/seors/reports/events_list.html?session_id=COP15

Brueggemann, Walter. (1977), *The Land: Place of Gift, Promise, and Challenge in Biblical Faith*, Minneapolis, MN: Fortress Press

Reuther, Rosemary Radford. *Gaia and God; An Ecofeminist Theology of Earth Healing.* (1992), Harper San Francisco: Harper Collins

Schaeder, G. *The Humanism of Martin Buber* (Detroit: Wayne State University Press, 1973)

The Accra Confession: *Covenanting for Justice in the Economy and the Earth,* (2004), World Alliance of Reformed Churches (now World Communion of Reformed Churches) http://warc.jalb.de/warcajsp/news_file/The_Accra_Confession_English.pdf

Watterson, Bill, *Calvin & Hobbes*. Universal Press Syndicate.

Wink, Walter. *Engaging the Powers*. (1992), Minneapolis: Augsburg Fortress

WCC *Statement on Just Finance and the Economy of Life*. (2009), Central Committee: World Council of Churches:

http://www.oikoumene.org/en/resources/documents/central-committee/geneva-2009/reports-and-documents/report-on-public-issues/statement-on-just-finance-and-the-economy-of-life.html

WCC *Statement on Eco-justice and Ecological Debt*. (2009): Central Committee: World Council of Churches, http://www.oikoumene.org/gr/resources/documents/central-committee/geneva-2009/reports-and-documents/report-on-public-issues/statement-on-eco-justice-and-ecological-debt.html

EDUCATION FOR CLIMATE CHANGE, ENVIRONMENTAL HEALTH AND ENVIRONMENTAL JUSTICE

P. NICOLOPOULOU-STAMATI,[1] L. HENS,[2*] I. MATIATOS,[1] C. KOTAMPASI,[1] P. STAMATIS[1] AND A.J. SASCO[3]

[1]Medical School, National and Kapodistrian University of Athens, Greece
[2]Vlaamse Instelling voor Technologisch Onderzoek (VITO), Mol, Belgium
[3]Université de Bordeaux, France
[*]To whom correspondence should be send at luchens51@gmail.com

Abstract

The climate change-health effects-environmental justice nexus is analyzed. The complex issue of climate change needs to be approached from an interdisciplinary point of view. The nature of the problem necessitates dealing with scientific uncertainty. The health effects caused by climate change are described and analyzed from a twofold inequalities point of view: health inequalities between rich and poor within countries, and inequalities between northern and southern countries. It is shown that although the emission of greenhouse gases is to a large extent caused by the industrialized countries, the effects, including the health effects, will mainly impact the South. On the other hand, the southern countries have the highest potential to respond to and offer sustainable energy solutions to counteract climate changes. These inequalities are at the basis of the call for environmental justice, of which climate justice is part. This movement calls for diversification of ecologists and their subject of study, more attention for urban ecology, more comprehensive human ecological analyses of complex environmental issues and more participation of stakeholders in the debate and the options for a solution. The movement advocates a more inclusive ecology targeted at management, socio-ecological restoration, and comprehensive policies. The fundamental aspects of complexity, inter-disciplinary approaches, uncertainty, and social and natural inequalities should be core issues in environmental health programs. Training on these issues for multidisciplinary groups of participants necessitates innovative approaches including self-directed, collaborative, and problem-oriented learning in which tacit knowledge is important. It is advocated that quality assessments of environmental health programs should take these elements into account.

1. Introduction

"Environmental justice" is a wide concept. It originated from the finding that workers in particular jobs such as the ones which involve exposure to asbestos, wood preservation products, and a wide series of solvents or pesticides, showed particular diseases. With the rise of concern for environmental quality, it was found that (poorer) people living closer to hazardous waste deposit sites, nuclear facilities, and airports are more exposed to physical (e.g. noise) and chemical (e.g. pesticides) pollutants. They have less access to clean energy, bio-food, and sustainable mobility. These differences in exposure explain in part why health and disease patterns differ from those of higher income groups. Often these people have to accept an elevated level of risk without compensation or other benefit. Moreover, inequalities which dovetail with cultural, ethnic, and educational differences also influence disease patterns [Muir and Booth (2012)].

More recently "environmental justice" has been defined as "the fair treatment and meaningful involvement of all people, regardless of race, colour, national origin, or income, with respect to the development, implementation, and enforcement of environmental laws, regulations and policies" [Bullard (1990)]. The focus on law and policies in combination with the equity principle indicates that the concept stems from the human sciences. It is therefore worthwhile to analyze to what extent it also coincides with environmental principles which originate in the natural sciences. This is particularly legitimate in view of the historical relation to health and facing the health problems that result from the current human footprint on the environment [Kissinger and Rees (2010)].

This paper focuses on the inequalities among the health effects that result from the changing climate. The analysis allows defining a set of characteristics of the environmental health debate that allows linkage with both environmental justice and environmental education. As such, the main aim of this paper is to define the climate change-environmental health-environmental justice-environmental education nexus. This approach facilitates identifying a framework for training students in environmental health.

2. Essentials of Climate Change

The discussion on climate change dovetails with the natural sciences. It starts from understanding the physics behind the greenhouse effect, which is a natural consequence of the composition of the atmosphere and the finding that the concentrations of gases which influence the heat trapping capacity of the atmosphere change. These phenomena are studied by physicists.

If the greenhouse effect is increasing, not only does temperature change, but also rain regimes, soil and atmospheric humidity, plus extreme weather events such as floods, hurricanes, heat waves and monsoons. Taken together, these are changes of the climate, the main area of study of climatologists.

Climate changes affect both the physical and the biological environment. The physical environment, because increased temperatures cause melting of the Arctic and mountainous ice masses. The resulting water thus produced, plus the expansion of the surface waters of the oceans as a result of the increasing temperature, generates increased sea-water levels, contributing to coastal erosion and inundation. This is the study area of the physical geographer.

Climate changes also affect the distribution of plants and animals in a given geographical space and therefore in ecosystems and biomes. These effects are studied by biologists and bio-geographers.

Inundations, stronger and more frequent hurricanes, impacts on air, water, and soil quality and ecosystems, all these affect human health. Some of them acutely and directly (like the number of deaths and injuries caused by floods or hurricanes), and some chronically and indirectly (like the increase in the incidence of malaria or dengue fever). It is unlikely that climate change will cause new diseases. Rather it will alter the incidence, geographical spread, and seasonality of existing health disorders. Table 1 lists the main health effects which are related to climate change. The table lists the impacts in five main groups:

(1) Effects which are the immediate consequences of heat waves. The heat waves of 2003 in Europe caused up to 70,000 deaths, mainly from respiratory and cardio-vascular diseases [Robine et al. (2008)]. Intense cold spells occur more and more frequently and contribute to increased mortality and frozen extremities.

(2) Aero-allergens as a consequence of ecosystem changes. Increased temperature and changes in humidity will cause alterations in the blossoming pattern of plants. A number of trees such as birches will blossom earlier, for longer periods, and produce more pollen. Moreover the ecosystemic changes result in larger distribution areas of plants with a high production of pollen such as birches, olives, Plantago and Artemisia species. This is the basis of an increased production of pollen. It might partially explain the increase of allergic conditions in Europe, such as the increase in hay fever, allergic eczema and atopic asthma. Also house mites are an important source of asthma and allergic reactions. Changing humidity patterns are considered as beneficial conditions for the propagation of house mites in northern Europe. On the other hand, causes of asthma and immunologic reactions are multiple and complex. Currently there is no conclusive epidemiological proof for these likely changes. Consequently the issue is a subject of scientific uncertainty [Huynen et al. (2008)].

(3) Changes in climate conditions will also affect the spread and transmission rates of vector-borne and rodent-borne diseases. Anopheles mosquitos which transmit malaria will breed in places once free of malaria. This applies also to tick-borne encephalitis, dengue fever, schistosomiasis, fascioliasis, alveolar echinococcosis, leishmaniasis, Lyme disease and hantavirus infections. Also other vector-borne diseases like the infectious hepatitis B virus (HBV) and hepatitis C virus (HCV) which are both risk factors for hepatocellular carcinoma (HCC) might become more widespread in the future. Moreover HBV is particularly a risk factor when it occurs together with the Aspegillusflavus toxin (Aflatoxin). Also the geographical spread of this mold, which is currently prevalent in sub-Saharan Africa, is sensitive to changes in temperature and humidity. Other virus-mediated cancers, such as bile duct cancer, which is caused by opisthorchisriverrin and is currently endemic in North-East Thailand, Laos and Cambodia, might spread over larger areas.

The spread of animal diseases, such as those caused by the blue-tongue virus and other viruses of worldwide spread, provides further evidence of climate change effects on the distribution of vector-borne diseases. Also other livestock viral diseases are climate-sensitive and at risk of spreading: east-coast fever, foot and mouth disease and Rift valley fever.

On the other hand there is no evidence at present for a climate change-related effect on influenza or avian flu. Moreover, just like allergens, vector-borne diseases too have multiple origins, among which changes in the climate are just one factor. [Costello et al. (2009)].

(4) Effects of extreme weather events such as cyclones, hurricanes, floods, storms, landslides or wildfires. These phenomena are expected to increase in number, severity and damage in a warmer planet. These events cause both direct and indirect health effects. Direct effects entail mortality and injuries. Indirect effects arise from the loss or pollution of drinking water and harvest losses. These lead to infections and malnutrition. In the long term, disaster-associated depression and anxiety may present serious mental-health problems.

(5) Clean water and good sanitary conditions are essential for health. The changing climate is likely to make access to clean water, decent sanitation and drainage even more complicated than it is now. An increasing number of people worldwide will experience water stress. Lack of access to drinking water and sanitation result in diarrhoea and other diseases caused by chemicals and biological agents. Standing water and poor drainage lead to more water-borne and vector-borne diseases.

Table 1 also shows to what extent the phenomenon can be considered as "scientifically proven". Moreover, the extent and the potential to mitigate the effect in a country of 14 million inhabitants such as the Netherlands are indicated. In particular these "extent" and "mitigation ability" columns show major variations among the countries and the continents. It is clear that malaria will mainly extend in areas bordering the current zones of incidence. Moreover, the mitigation potential of the dispersion of the dengue and malaria mosquitoes is only predictable within wide margins of variation. These climate change-induced health patterns are the field of interest of the environmental health expert.

Table 1. Health effects of climate changes in the Netherlands [after Huynen et al. (2008)]

	Extent (1)	Certainty Risk (2)	Mitigation ability (3)	Comment
Temperature-related mortality (heat and cold)	+	+	+	Cardio-vascular and airway related mortality
Aero-allergens (hay-fever; rhinitis; atopic asthma)	±	-	+	Pollen-related complaints
Vector-linked diseases (Malaria, Dengue fever, West Nile). Rodent and teak host-borne diseases (Lyme). Virus-caused cancer (bile duct)	±	+	-	Worldwide 80 million more people are threatened by malaria if the temperature increases on average by 4°C; 2.5 billion threatened by dengue
Floods, cyclones, storms, wildfires (mortality and injuries; breakdown of sanitation infrastructure)	-	+	+	In particular lowland areas, both coastal and inland, are threatened
Food and water transmitted diseases (Salmonellosis, Campylobacter, viruses, cyano-bacteria)	-	-	+	Diarrhoea, infections

(1) + High number (over 100 per 10 million inhabitants) of yearly cases from 2050 on
- Low number (less than 10 per 10 million inhabitants) of yearly casualties from 2050 on

(2) + The relationship between climate change and health has been scientifically proven.
- The relationship between climate change and health is uncertain.

(3) + It is possible to influence the health effect.
- Health effects can hardly be influenced by national policy.

Changes in climate and ecosystems do not only affect health. Also harvests are threatened (for example). A reduction in agricultural yield will affect the income of farmers and their families. Provided these effects last for a lengthy period and in the absence of policy-mediated or individual resilience, this will affect the macro-economics of counties, states and countries. The phenomenon is expected to be an important driver of migration. Although it is impossible to predict with certainty what climate change will mean for human migration, the number of climate change related-migrants that could exist by 2050 is estimated to be in the hundreds of millions [Costello et al. (2009)]. Driven by migration, climate change might be one of many factors influencing violence. These effects are studied in economics and sociology.

The complete chain of events is subject to policy and management decisions. Worldwide policy makers have not succeeded until now in establishing an effective policy to deal with climate change. At the conceptual policy-theory level climate change has fostered the establishment alongside a policy of mitigation (bring down greenhouse gas emissions) of a policy of adaptation (be prepared for the next flood or heat wave). Moreover an increasing number of experts are convinced that natural scientists do not properly communicate the climate policy message, and that more and better targeted awareness-raising, information, training, and communication might contribute to alleviating the problem [Almund et al. (2012)]. In contrast to the lack of a global policy, local authorities (e.g. decision-makers in local communities) have succeeded in implementing most interesting and effective policies to bring down their greenhouse-gas emission levels. Companies of a wide variety of sizes have established advanced management approaches to reduce their energy consumption and CO_2-emission levels [Leal Jr. et al. (2012)].

Figure 1 represents the different main aspects of climate change as a cascade of events. The physical changes in the atmosphere cause meteorological changes. In their turn, changes in temperature and humidity alter ecosystems and biomes. These cause changes in the spread of disease vectors, cause effects on agriculture and income, and have social effects. The whole system results in policy.

The figure equally shows a number of feedbacks. This might be illustrated by a few examples. A reduction of income for a family or a state affects expenditure on medical needs and social security. The occurrence of diseases alters ecosystems. Land use changes in agriculture affect greenhouse gas emissions.

Alongside the cascade effect caused by climate changes, figure 1 also illustrates basic characteristics of the problem:

(1) Multi- and inter-disciplinarity: understanding the cascade of effects and their feedbacks necessitates knowledge from physics, chemistry, biology ecology, geography, health sciences, socio-economics, ethics, philosophy and policy. Therefore the issue is highly multi-disciplinary. Moreover it is not sufficient to collect selected data from this wide array of scientific disciplines. Understanding climate changes entails the ability to link different elements to each other. In other words, to link chemical changes in the atmosphere with variations in and alterations of the weather, and these latter with the functioning of ecosystems. This interdependent understanding is referred to as "inter-disciplinarity". This is characterized by the fact that collecting and organizing items of information from different disciplines results in added value in terms of contents. It necessitates collaboration and interaction between diverse domains of thinking, combined with their own rationales. This paradigm should lead to "new thinking". Often the metaphor of a mosaic is used: the total information resulting from the mosaic picture is more than the sum of the individual stones [Marten (2001)].

This interdisciplinary character does not only apply to the science surrounding climate change. It also applies to environmental health, which combines environmental data (e.g. on transport of chemicals through the ecological system, or on exposure) with epidemiological or toxicological data and statistics, and with medical data from a wide range of specializations like oncology, internal medicine and pediatrics, to list only a few. Also these data are organized in frameworks which are specific to environmental health as health impact assessment or cause-effect relationships, which in turn provide a basis for preventive policy [International Council for Scientific Development (2006)].

(2) Complexity: The cascade of events represented in figure 1 is a simplification of reality. As shown in figure 2, climate change involves many more aspects. The figure shows how human activities such as industry, agriculture, tourism, transport and services contribute to deforestation, urbanization and land conversion, which all affect greenhouse gas emissions and the greenhouse effect. This latter affects,

besides the elements mentioned in figure 1, *inter alia* the circulation of currents in the oceans, famine and wildfires [Cherian (2012)]. In short, climate change events can hardly be explained through simple heuristic “if…then” logical constructs, but rather should be explained using more complex, neural models of reasoning.

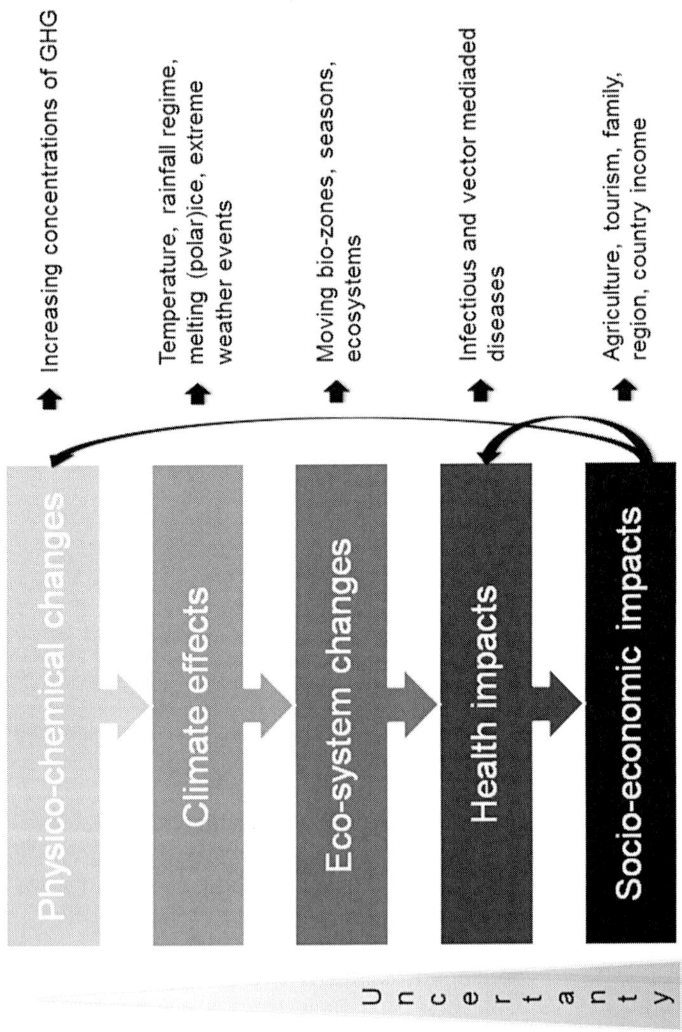

Figure 1. The cascade and feedbacks of effects caused by climate change

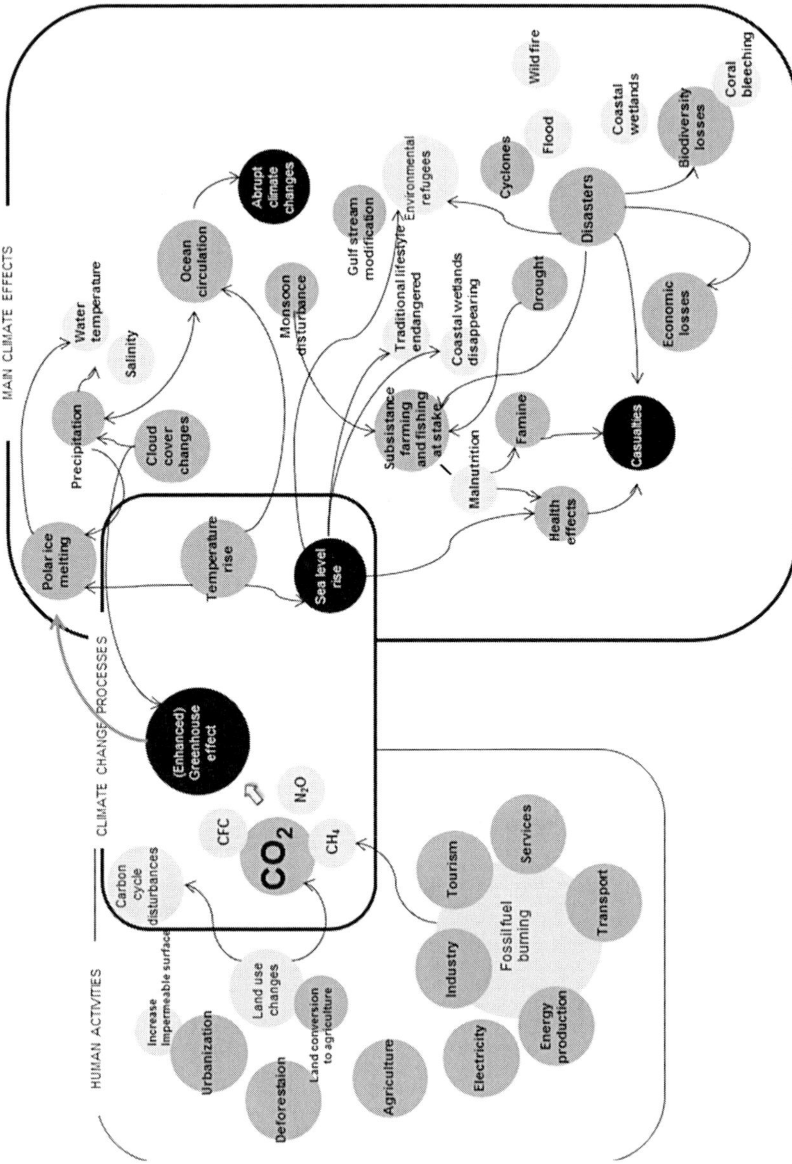

Figure 2. Climate change global processes, causes and effects [after Cherian (2012)]

(3) Uncertainty: The elements of the cascade of events shown in figure 1 are also characterized by different degrees of scientific (un-)certainty. The increase in greenhouse gases, the impact on temperature and extreme weather events, the melting of the polar land- ice, and sea-level rise have all been causally linked with the increase of greenhouse gases in the atmosphere and can be considered as "scientifically proven". In contrast the geographical spreading of tumour-associated viruses or the expected differences in incidence of malaria or dengue fever can only be forecasted with a more limited degree of scientific certainty. Consequently, the effects of these likely impacts upon expenditure on medical drugs and services, upon agriculture, and upon socio-economics at large are even more difficult to predict. As shown by the arrow at the left hand side of figure 1, scientific uncertainty increases as the one assumption flows from the other.

A particular aspect of the uncertainty linked with climate change is that the effects often appear in a discontinuous way. Thus small changes in a combination of temperature and humidity can induce major changes in the ecosystem. Ten years ago no model allowed predicting that the melting of montane and Arctic ice would proceed as fast as it has. We now know that minor changes in one or a few components of the system often trigger major effects further downwards in the cascade of events. We do not know however what these key points are and when they will appear.

Dealing with scientific uncertainty is inherent in the discussion on climate change and its health effects. In general, as shown in figure 3, uncertainty increases as the spatial, geographical scale and the time-scale increase. Predictive science performs best on small-scale problems which happen now. The larger the scales, the greater the uncertainty, and the more one enters post-normal science, where methods such as risk- and uncertainty-analysis and chaos theorems prevail.

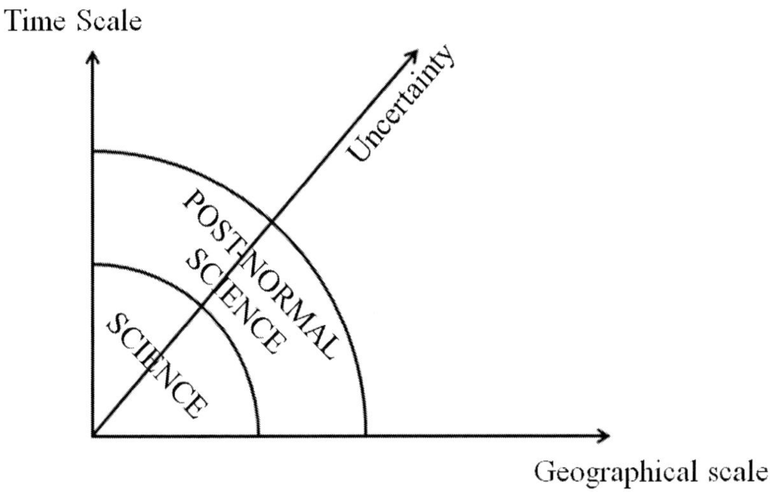

Figure 3. How scales influence uncertainty, science and post-normal science

3. Inequalities in Human Health

Human health is characterized by basic inequalities over space and over time. Disease patterns differ according to the period of life: prenatal, birth, childhood, adolescent, adult, or elderly person. They differ according to gender, population, environmental quality, diet, cultural factors and lifestyle. Worldwide, mothers and children are the populations at greatest risk of premature death and illness. Acute lymphocytic leukaemia (ALL) shows the highest incidence in children of two to three years of age, while prostate cancer occurs in ageing males. Colon cancer has the highest mortality in both men and women who live in industrialized, western countries. In Western Europe malignancies are a main cause of death, while in South-East Asia people die most often from lung diseases.

Human health is equally deeply influenced by development and poverty. For example, cancer, the second most common cause of death worldwide, was until recently considered a disease of westernized, industrialized countries. However, 63% of all cancer deaths (worldwide an estimated 7.6 million in 2008) were found in low- and middle-income countries, and this figure is expected to increase over time [Neira (2012)]. Moreover trends in mortality for a variety of cancers are increasing faster in Asia than in

Europe. More generally, low-income and disadvantaged groups are more exposed to avoidable risk factors and have less access to health services.

Environmental health does not escape these general health trends. Exposure to environmental toxicants is unevenly distributed over the population. Expensive villa housing is not located near industrial zones or areas of intense traffic. In the neighbourhood of European airports, plots for house-building are 10 to 20 percent cheaper as a result of the noise pollution. In developing countries, the section of the population with modest income has no access to modern kitchens with a high hygienic standard, and remains exposed to the exhausts of indoor coal or fuel stoves. On most waste-deposit sites in sub-Saharan Africa and India, the children of the poor collect recyclable materials.

Part of the differences in health patterns between industrialized and developing countries can also be explained by occupational exposure. Asbestos is the main occupational carcinogen. In 2004 it caused worldwide an estimated 41,000 lung cancer deaths, and 59,000 exposed people died from mesothelioma. While asbestos has been banned in the European Union since 1998, it is still abundantly used in developing countries, often without sufficient (or any) protection of the worker and his/her family. Analogous is the exposure to other carcinogens. Vinyl chloride monomer (VCM) causes DNA lesions and cancer among workers in the PVC plastic sector. While in most industrialized countries the autoclaves in which VCM is used are constructed in a way that limits exposure most significantly, the conditions in China still induce chromosome and DNA repair damage in the workers [Sun et al. (2012)].

Another aspect of health inequalities is attributed to climate change. WHO estimates that climate change caused a loss of 5.5 million disability-adjusted life years in 2000 – 84 percent of them in sub-Saharan Africa and East and South Asia. As temperature rises, the number of people exposed to malaria, dengue and other vector-bound diseases will increase, with the burden most pronounced in developing countries. The incidence of drought, projected to increase in the Sahel and elsewhere in the world, is strongly correlated with past meningitis epidemics in sub-Saharan Africa. Declining agricultural yields will increase malnutrition and consequently reduce people's resistance to illness. The burden of diarrhoea from climate change is projected to increase up to 5 percent by 2020 in countries with per capita incomes below $6000. Higher temperatures are expected to increase cardio-vascular diseases, especially in the tropics, but also in the higher-latitude countries [WB (2010)].

In general, adverse health outcomes are likely to be most pronounced in low-income countries and in poor people living in urban areas, elderly people, children, traditional societies, subsistence farmers, and coastal populations.

Exposure to these environmental factors explains shorter life-expectancies and differences in disease incidence. It is therefore not surprising that the health effects related to climate change also show differences which are in line with these patterns of inequality. The fourth report of the Intergovernmental Panel on Climate Change concluded that adverse health effects are much more likely in low-income countries and vulnerable subpopulations. These disparities may well increase during the decades to come, not only because of regional differences in the intensity of environmental changes (water stress, soil erosion, desertification), but also because of important differences in economic conditions, levels of social and human capital, political power and local environmental dependency [IPCC (2007); McMichael et al. (2008)].

Costello et al. (2009) describe five ways that link climate change and health in an inequitable way among countries:

(1) Food: Climate change threatens human health through its effect on under-nutrition and food security. Before the 2008 food crisis, more than 800 million people had calorie-deficient diets. The severity of malnutrition is most serious in sub-Saharan Africa and South Asia, where population growth is greatest. After the rise in food prices in 2008, an estimated 100 to 850 million more people may have suffered from hunger or food security. Currently more than 66 percent of the world population is malnourished [WHO (2000); FAO (2010)]. Although agricultural productivity might increase in some regions of the world as a result of climate change, increasingly severe droughts and floods, salinization, destructive hurricanes and monsoons, and soaring food prices will all aggravate food insecurity. The regions most likely to be adversely affected are those in developing countries which are already most vulnerable to food insecurity and malnutrition. Malnourishment inhibits the mental development of children, reduces human reproductivity, and is the prime cause of death in the world [WHO (2000)].

(2) Water and sanitation: Reliable access to drinking water of good quality and decent sanitary conditions are essential for health. In 2002, 21 percent of people living in developing countries did not have sustained access to a reliable drinking water source, and 51 percent did not have access to improved sanitation. Changing rainfall and temperature over the next decades are likely to make provision of clean water, good sanitation and drainage even more complicated than today.

More than one sixth of the world's population currently lives in glacial-fed water catchments which are vulnerable to climate change. Increased rates of glacial melting are predicted to result in great reductions of water availability.

Reduced river flows and increased water temperature will lead to declining water quality, less dissolved oxygen in the water, and increased microbiological activity. These effects will likely contribute to major health problems for vulnerable groups, especially during drought.

(3) Settlements: Over 50 percent of people worldwide now live in urbanized areas. And according to current projections, this will rise to 70 percent by 2050. Almost all of this growth will take place in developing regions, where authorities have difficulties in providing or facilitating urban infrastructure, where residents face more difficulties in paying for such services, and where resilience to natural disasters is limited. Rapid growth of urban slums and squatter settlements is the result. Coping with environmental pollution (especially solid and liquid waste and air pollution) and degradation has overwhelmed many cities in particular in the developing world. Close to 1 billion people (32 percent of the world's current urban population) live in slums in inequitable and life-threatening conditions, and are directly affected by both environmental threats and social crises [UN Habitat (2009)].

Within cities, climate change will negatively affect, besides access to water and sanitation, coastal flooding, landslides and related natural disasters. Hundreds of millions of people will be affected, who live mostly in the poorest countries. They will suffer the earliest and the most. They are the most vulnerable and have the least opportunities to adapt (e.g. to transit to low-carbon energy resources). The high cost of urban land and housing push the lowest-income urban populations into locations which are prone to natural hazards.

Those without access to electricity for cooking, lighting, refrigeration, and communication are exposed to adverse health risks. Poor people who have access to energy often pay a lot for limited amounts of energy which pollute the indoor air, in such a way that health risks are entailed.

(4) Extreme events: Weather-related disasters are exacerbated by climate change. These include heavy precipitation-events, and more intense and longer droughts, which all are on the increase across always wider areas. Since 1975, there has been a fourfold increase in the number of recorded natural disasters. While all continents report more natural disaster events, the rate of increase is highest for Africa, where a threefold increase has been experienced during the first decade of the new millennium. Human-made disasters have seen a tenfold increase from 1975 to 2006, with the greatest rates of increase in Asia and Africa. The low-elevation coastal zone (LECZ – less than 10 meters above sea level) is highly vulnerable to natural disasters. This zone accounts for 2 percent of the world's area, but contains 10 percent of its total population. In developing countries these areas develop faster than in industrialized countries. People here suffer disproportionately from the impacts of natural disasters. This is a function of the inability to manage pre- and post-disaster situations. In 1999, the US reported two to three times as many natural disasters as Bangladesh; yet, Bangladesh experienced 34 times more deaths [UN Habitat (2009)].

(5) Population and migration: The current rate of population increase is 1.2 percent, which means a doubling time of 58 years. Thus, the current world population of 7 billion is projected to double in less than 60 years [Pimentel (2012)]. This evolution will interface with climate change in ways that intensify other mechanisms, especially shelter, urbanization, food, and land and water accessibility. Changing climate will result in abandoning flooded and arid, inhospitable environments. The resulting migration will lead to serious health problems, both directly and indirectly. Overall, many of the most serious public health consequences of climate change will be experienced by the world's poorest nations, increasing global health inequalities. This is of particular concern as the capacity of these countries for meeting essential health-care needs, and coping with the aftermath of natural disasters and related changes that affect health, is more limited than elsewhere in the world. In particular, in these countries climate change will exacerbate the determinants of global illness and premature death.

4. Environmental Justice

In the United States environmental justice dovetails to an important extent with these occupational and environmental health inequalities. Asbestos workers suffering from lung diseases and mesothelioma, or polyvinyl chloride (PVC) workers exposed to the liver monovinyl chloride carcinogen, provide historical examples. The environmental justice movement further focuses on the environmental impacts on cities and urban areas. Cheaper and less well-maintained houses show poorer indoor air quality (higher concentrations of carbon monoxide, nitrogen oxides and volatile organic substances originating from indoor chemistry products), more pollution by biological agents like moles and mites, and less hygiene (resulting in more exposure to dust and pesticides). These exposures contribute to allergic and asthma symptoms, lung-cancer, chronic obstructive pulmonary disease, respiratory infections, cardiovascular morbidity and mortality, odour and irritation. Moreover better accessibility to green urban areas is associated with fewer visits to general practitioners, less consumption of medical drugs and less hospitalization [Werquin et al. (2005)].

People living near waste deposit sites face higher incidences of trisomy 21 progeny, and those living near waste incinerators are often exposed to higher concentrations of pollutants resulting from incomplete combustion such as dioxins [Nicolopoulou-Stamati et al. (2000)]. The disproportionate impacts upon exposed societal groups and their legal and policy impacts are part of the original drivers behind environmental justice.

Also climate change phenomena are characterized by social inequalities. Worldwide, people in the industrialized, OECD countries and in the oil producing OPEC countries consume per capita more oil than citizens of developing countries. Oil consumption in non-OECD and non-OPEC countries is however increasing. In 2005 energy consumption in non-OECD countries overtook that in OECD countries for the first time. This trend continues, but globally the relative differences in oil consumption between populous, low-income countries and the western world increase [Lesage et al. (2010)]. On the other hand, far more of the impacts are experienced more and more intensely in developing countries, which often have less means to invest in abatement technology, and consequently experience less resilience. For example, hurricanes prevail in the Caribbean and the Philippines. Lowlands such as large parts of Bangladesh, the Maldives, many low-lying Pacific islands (like Fiji, the

Marshall Islands and Tuvalu) and Central Vietnam are particularly prone to floods. Table 2 shows details of the area and the population in Bangladesh which are affected by different scenarios of sea-level rise. The data show that under the most likely assumption by 2050 the sea-water level will increase by 25 cm, and 4 percent of the country will be submerged, which forces 3.5 million people from the lowland areas to settle elsewhere in the country. If the sea-level should rise by 1 meter, the corresponding figures are 18.3 percent of the land, affecting one inhabitant out of 5, or 20 percent of the total population of 133 million people. This is most significant in a country which mainly depends on agriculture. These are figures to which no policy can formulate a sufficient reply.

The southern hemisphere is expected to face most of the costs of the climate change damage: some 75-80 percent. Important aspects of these include reductions in agricultural yield. Northern Brazil, Western Australia and the desert countries Namibia and Botswana are expected to face overall yield reductions by 2050 of between 20 and 50 percent (supposing the climate policy does not change). But also countries in the northern hemisphere such as Mauretania, Libya, Egypt, Saudi Arabia and the northern part of India face forecasted yield reductions of over 20 percent. For Canada, the Scandinavian countries and Russia, the models foresee an increase in agricultural yield of up to 10 percent in some areas [WB (2010)].

Also at an intra-country level, important differences can be noticed. Brazil may provide an example. Projections of changes in income and of poverty based on projected declines in agriculture by 2050 show that in particular the northern equatorial and sub-equatorial areas will face an increase in poverty of over 5 percent. In contrast, the already richer southern provinces including SaoPaulo, Rio de Janeiro, Minas Gerais and Santa Catharina are expected to show the largest income growth of the country [WB (2010)].

Overall the data point to a problem which is mainly caused by the excessive energy consumption in the industrialized countries of the North, while the main effects will be felt by the developing countries in the South. Consequently, the expected changes will increase the gap between North and South, between rich and poor. This uneven distribution of causes and consequences should be balanced by the mitigation potential and the resilience of these countries. The Namib, Sahara and Arabian deserts have a high potential to host industrial solar electricity plants. Also the solar energy potential of Central America, the Caribbean, and South Africa is significantly above average. More than 90 percent of the unexploited economically feasible potential of hydropower is in developing countries, primarily in sub-Saharan Africa, South and East Asia, and Latin America. Today, Africa uses only 8 percent of its hydropower potential. This means that not only most of the negative effects, but also of the long-term solutions are found in these developing countries. An inventive combination of these facts should significantly strengthen their position in international discussions.

Table 2. Affected area and population due to sea level rise in Bangladesh [Paul (2011)]

	Flooded area		Affected population	
Sea level rise (projected year)	Area (sq. km)	Percentage	Pop. (millions)	Percentage
10 cm (2020)	3 150	2.0	1.8	1.3
25 cm (2050)	6 300	4.0	3.5	2.5
50 cm (2075)	16 130	11.2	6.6	5.5
1 m (2100)	26 350	18.3	30.0	20.4

In view of this important climate change aspect in environmental justice, it is not surprising that a climate justice movement was developed as a branch of the wider discussion. The climate justice movement is a coalition of small farmers, indigenous people, unions and critical environmental NGOs, which are all concerned about climate change and current policy. More fundamentally they advocate a social model which is not predominantly driven by traditional, Keynesian economics and free market supply and demand issues. They challenge the current climate change-policy which uses "green" economy instruments such as emissions-trading rights and funding CO2-emitting projects in developing countries as ineffective. They do not agree with the "polluter pays" principle to the extent that it entails that as long as one pays, one can continue polluting. The alternative is a complete halt to carbon-dependent energy (including an immediate cessation of deep-sea and Arctic oil extraction) and the installation of effective redistribution of wealth.

5. Education for Climate Change, Health, and Environmental Justice

The analysis above shows that:

(i) The atmospheric chemical, physical and meteorological changes induce a cascade of effects which are often discontinuous and of which the appearance is often triggered by minor changes in the evolution of the problem.

(ii) The effects caused by climate change are multiple and complex. A cascade representation as in figure 1 is an over-simplification of reality. Effects of climate change are more accurately described as environmental syndromes (a structured whole of different environmental, health, and societal signs) models, or scenarios, as in the complex representation of figure 2.

(iii) In view of the manifold disciplines involved in understanding climate change and its effects, the problem and the study area relies on this wide range of scientific disciplines. However, real understanding is based on integrating the multi-disciplinary elements in an interdisciplinary framework. This latter should result in new frameworks of interpretation and new knowledge which adds to the sum of the individual units of information. This latter aspect defines the trans-disciplinary character of the discussion.

(iv) The geographical scale of the effects of climate change ranges from local (e.g. inundation; melting of mountain ice) to worldwide (sea-level rise; atmospheric increases in the concentration of greenhouse gases); the time scale on which the effects appear ranges from immediate (heat-waves and cold spells; changes in humidity) to long-term (changes in income, worldwide policy). The larger the scale of the effects, the longer their timeframe, and the more they are subject to forecast-triggering effects, the more they are affected by uncertainty.

These should be the core elements and central goals for training programs for environmental health professionals and human ecologists, who all have a vital contributory role in preventing and reducing the health effects of global environmental change. Further, in particular these training programs should focus on information, competencies and skills targeted to:

The competency for analyzing complexity and for suggesting networks and perspectives that provide solutions for the problems at stake. Central in this type of competency-development is dealing with interdependencies and interactions. It is inclusive, which means that it not only embraces approaches from the natural sciences, but also offers the relevant and necessary elements from the socio-economic and policy-legal sciences. The knowledge basis needs to entail:

(1) General elements including:

- The bio-physical-chemical background which is needed to understand e.g. the physics and the chemistry behind climate change and the physiology behind related health effects.
- The socio-economic, cultural, legal, and ethical-philosophical background which is at the basis of the environmental justice discussion.
- Relevant information from applied sciences such as engineering and medicine: the principles of water-treatment and soil-decontamination, although it is unnecessary to build a water treatment plant or to develop a flue-gas cleaning installation for a municipal waste incinerator. Analogously, while it is essential to understand the basic principles of e.g. toxicology and (pathologic) histology, it is not necessary to build up clinical experience with the associated diseases.
- Assessment of the data in a holistic context leading to applications in management and decision-making.

(2) Specific elements:
The number of applications of environmental health issues is steadily increasing. While 10 years ago only local inhabitants near an asbestos waste deposit site, or authorities facing an acute episode of environmental threats, searched advice on the health impacts, today more and more specific areas of environmental health experts emerge. Examples include:

- The health experts involved in disaster prevention, disaster alleviation, and post-disaster management.
- Those who perform environmental health risk-assessments for lending or insurance calculations.
- Those who advise on changing habits among aircraft or clinical personnel.

All this should lead to holistic views which are analytically well-defined, unbiased, and far from fuzzy and unstructured.

(i) The competency of dealing with uncertainty. From a fundamental point of view this necessitates dealing with the indeterminacy and the value-loaded character of systems (in contrast with the absolute deterministic nature of approaches in fundamental sciences on which their predictive and non-normative character is based). It is related to pro-active thinking using the available data and their margins of variation to establish actions for the future. It is supported by a proper set of methods: risk-analysis, chaos theory and related probability approaches which allow dealing with large-scale and long-term forecasts.

(ii) The competency to deal with large sets of information and knowledge. As a result of the multi-/inter-/trans-disciplinary nature of the environmental health discussion, access to a wide variety of scientific information sources and dealing with them in a systematic and targeted way are imperative. So, students need to learn how to deal with corpuses of knowledge which need to be classified, processed, assessed, discussed, and used in a context which allows conclusions.

Although incomplete, this is an ambitious learning program. Acquisition of these competences necessitates learning approaches and techniques which are different from these in traditional programs. They include:

- Self-directed learning in which students collect and organize information on a subject or a case, present it, and make it subject to discussion with their colleagues, supervisor and others. Of core importance is the active student input in this learning format, while the trainer guides and facilitates. Team learning is closely associated with this approach.
- Collaborative learning: To achieve the dcsired level of inter-disciplinarity a collaborative effort between the natural, human, and applied sciences is necessary. Moreover, learning the health effects of climate change is not only an academic activity. Working on environmental health issues should have added value for solving problems and acting in a preventive and proactive way. Students should learn dealing with and communicating the proposed solution.
- Problem-oriented learning: training on the environmental health-effects of climate change should be based on facts and empirical findings, on the analysis of cases and on evidence from situations. It makes use of publications in the published, international, scientific literature, but also of the "tacit", unpublished knowledge from practitioners in the area.

Part of the targeted inter-disciplinarity and of the team and collaborative learning atmosphere are determined by the participants to these programs. This group should be inter-disciplinary in such a way that different basic training capacities are available. Moreover the group should be multi-professional so that experiences from different areas of application should be present and be contributed to the discussions.

Both people who are directly involved in dealing with environmental health issues and those who are indirectly involved should participate in the multi-disciplinary and multi-functional group. This includes people who have a direct need of information, competences and skills on environmental health (e.g. teachers, consultants, researchers, policy advisors and decision-makers), and health experts (e.g. medical doctors, nurses, public health workers). Also the stakeholders of civil society interested in environmental health (e.g. representatives of unions, employers, farmers, NGOs) are directly involved interest-groups. Training events on environmental health effects, in particular those of climate change, at least in part, target also the public at large. Alleviating environmental health is indeed a matter of changing lifestyles and reducing the overconsumption of natural resources in western, industrialized societies, but also to an increasing extent in the upper societal layers of the society of (fast-) developing countries.

Indirectly involved audience groups include persons who have a profession in which they face environmental health issues on an occasional basis. Nevertheless they might need quite an in-depth understanding of the issues at stake. Examples include lawyers dealing with court cases on environment and health, insurance and corporate social responsibility fund-managers, and environmental journalists. More information and skills for these groups might erode or remove misconceptions and bias on environmental health.

6. Discussion

This paper has analyzed climate change as a sequence of inter-dependent and logically connected events. It puts emphasis on those environmental health effects which are often linked to the spread of vector-bound diseases, extreme weather-events, and changing patterns of pollution. These are intertwined with under-nutrition and food insecurity, declining water and sanitation facilities, urbanization, and population growth and migration. The analysis illustrates that understanding the health effects of climate change necessitates an interdisciplinary approach which is prepared to deal with complexity and uncertainty.

The health burden is unevenly distributed over the population. This applies to disease and mortality in general, but it equally applies to the health effects of climate change. These are acute mortality and morbidity effects which accompany floods, heat-waves and cold spells, monsoons and hurricanes. They result in an increasing burden from malnutrition, diarrhoea, cardiovascular and infectious diseases. Among the drivers of these inequalities are socio-economic factors: poor people experience higher exposure as a result of environmental conditions and spend less on medical care. Inequalities related to climate change also apply to the relation between industrialized and developing countries: while the increased greenhouse gas emissions are mainly caused by industrialized countries, most of the effects, including the socio-economic ones, are localized in developing countries. As such, climate change will have its most significant health effects on those who have the least access to world resources and who have least contributed to its cause. Climate change is expected to increase health inequality especially through negative effects on the social determinants of health in the poorest communities. On the other hand, the developing countries also show the highest mitigation potential for long-term climate change solutions.

Health, socio-economic, and involuntary exposure inequalities are the main drivers of environmental justice. This latter calls for a more inclusive, human ecological approach which is not only based on a careful inter-disciplinary analysis of the situation, but also integrates management, remediation, and socio-ecological restoration programs, addressing the inequalities with which environmental justice dovetails. Management of the health effects of climate change will require inputs from all sectors of academia, government and civil society. It will necessitate new ways of collaboration which till now are still unexplored.

This analysis equally indicates the main lines along which solutions for mitigating and preventing health effects need to be established:

(i) Positive effects will only be reached by lifting hundreds of millions out of poverty. Reduce malnutrition and act on food security, improve access to drinking water and sanitation, reconsider urban planning and urbanization, prepare for extreme weather events, and act on environmentally induced migration and insecurity.

(ii) Stopping greenhouse gas emissions. Transit fast to a low-carbon energy system. Mitigate the methane emissions from rice paddies and other sources. Adopt a policy-making attitude which allows more drastic international measures than have been established during the past 20 years.

(iii) Establishing a policy which prevents local health impacts of climate change. Plan and exercise to be prepared for the health impacts of the next weather disaster in the region.

(iv) While pushing on policy action, the health sector should minimize greenhouse gas emissions from its own infrastructure. Hospitals, maternity units, organizations providing shelter for elderly people, mental and physically disabled people, and medical and nursing schools should minimize greenhouse gas emissions from their own infrastructure.

Awareness of and training about environmental health risks have an important role in strengthening greenhouse-gas mitigation debates and targets. Environmental health professionals have a core role in preventing and reducing the health effects of climate change. Training for environmental health should take into account environmental justice. This necessitates specific competences such as analyzing multiple and complex networks, dealing with uncertainty, and proactive thinking. It is possible to realize this using methods of self- and team-directed learning, collaborative learning, and problem-oriented learning. In particular these latter approaches allow combining published scientific information with tacit knowledge.

The consequences of this analysis are multiple:

(1) There is a need for a concerted effort for more research and understanding of the links and feedbacks which are summarized in figure 1. The relationship between health effects and the socio-economics of a family, a county, or a state is often more driven by intuition and perception than by a careful analysis of measured data.

(2) The nexus of climate change, health effects, and environmental justice is complex and difficult to manage. Not only more in-depth understanding, but also greater involvement of the wide variety of civil society stakeholders in the debate, might provide more significant results than the ones international climate change diplomacy generates today. Overall, a

more inclusive approach is a major challenge for environmental stewardship.

(3) Training about the health effects of climate change necessitates specific approaches to stimulate specific competencies. This can only partially be realized by integrating this information within medical and nursing curricula. More in-depth training programs which cover the complexity of the issues at stake are needed. It is important not only to integrate the competences which are specifically targeted to environmental health training into the programs, but also into the terms of reference of the program assessment. These new competences are the heart of new professions and of concerns in existing professions, which are increasingly recognized as essential in delivering environmental health-care.
(4) Unfortunately training and policy alone will not be able to solve the inequitable effects of climate change on health. Nor will it be sufficient to set out strategies to help prevent and lessen the harm. An integrated response, which is most vital in this context, will also necessitate societal, lifestyle and cultural changes to profoundly in-built attitudes in our societies.

References

Almund P., Homann Jespersen P., Riis S.V. (2012), *Rethinking climate change research*. Clean technology, culture, communication. Ashgate, Surrey, UK, pp. 334.

Brulle R.J., Pellow D.N. (2006), *Environmental justice: human health and environmental inequalities*. Ann. Rev. Public Health 27, 103-124.

Bullard R. (1990), *Dumping in Dixie*. Westview Press, Boulder Co., pp. 287.

Cherian A. (2012), Grappling with the global climate challenge. In *Global environmental issues* F. Harris Ed. pp. 65-86.Wiley-Blackwell, Chichester, UK.

Costello A., Abbas M., Allen A., Ball S., et al. (2009), *Managing the health effects of climate change*. Lancet 373, 1693-1733.

FAO – Food and Agriculture Organization (2010) , *The state of food insecurity in the world*. Rome, Italy. http://www.fao.org/docrep/013/i1683e/i1683/i1683e.pdf.

Huynen M.M.T.E., de Hollander A.E.M., Martens P., Mackenbach J.P. (2008), *Mondiale milieuveranderingen en volksgezondheid. Stand van de kennis*. Bilthoven, RIVM, the Netherlands, pp.94. (In Dutch).

ICSD-International Council for Scientific Development (2006), *Science without borders*. Transactions of the International Academy of Science. Health and Environment. ICDS/IAS, Innsbruck.2, pp.524.

IPCC-International Panel on Climate Change 2006 (2007), *Impacts, adaptation and vulnerability*. Cambridge University Press, Cambridge, UK, pp. 286.

Kissinger M., Rees W.E. (2010), *Exporting natural capital: the foreign eco-footprint on Costa Rica and implications for sustainability*. Environment, Development and Sustainability 12, 547-560.

Leal Jr. I.C., de Almada Garcia P.A., de Almeida D'Agosto (2012), *A data envelopment analysis approach to choose transport modes based on eco-efficiency*. Environment, Development and Sustainability 14, 767-782.

Lesage D., Van de Graaf T., Westphal K. (2010), *Global energy governance in a multipolar world*. Ashgate, Farnham, UK, pp. 217.

Marten G.G. (2001), *Human ecology. Basic concepts for sustainable development*. Earthscan Publications Ltd. London, UK, pp. 238.

McMichael A.J., Friel S., Nyong A., Corvalan C. (2008), *Global environmental change and health: impacts, inequalities, and the health sector*. BMJ336, 191-194.

Muir B.R., Booth A.L. (2012), *An environmental justice analysis of Caribou recovery planning, protection of Indigenous culture, and coal mining development in northeast British Columbia, Canada*. Environment, Development and Sustainability 14, 455-476.

Neira M.P. (2012), *Environmental and occupational determinants of cancer: interventions for primary prevention*. Program and abstracts book of the 7th Princess Chulabhorn International Science Congress on "Cancer: from basic research to cure". Bangkok, Thailand. P. 59.

Nicolopoulou-Stamati P., Hens L., Howard C.V. (2000), *Health impacts of waste management policies*. Environmental Science and Technology series. Kluwer, Dordrecht, the Netherlands, pp. 326.

Paul B.K. (2011), *Environmental hazards and disasters. Contexts, perspectives and management*. Wiley-Blackwell Chichester, UK, pp. 322.

Pimentel D. (2012), *World overpopulation*. Environment, Development and Sustainability 14, 151-152.

Raman Q. (2013), *Health impacts from asbestos in developing countries*. Environment, Development and Sustainability (in preparation).

Robine J. M., Cheung S.J.K., Le Roy S., et al. (2008), *Death toll exceeded 70000 in Europe during the summer of 2003*. C. R. Biol. 331, 171-178.

Sun Y., Jiao J., Xia Z-L. (2012), *Chromosome damage and the damage repair capacity in Chinese VCM-exposed workers*. Program and abstracts book of the 7th Princess Chulabhorn International Science Congress on "Cancer: from basic research to cure". Bangkok, Thailand, pp. 239-240.

UN Habitat (2009), *Planning sustainable cities*. Earthscan, London, UK, pp. 306.

WB - World Bank (2000, 2009), mentioned by Burandt S., Barth M. (2009) *Learning settings to face climate changes*. J. Cleaner Production 18, 1-15.

—. (2010), *Development and change*. World development report 2010. Washington DC, pp. 417.

Werquin A.C., Duhem B., Lindholm G., Oppermann B., Pauleit S., Tjallingii S. Eds. (2005), *Green structure and urban planning*. Office for Official Publications of the European Communities. Luxemburg, pp. 438.

WHO – World Health Organization (2000), *Nutrition for health and development: A global agenda for combating malnutrition*. Geneva, Switzerland. http://whqlibdoc.who.int/hq/2000/WHO_NDH_00.6.pdf.

OF TSUNAMIS AND CLIMATE CHANGE: THE NEED TO RESETTLE

EBERHARD WEBER

The University of the South Pacific,
Laucala Campus, Suva, Fiji Islands
E-mail: Eberhard.weber@usp.ac.fj

Abstract

Next to hunger, poverty, war and violations of human rights the degradation, pollution and destruction of the natural environment as well as natural hazards have become major reasons for people to leave the places where they lived previously. Environmentally-induced migration is as difficult to define as environmentally displaced people. Displacement is often the result of many factors, of which the change in environmental quality can be one. Still there is little doubt that destruction and degradation of the environment can become pressures for individuals or whole groups to go elsewhere.

In Samoa and the Solomon Islands people left their villages after they were devastated by tsunamis in 2007 and 2009. In both cases they settled just a few kilometers away on higher ground. In Samoa resettlement started right after the tsunami. It received strong support from the Samoan Government, many NGOs and international organizations. Houses and infrastructure were in very good shape a few weeks after the tsunami of September 2009. The people of the island of Ghizo in the Solomon Islands on the other hand were forgotten soon after the event. Today they even fear that government authorities might ask them to move back to the coastal area, as the land they now occupy is government land and officially they are seen as squatters.

1. Introduction

Climate Change has become an important discourse at the beginning of the 21st century. For long scientists saw climate change mainly from a perspective of mitigation: what can be done to prevent climate change? Once however the scientific community acknowledged that climate change is real, the perspective, the challenges to take up have changed fundamentally. Today's major concerns are: how can societies adjust, adapt to or at least cope with the impacts arising from climate change? What can poor people and countries in particular do to adapt to climate change, or to recover from the impacts of a changing climate? Closely related to the discourse on climate change is the question of what impact a changing climate might have on internal and international migration. Fears about environmentally-induced migration put island countries in the limelight as they might disappear in a few decades, if sea levels rise as predicted.

It is obvious that changes in environmental quality can make people leave their homes. Still the terminology "environmental refugees" is rejected by many. Legal frameworks about refugees don't accept environmental reasons when granting refugee status. The Geneva Convention of 1951, for example, does not recognize the destruction of the environment as a reason to accept people as refugees. Although Burson (2010) discusses "Pathways to Recognition" and also Bauer (2010) reports on national legislation (e.g. in Sweden) that includes refugee recognition for environmental refugees, right now it is rather unlikely that those who become homeless because of environmental change will ever be given refugee status.

There is little agreement about basic aspects of the relationship between migration and the environment. It therefore does not come as a surprise when the dimension of the challenge, the figures of how many people are presently on the move because deteriorating environmental conditions or natural hazards, are equally heavily disputed.

There is little doubt that Pacific Island countries are much exposed to the impacts of climate change (including natural hazards). There is also little doubt that many Pacific Island countries have been a source region for a large numbers of emigrants. The question is, whether there are indications that climate change and migration in the Pacific have been related in the past, and whether with respect to the future one can expect large numbers of people from Pacific Island countries to be driven away from their homes.

2. Climate Change and Migration

Many publications have highlighted that climate change and its impacts have the potential to trigger migration flows in dimensions unknown so far. Entire states might disappear, like the Marshall Islands, Kiribati, Tokelau, and Tuvalu. Long before atolls start to get inundated, scientists and activists pretend to know what the world will look like in a few decades, when this might happen. Migration is the last resort, when other attempts to cope and better to adapt to changing environments have failed. Climate change also might not be the major reason why people leave. People might have other reasons to get away from their tiny atolls, some with population densities of more than 10,000 people per km^2, amongst the most densely populated places on this planet. The idyllic blue-greenish lagoons have *coli* bacteria counts so high that people should not get into the water. People might want to leave because atoll environments are harsh and it is extremely difficult to live there. They might want to leave to get a better job, better health or better educational facilities somewhere else. They might want to leave to get a better life elsewhere. The prospect of getting inundated in a few decades will doubtless intensify their willingness and efforts to move away.

Migration because of climate change is represented as a threat to the free world. Predictions of millions that possibly could flood developed societies and put pressures on ailing social support systems create fears amongst people and politicians alike.

3. Hazards, Climate Change and Migration in Pacific Island Countries

Small Island States are amongst those countries that are most exposed to various impacts of climate change. They "have characteristics which make them especially vulnerable to the effects of climate change, sea-level rise and extreme events (IPCC 2007, 15, IPCC 2012, 184, see also: UNFCCC, 1992). Their high vulnerability is intensified by low adaptation capacities and capabilities. Terry and Goff (2012) reflected on the vulnerability of Asia–Pacific islands to natural hazards. The paper starts with highlighting the physical and demographic differences between islands, but it doesn't go further, missing the opportunity to clearly carve out differences between islands in Asia and Small Island Developing States (SIDSs) in the Pacific Islands region when it comes to natural hazards. Briguglio (1995) highlighted some differences when he stressed that "Many small island developing states (SIDS) face special disadvantages associated with small

Pacific Islands and Territories

Country / Territory	Sub-Region	Population (2010)	Population (around 1960)	increase since 1960 (%)	Land Area (km²)	Population Density (per/km²)	EEZ (km²)	Atolls / Coral Islands	Raised islands	volcanic Islands	total
Cook Islands	Polynesia	15,708	18,378[a]	(15)	237	66	2,200,000	7	0	8	15
Niue	Polynesia	1,479	4,864[a]	(71)	260	6	390,000	0	1	0	1
Samoa	Polynesia	183,123	114,427[a]	61	2,930	62	120,000	0	0	2 + 8 islets	2
Tonga	Polynesia	103,365	56,838[c]	76	650	159	700,000	a few	>100	a few	~170
Tuvalu	Polynesia	11,149	5,444[k]	111	26	429	757,000	9	0	0	9
Federated States of Micronesia	Micronesia	111,364	39,284[b]	176	702	159	2,600,000	~600	0	>10	607
Kiribati	Micronesia	100,835	47,735[e]	118	810	124	3,500,000	32	1	0	33
Marshall Islands	Micronesia	54,439	13,928[b]	280	181	301	2,100,000	34	0	0	34
Nauru	Micronesia	9,976	4,613[a]	119	21	475	320,000	0	1	0	1
Palau	Micronesia	20,518	9,344[e]	127	487	42	600,900	<300		>5	340
Fiji	Melanesia	847,793	345,737[c]	134	18,300	46	1,260,000	2	>10	<100	~320
Papua New Guinea	Melanesia	6,744,955	2,184,986[h]	237	462,000	15	3,100,000	a few	a few	>600	>600
Solomon Islands	Melanesia	549,574	124,076[g]	336	28,000	20	1,600,000	a few	a few	>900	>990
Vanuatu	Melanesia	245,036	78,088[i]	249	12,190	20	680,000	0	0	82	82
American Samoa	Polynesia	65,896	20,051[f]	229	199	331	390,000	2	0	5	7
French Polynesia	Polynesia	268,767	84,551[d]	227	3,660	73	4,500,000	~80	a few	~40	130
Tokelau	Polynesia	1,165	1,870[a]	(38)	12	97	319,000	3	0	0	3
Wallis & Futuma	Polynesia	13,256	8,546[m]	67	143	93	300,000	0	0	2	2
Guam (USA)	Micronesia	187,140	67,044[l]	179	549	341	218,000	0	0	1	1
Northern Mariana Islands	Micronesia	63,072	8,290[b]	635	457	138	777,000	0	0	15	15
New Caledonia (France)	Melanesia	254,525	86,519[b]	187	18,600	14	1,740,000	0	0	7	7

[a]1961, [b]1962, [g]1959, [k]1963; [c]1958, [e]1963, [h]1966, [m]1969; [f]1956, [i]1960, [l]1967

(shaded) theatended by Sea-level rise; threatened by water shortage

Source: for Population Data: SPC Population Data 2010 and Time series from 1900; for island size, EEZ and island type: Pacific Islands Yearbook, 1989 size, insularity, remoteness and proneness to natural disasters. These factors render the economies of these states very vulnerable to forces outside their control – a condition which sometimes threatens their economic viability" (ibid, 1615).

There are a number of physical hazards Pacific Islands are exposed to. Situated in the centre of the Pacific Ring of Fire as they are, geological hazards like earthquakes, tsunamis and volcanic eruption play a significant role. Pacific Island countries are also exposed to hazards that are of meteorological background, such as flooding, tropical storms (cyclones and hurricanes) and drought (Asia Pacific Disaster Report 2010). Sea-level rise as a slow-onset, low-intensity hazard can make whole atoll island states disappear. Before this happens fresh-water sources are destroyed, together with land resources that are anyway scarce. Such hazards and risks, environmental, economic and social, plus climate change and its impacts, are major threats to development achievements (Asia-Pacific Disaster Report 2010). Within a few decades these island states could become uninhabitable. Holland (2008) highlights that "it is a long accepted fact that the Pacific is one of the most natural disaster prone regions in the world". (ibid, 1).

Given a closer view of disaster events in Pacific Island countries, the particular character of disasters becomes apparent. Compared to the regions of Asia the Pacific Islands (incl. Australia and New Zealand) have by far the lowest occurrence of disasters, the lowest numbers of victims as well as the lowest numbers of people affected. The number of people affected and the damaged caused is mainly concentrated on Australia and New Zealand, so that Pacific Island countries appear relatively little exposed.

Table 2 Disaster events in Asia-Pacific and impacts by sub-region, 1980-2009

Region	Events	Deaths	People affected (millions)	Damage ($ millions)
East & NE Asia	908	162,804	2,567,214	578,602
North &	297	34,644	17,231	15,636
South & SW	1,283	566,423	1,914,696	141,506
South East Asia	1,069	394,687	272,777	48,220
Pacific	406	5,425	19,126	39,078
Australia & New	197	978	16,832	36,252
Pacific Islands	209	4,447	2,294	2,826

Note: Damage and loss reported in $millions at 2005 constant prices

Another way of looking at disasters in small island states reveals very high economic and social vulnerabilities relative to the small physical size and populations of affected countries. Relative damage as a share of affected population and economic performance is often much higher than the biggest disasters in history. Disasters in SIDSs can have consequences that might threaten the very existence of countries. The tiny island of Niue in the Pacific turned from a food-exporting country into a food-dependent country after a cyclone in 1990. A cyclone that hit the country in 2004 had an even bigger impact on its agriculture (Wade, 2005). In Samoa a cyclone in 1991 caused damages equivalent to 230 percent of the country's real 2004 GDP (World Bank 2006). Between 1950 and 2004 more than 10 percent of Fiji's population had been affected on average in a disaster year and the damage was close to 8 percent of the GDP on average. Figures for Samoa, Tonga and Vanuatu are even higher, where on average more than 40 percent of population were affected in a disaster year, and related damage took a major share of the countries' GDP.

Table 3 Estimated economic and social impact of disasters in selected Pacific Island countries (1950-2004)

			Average population affected (%)		Average impact on GDP (%)	
Country	No of Disasters reported	Total reported losses in 2004 (US$m)	In disaster years	In all years	In disaster years	In all years
Fiji	38	1,175	10.8	5.1	7.7	2.7
Samoa	123	743	42.2	6.1	45.6	6.6
Vanuatu	37	384	15.5	4.5	30	4.4
Tonga	16	171	42	5.3	14.2	1.8

Source: EM-DAT, World Bank 2006, GDP and population estimates from World Development Indicators. All economic data converted to 2004 values

The most expensive disaster in history had been the 2011 Tōhoku earthquake and tsunami in Japan with an estimated damage of US$210 billion (EM-DAT 2012), about 3.5 percent of Japan's GDP in 2011. Hurricane Katrina striking the southern states of the USA in 2005 caused damage in the dimension of 125 billion US$ (EM-DAT 2012), around 1 percent of the GDP of the United States in that year. Hurricane Katrina affected some 500,000 people, or less than 0.2 percent of the population of the USA.

Since 1980 disasters in the Pacific Islands have directly affected about 3.5 million people and led to about 4,500 deaths, which seems to be modest compared to the billions of people affected in Asia and the hundreds of thousands killed in such events (see table 2), but one needs to consider the small population size of Pacific Island countries to get a better understanding of how a disaster can dominate the society and economy of a Pacific Island country for long time. The tsunami of September 2009 killed 143 people in Samoa. As a share of Samoa's population this disaster was bigger than the 2004 tsunami in the Indian Ocean for most of the countries most affected.

Table 4 Casualties of the 2004 Indian Ocean tsunami in comparison to Samoa tsunami

Country	People killed	Population	ration (people killed/'000 people)
India	16,389	1,140,000,000	0.014
Indonesia	165,708	227,300,000	0.729
Sri	35,399	19,800,000	1.788
Thailand	8,345	66,600,000	0.125
Samoa	143	183,000	0.781

Source: EM-DAT: The OFDA/CRED International Disaster Database, *Population Data: ESCAP database

4. Coping, Adapting, Recovering – Community Resilience and Migration

In the Caribbean and Pacific Islands more than 50 percent of the population lives within 1.5 km to the shore. Major infrastructure and economic activities are in this area serving communication and transportation needs. Small islands already today have enormous challenges arising from scarcity of land for agriculture and other purposes. Erosion as a result of sea-level rise will further make land resources shrink and degrade the quality of remaining areas, causing negative impacts on agriculture and food security (IPCC 2007).

Risk management and risk reduction at community levels should be given highest priority when preparing societies for a more insecure future. Migration is one of many choices people have to adapt to adverse conditions. Most likely it is one of the last resorts, when other adaptation strategies have been tried and didn't work. As a reflection and adjustment to their environment people have developed skills, knowledge and management systems that are especially important to live with risks arising from the environment. For example, in their economic activities poor subsistence farmers usually give highest priority to reducing risk by diversifying production. Also extensive networks that provide mutual assistance and support can help to mitigate adverse outcomes arising from shocks, threats and environmental risks. Societies in the Pacific Islands have undergone rapid changes in the past one-hundred years. Individualization, commercialization and urbanization have weakened coping mechanisms that were based on community solidarity, subsistence production and exchange in rural, small communities. Already then as well as today people were agents of their destinies. People are not victims of natural hazards, nor of social structures. Social structure is both the medium and the outcome of social action, of group action. In this tradition Pacific Islanders seem better equipped to take up the challenges of the future as organized groups rather than individually.

Campbell's report on traditional disaster management practices (2006) brings together a huge number of relevant publications. More than that, it also provides a structure for what to look at when investigating traditional ways to cope with and adapt to adverse events such as natural hazards. What Campbell calls 'Agricultural Resilience' (Campbell, 2006, 15) increases crop and location diversity to minimize or to spread risk, a principle many subsistence economies and even small commercial agricultural systems know, and where livelihood security is more important than profit maximization. Mutual help as a strong feature and belonging to social as well as spatial networks was constantly reinforced through rituals that strengthened reciprocal relationships. Thaman (1982ab, 1979) stresses the need to preserve the 'time-tested, socially and environmentally suited Pacific Island food systems' (Thaman, 1982a:109). Societies were self-sufficient and in a state of subsistence affluence (ibid.:109). Local communities are highly resilient to natural hazards. External intervention (e.g. in the form of inappropriate disaster relief) can weaken this resilience, undermining community solidarity, or in today's development language: people's 'social capital' (Thaman, 1982b).

5. Migration in the Pacific Islands

"Migration involves a search for well-being and security, but is not guaranteed to bring either" (Gasper 2011, 259). This is also the major reason why people of the Pacific Islands' migrate; this is their motivation and their experience of migration. This is particularly true in a time when more and more people realize that they might need to abandon the safety of their homes and prepare themselves (or their children) for living elsewhere because of climate change. Sea-level rise and other aspects of climate change are changes in the environment. At the same time they are social events: they create spaces of consideration, of discourses, of fear and assumption about an unsure (and rarely admitted) unknown future. People's "concerns" often have been hijacked by scientific, technocentric, masculine, institutionalised and Western perspectives, that "marginalize(s) oceanic perspectives and ways of being among inhabitants of small islands in the Pacific" (Farbotko 2000, 2).

The islands of the Pacific have a rather mixed history where the mobility and migration of their people are concerned. It was one of the last parts of our planet that to be settled, but when this started some 3,000 years ago (Bedford et al. 2007, Nunn 2007) the first settlers probably mastered the most sophisticated navigational skills available anywhere on this planet at that time. Today many citizens of Pacific Islands no longer live in their home countries; in the countries they hold citizenship. There are countries where more than 90 percent of the citizens have permanently migrated, mainly to metropolitan countries at the rim of the Pacific Islands. However there always also had been an exchange of goods and people within the Pacific Islands regions which created pluralistic, multi-cultural, transnational societies that actually benefited much from the mixing of people. An important economic feature of this out-going characteristic of many Pacific Island societies is that some states draw more than half of their incomes from migrants' remittances (Ware 2005).

As in many other parts of the world it was the colonial period that had huge demographic consequences in the Pacific Islands. In the efforts to recruit labourers for colonial plantations and mines kidnapping and abduction, in the Pacific Islands generally known as *black-birding*, was widely used in the middle of the 19th century. Later the colonial powers brought workers mainly from Asia to the Pacific Islands. They also resettled whole islands, often to curtail criticism from Pacific Islanders when they experienced what colonial exploitation did to societies, but also to the environment of their islands. The people of Banaba, a once phosphate-rich island that today belongs to Kiribati, were first resettled by the Japanese during the Second World War, and then brought to Rabi in Fiji by the British just after the war. They were forced to leave their home islands as their environment was slowly destroyed by mining. Or consider the violence that was done to people in connection with nuclear testing in the Pacific. It started when the USA began nuclear tests in the Marshall Islands, which had been given to the USA as a trust territory by the United Nations. Similar forced resettlements happened in French Polynesia and also Kiritimati, where the French and the British respectively conducted nuclear tests, destroying the environments and societies of these islands.

There are also instances of community-driven relocation supported by colonial or state authorities (e.g. relocation of people from Vaitupu Atoll, Tuvalu, to Kioa in Fiji, starting in 1947; relocation of people from the Phoenix Islands, Kiribati, to various locations in the Solomon Islands in the 1950s; relocation of Samoan villages after the tsunami of 2009; relocation of villages in Kadavu,

Fiji Islands after a tsunami in the 1950s). Finally there are cases of relocation of villages which were disapproved by governments (e.g. relocation of a few villages on Ghizo Island, Solomon Islands, after a tsunami in April 2007). The effort of the government is to bring people back to their original places.

Pacific Islands leaders should consider challenges arising from climate and environmental change as a huge opportunity to strengthen solidarity amongst themselves, an opportunity to enhance a Pan-Pacific consciousness, to collaborate and help each other. Such renewed and re-strengthened Pacific identity actually can build on historical bounds, and is urgently required not only to master the challenges that come along with climate change, but also those that arise from globalization processes. (These changes are not necessarily favuorable to small island countries.)

Potential relocations of entire island states in the Pacific have made many leaders express the view that they don't see themselves and their people as environmental refugees that seek new homes in an alien environment. The President of Kiribati for example doesn't leave any doubt that he is aware that a time most probably will come when the 32 atolls of his country are no longer suitable to provide a home to the people of Kiribati. President Tong has started to find alternatives for the more than 100,000 I-Kiribati.

For people in the Pacific Islands it seems to be an unpopular thought to give up homes and country, and start anew elsewhere. People in the Pacific highlight that social networks will be destroyed through this, and that people's identity and culture are at risk of drowning, along with the tiny lands they have lived in for generations.

Pacific Islanders have been amongst the most mobile peoples over the centuries. They have strongholds around the Pacific Rim. Take the tiny country of Niue, which today has a population of just a bit more than 1,500. Another 20,000 live mainly in New Zealand, where they have citizenship by birth. Or the people of the Cook Islands, also endowed with New Zealand citizenship. Some 16,000 people live at home, while more than 60,000 live mainly in New Zealand. Even bigger countries with no dual citizenship with one of the metropolitan countries in their vicinity have meaningful sections of their population outside their territories. Samoa with a present population of around 190,000 has another 300,000 people living overseas. Similarly Tonga, with a bit more than 100,000 people back home and roughly the same numbers outside the country. Even Fiji, until recently not perceived as an important source country of migrants, has lost around 120,000 citizens to migration since 1987. A huge number, considering that the country in 2012 has a population of slightly less than 900,000. Ware (2005) suggests that emigration has acted somehow like a safety-valve, mitigating potential internal conflicts that might arise due to population pressures. Conflicts that societies have with well-educated youths but not the jobs they are looking for, conflicts that arise when expectations of the future are not fulfilled and that then lead to frustration and violence. “One adaptive strategy is simply to leave” (Ware 2005, 443).

Around 3 million people lived in 2011 in the Pacific Islands (excluding Papua New Guinea). Campbell (2009) estimates that within the next 40 years 665,000 to 1,725,000 people in Pacific Island countries won’t be able to stay where they used to live. If we take the maximum number, it would mean that on an average every year some 43,000 people would get displaced, most of them towards the end of the 40 years period. A fraction of this number would be international displacements, mainly those who live in atoll countries where no safe grounds within the country exist to resettle affected populations. This would be 320,000 people, or 8,000 on average per year for the next 40 years. Considering that Australia and New Zealand annually absorb some 200,000 immigrants, the challenge to take those who cannot remain in their home countries because of environmental deteriorations seems to be manageable, especially if one considers that there are still decades until the “evacuation” needs to be completed.

Solutions to challenges around environmental change and migration should be as close as possible to the places where people live. In ideal situations migration / resettlement should be avoided and where it becomes necessary it should keep people as close to their original residencc as possible. People’s foremost motivation to avoid difficult situations of all sorts can be broadly described as their wish for security, the avoidance of risks, exposure to shocks and stress they might not be able to withstand. Preference should be given to measures that achieve this in the places where people live. In many instances it should be possible to resettle people within their own customary land. Examples of this are small-scale resettlements in Samoa and the Solomon Islands after tsunami events, where whole villages relocated about 1km away from the coastline to safer areas. Hopefully large numbers of people, especially in the high islands of Melanesia, can be resettled very close to where their homes are right now.

Table 5 Pacific Island countries and territories and migration

Country	Sub-Region	Population (2011)	Population (around 1960)	Net Immigration Rate (per 1000)	Total in NZ, Aus., USA*	Percentage of total ethnic population in NZ, Aus, USA*	Pacific Access Category (PAC) / Samoa Quota (SQ) (annual quota)	Unskilled Labour Schemes	Migration Opportunities
Cook Islands	Polynesia	15,576	18,378[f]	(21.7)	60,355*	82.0%*			Full access to NZ
Niue	Polynesia	1,446	4,864[f]	(43.2)	20,548*	93.0%*			Full access to NZ
Samoa	Polynesia	183,617	114,427[f]	(17.6)	299,962*	61.8%*	1100 (SQ)	RSE, PSWPS2	
Tonga	Polynesia	103,682	56,383[a]	(15.1)	107,164*	52%*	250	RSE, PSWPS1	
Tuvalu	Polynesia	11,206	5,444[h]	(5.2)	3,075*	24.0%*	75	RSE, PSWPS2	Special Deal with German mercant marine
Federated States of Micronesia	Miconesia	102,360	39,284[e]	(5.9)					Full access to USA
Kiribati	Miconesia	102,697	43,336[h]	0.0	482*	0.5%*	75	RSE, PSWPS1	Special Deal with German / Japanese merchant marine
Marshall Islands	Miconesia	54,999	13,928[e]	(20.0)					Full access to USA
Nauru	Miconesia	10,185	4,613[f]	(0.2)				PSWPS2	
Palau	Miconesia	20,643	9,344[e]	8.1					
Fiji Islands	Melanesia	851,745	345,737[a]	(3.7)	123,473*	12.7%*	**	RSE**	
Papua New Guinea	Melanesia	6,888,297	2,184,986[i]	0.0	13,798*	0.2%*		PSWPS1	
Solomon Islands	Melanesia	553,254	124,076[g]	0.0	1,117*	0.2%*		PSWPS2	
Vanuatu	Melanesia	251,784	78,088[k]	0.0	512*	0.1%*		RSE, PSWPS1	
Territory				total	630,486				
American Samoa	Polynesia	66,692	20,051[e]	2.7					Full access to USA
French Polynesia	Polynesia	271,831	84,551[g]	0.0					Full access to France (incl. French Pacific territories)
Tokelau	Polynesia	1,162	1,870[f]	(24.9)					Full access to NZ
Wallis & Futuma	Polynesia	13,193	8,546[m]	(8.7)					Full access to France (incl. French Pacific territories)
Guam	Miconesia	192,090	67,044[e]	(10.8)					Full access to USA
Northern Mariana Islands	Miconesia	63,517	8,290[e]	33.5					Full access to USA
New Caledonia	Melanesia	252,331	86,519[e]	2.4					Full access to France (incl. French Pacific territories)

[a]1956 [b]1957 [c]1958 [d]1959 [e]1960 [f]1961 [g]1962 [h]1963 [i]1966 k1967 [l]1968 [m]1969

Sources: for Population Data: SPC Population Data 2011 and Time Series from 1900; Migration rates and opportunities: Ware 2005, information population in NZ, Austr., USA: Hayes 2010) (except *: Hayes 2010); n.a. (not available). New Zealand Government announced the "Recognized Seasonal Employer" (RSE) policy in October 2006 and it began in April 2007. Australia launched the "Pacific Island Seasonal Workers Scheme" (PSWPS1) in Aug 2008, in September 2011 more countries were added (PSWPS2)

New Zealand / Tuvalu guest-worker scheme of 1986 allowed up to 80 workers from Tuvalu to be employed in NZ from one to three years (Simati 2009)
In 2002 New Zealand started the Pacific Access Category (PAC) which allowed 75 Tuvaluans each year to work and live permanently in New Zealand (Simati 2009, 44)

Pacific Islanders and societies at the rim of the Pacific have a long tradition of migration, including forced migration. In the second half of the 19th century some 55,000 to 62,500 Pacific Islanders were taken to Australia by *black-birding* ships to work in Queensland's sugar plantations (Flanagan et al. 2003). Pacific Islanders were instrumental in the development of Australia and also New Zealand.

Stahl and Appleyard (2007) highlight the development aspects of New Zealand and Australian programs to grant permanent access to people from Pacific Island countries for work purposes. New Zealand is granting citizenship to the people of the Cook Islands, Niue and Tokelau. The country also has a quota for Samoan citizen, the Samoa Quota (SQ), under which since 1970 1,100 Samoan nationals annually can become permanent residents. The Pacific Access Category (PAC) of 2002 allows 250 persons from Fiji, 250 from Tonga, 75 from Kiribati and 75 from Tuvalu to become permanent citizens, if they meet particular requirements. Under the SQ and PAC people from Pacific Island countries can enter New Zealand who under the general programs for skilled immigrants won't qualify as their skills levels are often too low to qualify for such programs (ibid., iv). These permanent migration schemes are supplemented by temporary unskilled labour schemes, such as the Recognized Seasonal Employer (RSE) policy that New Zealand introduced in 2007 and a similar policy Australia launched a year later. Under the RSE policy at present up to 8,000 persons from (in order of numbers) Vanuatu, Tonga, Samoa, the Solomon Islands, Tuvalu and Kiribati can come to New Zealand for seasonal agricultural work. The Australian scheme allows people from Papua New Guinea, Vanuatu, Kiribati, Tonga, Samoa, the Solomon Islands and Tuvalu to enter Australia for seasonal unskilled employment, mainly in the agricultural sector. Others economic sectors have already shown interest in employing temporary workers from the Pacific such as the New Zealand dairy and meat processing industries (Labour & Immigration Research Centre 2012).

6. Conclusion

It is difficult to give even a rough estimation of what number of (additional) migrants to expect in decades ahead because of climate change. A big challenge lies in the opportunities people have to move away from adverse environmental conditions. Not many people in Kiribati can afford a flight ticket that would bring them to Fiji. Even if they could, they still need to show funds such that they are able to sustain themselves, and a return ticket or proof of onward travel, before they can make it on board a plane. An important question therefore is to determine who migrates, to where and when. In particular, “in the face of slow environmental change those who are able to move – those with money, social networks and alternative livelihoods – may tend to migrate independently. The vulnerable poor, those with no capacity to move when environments deteriorate, the very young and the elderly, may be left behind or forced to resettle later” (Warner and Laczko 2008: 60). Few others even consider the possibility of people left behind in adverse environments. This seems not to be unlikely: people in the so-called developed world bother little when people (already today) live under extremely adverse environmental conditions, at least not so much that they would like to have these people evacuated to safer locations. The evacuation (Kelman 2008) of “sinking” islands in the Pacific won't happen. It is more likely than anything else that the last plane and ship will have left Kiribati and people are still there, not forgotten, but never intended to be brought to safety, either socially or environmentally.

The majority of Pacific Island emigrants have left their home countries not because of environmental concerns, nor as refugees, but as free migrants to establish a new, a better live elsewhere. For Pacific Islanders culture and cultural identity are very important aspects. Hundreds of thousands of emigrants have shown that there is no contradiction between cultural identity and emigration that is big enough to prevent migration from happening. Still, there are the challenges of unknown consequences of what will happen, once whole home countries are physically no more present, when the land has been swallowed by a rising ocean and when the institutions of the state have switched off the lights in the parliaments and corridors of the administrative buildings. Will these people then become stateless, or do they automatically become citizens of New Zealand, Australia, the USA, Fiji or other Pacific Island countries? Who will govern them, who will feel responsible when they need support, which to provide until then was the responsibility of their governments and administrations?

References

Asia-Pacific Disaster Report (2010), Protecting Development Gains, Reducing Disaster Vulnerability and Building Resilience in Asia and the Pacific, ESCAP / UNISDR, Bangkok, Thailand

Bauer, S. (2010), “Klimaflüchtlinge” nach Kopenhagen. Rechtliches Konzept, politische Folgen, normative Überlegungen. Stuttgart: Brot für die Welt.

Bedford, S., Sand, C. and Connaughton, S (eds) (2007), Oceanic Explorations: Lapita and Western Pacific Settlement

Briguglio, L. (1995), "Small Island Developing States and Their Economic Vulnerabilities". World Development, vol. 23, issue 9 (September), pp. 1615-1632

Burson, B (2008), 'Environmentally induced displacement and the 1951 Refugee Convention: Pathways to recognition.' Paper presented at United Nations University Institute for Environment and Human Security and International Organisation for Migration Environment Conference, Forced Migration and Social Vulnerability, Bonn, October

Campbell, J (2006), Traditional Disaster Reduction in Pacific Island Communities. GNS Science Report 2006/038. Lower Hutt: Institute of Geological and Nuclear Sciences (GNS Science).

—. (2009), Climate change and migration: Community relocation in the Pacific, paper presented at the Regional Symposium on Population and Development in the Pacific Islands held at The University of the South Pacific, 23–25 November, 2009

Farbotko, C. (2000), Migration, adaptation and indigenous mobilities and spaces in the Pacific. Paper presented at the ICARUS workshop on Climate Change, Vulnerability and Adaptation, 11-13 February 2010, at the University of Illinois, Urbana Champaign, http://www.icarus.info/wp-content/uploads/2010/06/Carol-Farbotko.doc

Flanagan, T., Wilkie, M. and Iuliano, S. (2003), Australian South Sea Islanders A century of race discrimination under Australian law: The role of South Sea Islanders in Australia's economic development, http://www.hreoc.gov.au/racial_discrimination/forum/Erace/south_sea.html.

Gasper, D. (2011), International Migration, Well-being and Transnational Ethics. In: Truong, T.-D. and Gasper, D. (eds.) (2011a) Transnational migration and human security, Springer Heidelberg, Dordrecht, London, New York, 259 – 271.

Holland, P. (2008), Disaster risk management in the Pacific: economic analysis and advocacy, Paper presented at Inaugural Pacific Regional Disaster Risk Management Meeting for Pacific CEOs of Finance/Planning and Disaster Management, 24th – 25th July 2008. Nadi, Fiji, South Pacific Applied Geoscience Commission.

IPCC (2007), Climate Change 2007: Impacts, Adaptation and Vulnerability. Contribution of Working Group II to the Fourth Assessment Report of the Intergovernmental Panel on Climate Change, M.L. Parry, O.F. Canziani, J.P. Palutikof, P.J. van der Linden and C.E. Hanson, Eds., Cambridge University Press, Cambridge, UK

— (2012), Managing the Risks of Extreme Events and Disasters to Advance Climate Change Adaptation. A Special Report of Working Groups I and II of the Intergovernmental Panel on Climate Change [Field, C.B., V. Barros, T.F. Stocker, D. Qin, D.J. Dokken, K.L. Ebi, M.D. Mastrandrea, K.J. Mach, G.-K. Plattner, S.K. Allen, M. Tignor, and P.M. Midgley (eds.)]. Cambridge University Press, Cambridge, UK, and New York, NY, USA, 582 pp.

Labour & Immigration Research Centre (2012), Population Movement in the Pacific: A Perspective on Future Prospects, Wellington.

Nunn, P. (2007), Climate, Environment and Society in the Pacific during the last millennium, Elsevier, Amsterdam

Stahl, C. W. and Appleyard, R.T. (2007), Migration and Development in the Pacific Islands: Lessons from the New Zealand Experience.

Thaman, R.R. (1982a), Deterioration of traditional food systems, increasing malnutrition and food dependency in the Pacific Islands. Journal of Food and Nutrition 39, 109–121.

—. (1982b), Hurricane Isaac and Tonga: A natural or cultural disaster? Review, 3(8), 22–35. School of Social and Economic Development, USP.

Thaman, R.R., Meleisea, M. & Makasiale, J. (1979), Agricultural diversity and traditional knowledge as insurance against natural disasters. South Pacific Bureau for Economic Cooperation Report (79) 25: 75–96.

Terry, J.P. and Goff, J.R. (2012), (Eds) The special vulnerability of Asia–Pacific islands to natural hazards. In: Natural hazards in the Asia–Pacific Region: Recent Advances and Emerging Concepts. Geological Society of London, special publication no.361

United Nations (1992), United Nations Framework Convention on Climate Change. http://unfccc.int/resource/docs/convkp/conveng.pdf.

Ware, H. (2005), Demography, migration and conflict in the Pacific. In: Journal of Peace Research, vol. 42, no. 4, 435 – 454.

Warner, K & Laczko, F (2008), A global research agenda, Climate change and displacement. In: *Forced Migration Review*, vol. 31, pp. 59-60.

CLIMATE JUSTICE AND THE PACIFIC CONFERENCE OF CHURCHES: MOVING THE RELOCATION AGENDA IN THE PACIFIC

JULIA B. EDWARDS

Pacific Conference of Churches
Box 208, Suva, Fiji Islands
Tel: +679 3311 277
Email: edwards@pcc.org.fj

Abstract

The Pacific is a region extremely susceptible to the effects of climate change, especially sea-level rise. It is also a region where the Church is very much respected and highly influential at all levels of society.

This paper examines the role of the ecumenical organisation, Pacific Conference of Churches (PCC) in its advocacy and accompaniment of people facing displacement from sea-level rise. Climate-induced relocation, still a relatively new approach, is a climate-change adaptation of last resort; as a result, to date, the relocation policy agenda within the Pacific has been slow to gain pace.

Hearing the cries of its member churches, PCC has been encouraged to promote the relocation issue at every level. Locally, PCC accompanies, both physically as well as spiritually, communities who have no alternative other than to move. PCC listens to and learns from their experiences, and, in turn, hopes to provide advice and hope to others facing similar circumstances and uncertainties. Specific PCC-accompaniment case studies are highlighted.

Nationally, PCC assists church leaders to prepare, send and receive their members displaced by climate change, and, also to draw attention to the relocation issue with key stakeholders, such as governments and other engaged groups. Internationally, PCC works consistently to raise awareness of its people's plight to the wider global audience.

Relocation within the Pacific is happening today; moving the relocation agenda is slow, but steps are being taken to raise the issue to a higher level, so that the dignity and rights of those affected are upheld.

1. Introduction

Climate change is one of the most pressing issues that the people of the Pacific face today, and the region, made up of nearly 30,000 islands, is especially vulnerable to slow-onset sea-level rise (Stern, 2006). Relocation of at-risk island communities is the ultimate and inevitable consequence of the rising sea levels; not only is the long-term habitability of atolls and islands under threat, but communities are already being forced to move because of changes to their home environments.

Any proposed relocation in response to the impacts of climate change requires extensive planning. Important considerations include how to manage the relocation process effectively so that issues are pre-empted, conflicts avoided and opportunities for those to be resettled maximised. Deliberations, however, must extend beyond mere economics, logistics and the political will of governments; the ethical, spiritual and emotional needs of both the displaced people and receiving host communities must be recognised and attended to.

The Church has much to offer in any resettlement scheme. It is ideally positioned to assemble and maintain networks of people, to partner key stakeholders, such as regional governments and other engaged groups, and to consult more widely with technical, economic and scientific expertise; yet the greatest advantage of the Church remains its direct involvement, both spiritually and physically, with its community. In the Pacific no other organisation is as close to the people affected nor as influential and independent in its advocacy role. As such, the Church has a considerable responsibility to ensure the rights and dignity of all those afflicted are maintained.

The specific aims of this paper are to:
Present a conceptual framework for the involvement of the Church in climate-induced accompaniment, and critically examine the role and responsibilities of a regional ecumenical body, the Pacific Conference of Churches (PCC), in the accompaniment and advocacy of climate-induced displaced people.

The remainder of this paper is structured as follows: sea-level rise in the Pacific will be discussed briefly in the next section, before the issue of relocation is examined in the context of the climate-induced movement of people. The advantages of the contribution of the Church to climate justice are discussed, and the historic involvement of the PCC highlighted. Current-day accompaniment and advocacy of PCC is illustrated, and reference to two relocation case studies from Fiji presented.

2. Sea-level rise in the Pacific

Unlike people living in other parts of the world, Pacific Islanders live on small parcels of land "surrounded by the wealth and menace – of the sea (Garrett, 1982, p.xi). With its numerous low-lying islands and high concentrations of people in low-elevation coastal areas, the Pacific region is particularly vulnerable to the consequences of climate change.

Relocation Case Study A: VUNIDOGOLOA village, Vanua Levu, Fiji

Village: 'Methodist' village of 26 houses;
Population: 122;
Location: On the shores of Natewa Bay, Fiji.

Relocation decision: Villagers formally approached Fiji government in 2007 for assistance with the relocation of the village.

Entire village is flooded when high tide coincides with heavy rain.

Relocation site: 2 km inland and uphill from original village, on land owned by the villagers. They have renamed the site *Kenani* (Caanan, the Promised Land): *"After many years our prayers have finally been answered. God has allocated a special place for us."* Vunidogoloa headman, Sailosi Ramatu.

Government support: Technical assistance and advice.

PCC involvement: Government approached the PCC to run a climate change and relocation workshop with the villagers.

Coastal erosion, Vunidogoloa

Latest research indicates that global sea levels are rising at 3.2mm per year since 1990 (60% faster than UN projections) (Ramstorf, Foster and Cazenave, 2012). In the Pacific, the rate is higher. From 1993 to 2009, in the western Pacific, the average rate of rise was up to three times the global average. Owing to the influence of El Niño and La Niña events, the rate is not necessarily representative of a longer time span, and while there is no agreement on future sea-level rise across the region, by the end of the 21st Century, a rise in sea levels of between 18 to 80 cm is predicted (Australian Bureau of Meteorology and CSIRO, 2011).

Likely impacts of sea-level rise will include:

- New or accelerated coastal erosion;
- More extensive coastal inundation and higher levels of sea flooding;
- Increases in the landward reach of sea waves and storm-surges;
- Seawater intrusion into surface waters and fresh groundwater lenses;
- Further encroachment of tidal waters into estuaries and coastal river systems; and,
- Higher nearshore, lagoonal and groundwater temperatures (Hays et al., 2003).

The viability of coastal communities is threatened. The interconnected impacts of climate change will pose a threat to the sustainability of agriculture and food security in the islands (Mimura et al. 2007), making conditions more difficult for communities and stress levels likely to increase for individuals (Weir and Virani, 2011).

The risk of inundation and flooding may be further intensified by social factors. Coastal erosion is often exacerbated by human activities such as the removal of coral reefs and mangroves that buffer storm surge impacts on coastlines (McLeod et al., 2010); furthermore, in the last few decades people have migrated from outer islands to principal cities and other low-lying coastal areas of the South Pacific, often as a result of economic difficulties. Such population movements create additional population stresses on potentially vulnerable areas (Mimura, 1999).

3. Relocation as a climate-change adaptation

Globally, millions of people will be forced to migrate because of sea-level rise, extreme weather events and water scarcity (Myers and Kent, 1995; Christian Aid, 2007), and while global estimates of those likely to be affected vary widely, substantial numbers of people will be forced to move in the future (Black, 2001).

Relocation, however, should be viewed as a strategy of last resort (de Sherbinin, 2010). Climate-induced relocation is an extremely complex process (Campbell, 2005; Corlett, 2008), and, as such, it is never a decision to be taken lightly. From experiences of past forced relocations, the evidence suggests that people affected by climate-induced displacement are potentially exposed to severe economic, social and environmental hardship. Communities that move in anticipation of climate change may also expose themselves to greater vulnerability from the move than they experienced previously (Barnett and Webber, 2009).

In the Pacific the situation is especially acute. Several low-lying nations, such as Kiribati, Tuvalu and the Marshall Islands, are threatened with complete submersion over the coming years, and their populations will be forced to flee elsewhere (Callick, 2009). Should entire states disappear in the next 40 years, as predicted (Weir and Virani, 2011), the retention of identity and culture of the displaced inhabitants will be a significant issue. Equally, how to address sovereignty and a government-in-exile are pressing concerns that need to be dealt with before 'deterritorialised' states become a reality (Yamamoto and Esteban, 2010). Yet it is the prospect of tens of thousands of displaced people in the region that is a major motivator for the PCC.

Most research work in the area of climate-induced resettlement concentrates on future resettlements (Maharaj, 2001; Campbell, 2005). However, there are examples from across the Pacific and elsewhere where relocation is on-going at present, and few nations can afford to be complacent about the issue. Thus far, the main relocation challenge in the region has been finding access to alternative land for those already forced to move (Campbell, 2010). The availability of land for displaced people globally will take on even more precedence in the future, but it will be a particularly sensitive issue in the Pacific.

4. Land and ownership issues in the Pacific

Pacific islanders have a very special relationship with the land. Land is more than an economic resource to the Pacific people; islanders possess an instinctive and spiritual attachment to the land (*vanua*) (Campbell, 2010). They receive their identity from the land. They belong to the land. Relocation challenges the relationship both for the displaced and the receiving host communities.

Ownership of land in the Pacific, broadly classified as customary, public or freehold, further complicates the already difficult issue of relocation. The vast majority of land, over 80% in most Pacific countries, is under customary authority (AusAID, 2008), and usually acquired by inheritance (Crocombe, 1995). Individuals and clans just do not have the ability to give land away, and virtually every tract of land has a claim to it by someone.

Acquiring alternative land for those displaced is a primary obstacle in any relocation. Even when land is available, those communities forced to move will have to leave their ancestral homes and in the process potentially lose their identity and became a displaced, landless people.

5. Role of the Church in Pacific community life

Owing to the geographical vastness of the Pacific, there is no one ecclesiological tradition across the region (Pascoe, p.343), yet “for most Pacific people, Christianity is the religion which helps define their world, their culture and their identity” (Davidson, 2004, pp.135).

The Church, an integral and important part of community life, is found in almost every Pacific Island settlement. Church leaders are community leaders and are highly respected within Pacific Island society. Ministers, however, are also receptive to the everyday concerns of local people, as living among their members they experience the same daily living conditions and physical environment as the rest of the community. Church leaders, therefore, have gained the trust of church-goers, and are ideally positioned to lead communities during times of hardship. A spiritually-strong community is one that is more resilient and better able to respond positively to adversity, and in the event of a sudden disaster, a community-based minister is able to reach out and mobilise the skills present in the wide population. Across the Pacific the Church, as a regional body, has extensive networks of communications, making it an ideal organisation to advocate and accompany Pacific Islanders in their struggle against climate change.

While the Church is very well established, and a cohesive force, recently the issue of climate change has also unified and drawn Pacific Islanders together into new networks of geographical, demographic and political global interconnectedness (Ryle, 2010).

6. Pacific Conference of Churches: Background

NARIKOSO village, Ono Island, Fiji

Village: 27 houses;
Population: 109;
Location: coastal settlement in Ono Island, southern Fiji.

Relocation decision: Villagers approached the Prime Minister, Frank Bainimarama, directly in 2011, requesting assistance. At each high tide parts of the village are regularly inundated.

Relocation site: Land adjacent to Narikoso, and owned by two of the village clans, has been being lowered to create a raised, tiered platform for the new village.

Government assistance: The government provided finance and a team of engineering experts to lower and stabilise the elevated new site.

PCC involvement: A representative of the Provincial Council approached PCC to accompany the villagers in the early planning stages of the relocation.

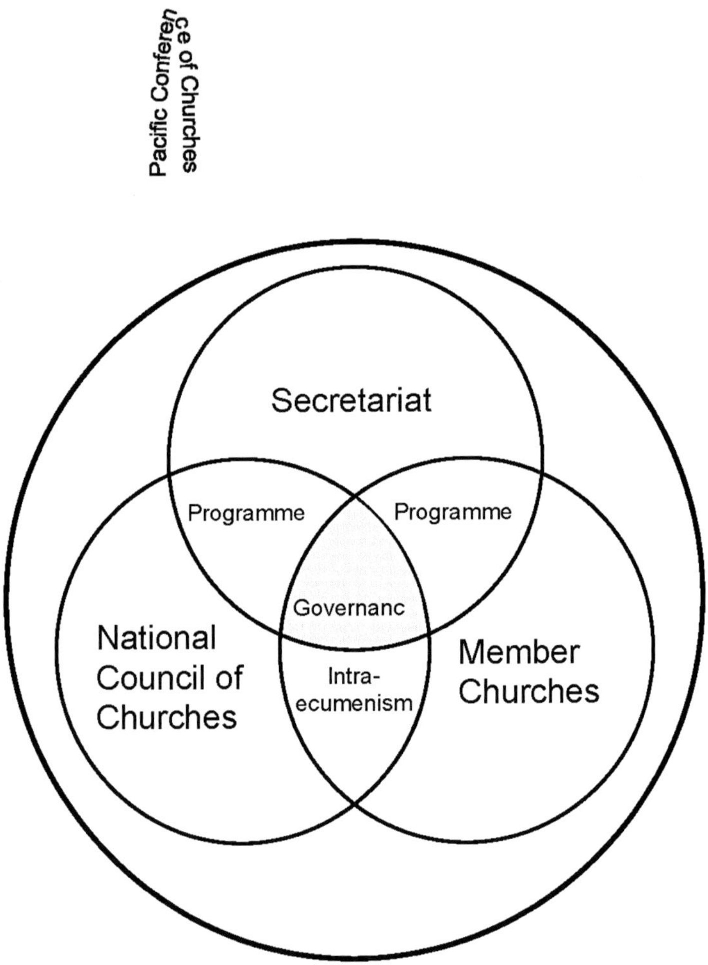

Figure 1: The interrelationship between PCC, Member Churches and National Councils of Churches in relation to the justice and equity programmes of PCC.

The Pacific Conference of Churches (PCC) is an active Regional Ecumenical Organisation (REO) that represents churches at all levels in the Pacific. Established following an inaugural meeting at Malua Theological Seminary, Samoa in 1961, the PCC has grown from relatively humble beginnings to a network of 26 member churches and 6 national councils of churches (PCC, 2007). Today, PCC represents the largest geographical region of the World Council of Churches and has a coverage extending from Papua New Guinea in the west to Tahiti in the east and from the Marshall Islands in the north to New Caledonia in the south, and, as such, is the leading community-focused agency in the region.

The work of PCC reaches beyond the spiritual wellbeing of its 6.5 million members. The secretariat of PCC have been mandated by its member churches and national council of churches to tackle issues of social justice and environmental stewardship through its various programmes (Figure 1). At the same time, the PCC seeks to promote the active participation of its members, to scrutinise government (in)action and ensure that the voices of the voiceless are heard.

7. PCC, Prophetic Activism and Climate Change

During the 1970s and 1980s the PCC focused on issues of regional concern to the Pacific, such as nuclear testing and self-determination (Siwatibau and Williams, 1982; Sidal, 2005). More recently, key programme areas have included good governance and leadership, human rights, globalisation and trade, and the empowerment of women and youth. The concerns of member churches about climate change have also been taken up.

The PCC is deeply aware of both the sensitive role it must play in providing physical and spiritual support to those affected by climate change, and also the enormity of the task it and other organisations face in highlighting the needs of climate-displaced people to the international political arena. Campaign work by PCC on climate change is not new, however.

In 1997, at the 7th Pacific Conference of Churches Assembly in Tahiti, a statement was released on climate change, and five years later, at the 8th PCC Assembly in Cook Islands, climate change at the church level became a focus.

Then in 2004 the World Council of Churches statement, called the *Otin Taai* Declaration, established a platform for church involvement. The statement began:

> "Throughout the Bible from the Genesis creation story onward, we learn about God's love for the earth and all its creatures including humanity. The Biblical understanding of the wholeness and inter-relatedness of all

> creation has some similarities to the traditional Pacific teachings about the land known as Vanua/Fonua/Whenua/Enua and the ocean referred to as Moana. The implications of this vision include the need for us humans to live with respect and humility within God's creation. Responding to God's love for creation, we are called to care for the earth and limit destructive activities such as those that contribute to climate change.
>
> Being good stewards or custodians leads to a reciprocal relationship between people and the earth. We are dependent on the earth and must take care of it. If we do so, the land and oceans will yield bounty sufficient for all. Conversely, if human societies damage the earth, people suffer."

As many other Pacific organisations were working in the area of climate change at the time, the PCC sought to focus on an aspect where the church could make a distinctive contribution; the 9th General Assembly, held in Pago Pago in 2007, highlighted for the first time the issue of the resettlement of Pacific people. Resolution No. 71 states:

> "PCC, in collaboration with member churches and NCCs, and, where appropriate, with government, advocates for a regional immigration policy giving citizens of countries most affected by climate change, especially sea-level rise, rights to resettlement in other Pacific Island nations or Pacific regional countries of their choice" (PCC, 2007).

Preparing people in both the sending and receiving communities for relocation needs a strong input of Christian responsibility to care for 'the other'. Acknowledging this reality, two years later the statement of the Pacific Church Leaders, the *Moana* Declaration, urged respect and protection for displaced people, and the development of plans to ensure that the rights of displaced people and receiving communities are upheld at all levels.

8. Regional Ecumenical Organisation (REO) Action on Climate Change: A Conceptual Framework

Based on the work of the PCC, figure 2 illustrates the approach an REO can undertake to advocate for, and to accompany, climate-induced displaced people. Developing good practice in reaching out to communities and engaging with government and other stakeholders are important parts of any advocacy and accompaniment strategy, and an REO is ideally placed to develop a regional relocation strategy that reflects the priority concerns of its membership.

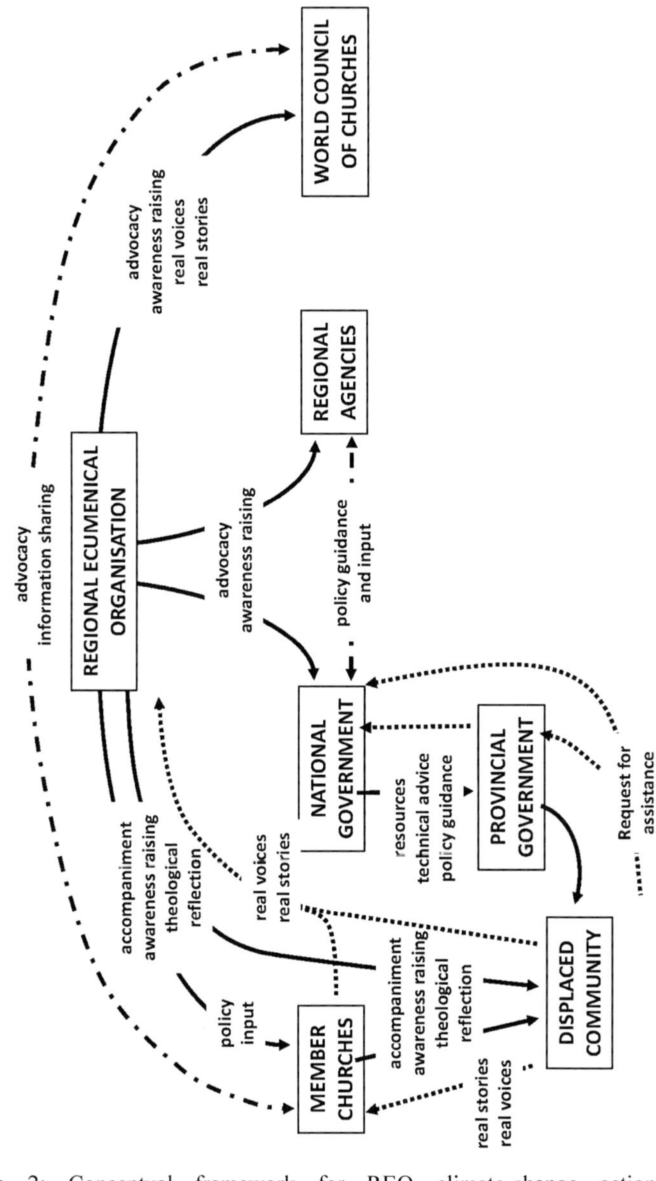

Figure 2: Conceptual framework for REO climate-change action, and accompaniment of climate-induced displaced people

As the impacts of climate change are always changing, it is important to maintain communications with those affected at the grass-roots level and the local church offers such a communications channel that assists in overcoming the issue of distance and population dispersement.

An REO can use its established networks of churches to disseminate information and also to learn about concerns affecting its local membership. Equally, local churches can receive climate change-related information from the headquarters of national churches and thus be empowered to take positive action at the community level. A suitably-trained and engaged minister on the ground can be the target recipient of the information received from either an REO, the national church or other stakeholders, and in turn can become the agent to relay first-hand information back. The clergy can also instill among their membership the responsibility to care for the the natural environment and each other.

Sometimes direct physical accompaniment of an affected community may be requested of an REO or representatives from the headquarters of a member church. People may wish to recall their experiences of local climate change face-to-face, and also to show the observed changes that affect their daily life. An REO, and especially the leaders of a national church, should be prepared to approach any community facing climate-induced displacement when news of their imminent relocation first becomes public knowledge, as the local members will gain great strength from such a visit.

During the visit, it is important that communities are aware that the changes that they recall are the result of human activity and not the work (wrath) of God. Understanding the climate-change issue is critical for people to be able to help themselves. Any climate-science input needs to be grounded in theological discussions related to the issue of climate equity and justice. Key communications should convey that those who have contributed least to the situation are the people who suffer the most.

For example, the Pacific region, because of its small population, restricted industrial activity and underdeveloped energy services, contributes less than 0.1% to global greenhouse gas emissions (Asian Development Bank, 2010), yet Pacific Islanders are among the first to experience the negative effects of climate change.

Emphasis should also be given to collective responsibility, as all people have a duty to care for the earth, God's good creation, and also to help those most in need within local communities and elsewhere.

An important role of an REO is to provide Christian hope to those threatened with relocation. Sometimes hope can be through tangible activities to empower people, such as the presentation of alternative adaptation options and disaster preparedness knowledge, so that communities can minimise the effects of climate change, or an REO can highlight to a community the benefits of acquiring new and transferable skills for those likely to be relocated to a new area. On other occasions, hope might be in the form of empathy, understanding the fears and anxieties of people and being present with the community.

At all times accompaniment must be grounded in acts of listening to the experiences and concerns of those affected, learning from past and present resettlement initiatives and conveying empathy and understanding for the difficulties being faced, empathy that translates into policy advice for church and government leaders.

An REO should be watchful of government (in)action in respect to climate change, and work closely with governmental and regional representatives to highlight local needs, to share information and to challenge injustices. The PCC has representation in national government climate-change teams, makes submissions at national and regional policy platforms, and advocates for the rights of climate-induced displaced people in the run-up to and during the annual UN climate talks, the Conferences of the Parties.

9. Conclusions

For any regional ecumenical organisation, the most important components of relocation accompaniment are to preempt a need and not wait until urgent action is required, to listen to the concerns of those affected, to endeavour to alleviate their fears by providing hope and to physically as well as spiritually accompany those requiring resettlement on their journeys.

It is hoped that the Pacific as a region will lead the world in formulating a region-wide relocation strategy that will respect the rights and dignity of all those affected and will have involved regional NGOs, government agencies and other key stakeholders in its development.

References

Asian Development Bank (2010), Responding to Climate Change in the Pacific: Moving from Strategy to Action. Manila: ADB.

AusAID (2008), Making Land Work. Volume 1: Reconciling Customary Land and Development in the Pacific. Canberra: Australian Agency for International Development, 3.

Australian Bureau of Meteorology and CSIRO (2011): Climate Change in the Pacific: Scientific Assessment and New Research. Volume 1: Regional Overview.

Black, R. (2001), Environmental refugees: myth or reality? New Issues in Refugee Research Working Paper 34, University of Sussex, UK.

Callick, R. (2009), Don't desert us, say sinking Pacific islands, The Australian, 30 July 2009, (Retrieved on 1st January 2013 from <http://www.theaustralian.com.au/news/dont-desert-us-say-sinking-pacific-islands/story-e6frg6n6-1225756097220>).

Campbell, J. (2005), Community relocation as an option for adaptation to the effects of climate change and climate variability in Pacific Island Countries (PICs). Final report for APN project 2005-14-NSY-Campbell, Asia-Pacific Network for Global Change Research.

—. (2010), Climate-induced Community Relocation in the Pacific: The Meaning and Importance of Land, in: Climate Change and Displacement: Multidisciplinary Perspectives. McAdam, J. (Ed.), Oxford and Portland, Oregon: Hart Publishing, pp. 68-71.

Christian Aid, (2007), Human tide: the real migration crisis. Christian Aid report, May 2007, London.

Corlett, D. (2008) Stormy Weather: The Challenge of Climate Change and Displacement. University of New South Wales Press Ltd.: Sydney.

Crocombe, R. (1995), Overview, in: Customary land tenure and sustainable development: Complementarity or conflict? Crocombe, R. (Ed.), Noumea and Suva: South Pacific Commission and University of South Pacific, pp.5-21.

Davidson, A.K. (2004) "The Pacific is no longer a mission field?" Conversion in the South Pacific in the Twentieth Century. In: D.M. Lewis (Ed.), Christianity Reborn: The Global Expansion of Evangelicalism in the Twentieth Century, Michigan: Wm B. Eerdmans, pp.133-153.

Garrett, J. (1982), To Live Among the Stars: Christian Origins in Oceania. Geneva and Suva: World Council of Churches and the University of the South Pacific, p.xi.

Hay, J.E., N. Mimura, J. Campbell, S. Fifita, K. Koshy, R.F. McLean, T. Nakalevu, P. Nunn and N. De Wet (2003), Climate Variability and Change and Sea-level Rise in the Pacific Islands Region: A Resource Book for Policy and decision makers, educators and other stakeholders, Japan Ministry of the Environment and South Pacific Regional Environment.

Maharaj, R. (2001), Pacific Islands at Risk: Foreshore Development and their vulnerability and implications for adaptation strategies to climate change, SOPAC Report, Suva Fiji.

Mcleod, E., J.Hinkel, A.T.Vafeidis, R.J.Nicholls, N.Harvey and R.Salm (2010), Sea-level Rise Vulnerability in the countries of the Coral Triangle, Sustainability Science, 5, pp.207–222.

Mimura, N. (1999), Vulnerability of island countries in the South Pacific to sea level rise and climate change, Climate Research, 12, pp.137–143.

Mimura, N., L. Nurse, R. F. McLean, J. Agard, L. Briguglio, P. Lefale, R. Payet, and G. Sem (2007), Small islands, Chapter 16, Climate Change 2007: Impacts, Adaptation and Vulnerability, in Contribution of Working Group II to the Fourth Assessment Report of the Intergovernmental Panel on Climate Change.

Myers, N., and J. Kent (1995), Environmental Exodus: An Emergent Crisis in the Global Arena, Climate Institute, Washington DC.

Pascoe, D. (2008), The Church in an Oceanic Perspective. In: Mannion, G. and L.Mudge (Eds.) The Routledge Companion to the Christian Church, New York and Abingdon, Oxon: Routledge, pp.342-363.

PCC (2007), Atua, Empower us to be Liberating Communities through a Shared Spirituality of Ecumenism in the Pacific. PCC Strategic Plan 2008-2012 (unpublished), Suva: PCC, pp.4.

—. (2007), Resolutions of the 9th General Assembly of the Pacific Conference of Churches (PCC), (unpublished), Kanaana Fou Theological Seminary, Pago Pago, American Samoa, 2-8th August.

Ramstorf, S., G. Foster, A. Cazenave (2012), Comparing climate projections to observations up to 2011, Environmental Research Letters 7, pp.1-6.

Ryle, J. (2010), My God, my Land: Interwoven Paths of Christianity and Tradition in Fiji, Ashgate Publishing Limited: Farnham.

Sidal, M. (2005), The Legacy of Nuclear Testing in the Pacific. Report of a Workshop of the PCC held in Kiribati, 7-9th February, Suva, Fiji: PCC.

Siwatibau, S. and B.D. Williams (1982), A Call to a New Exodus: An Anti-nuclear Primer for Pacific People – An Introduction of Science,

Health and Environmental Topics with Social, Economic and Political and Human Values Questions. Suva, Fiji: PCC.

Weir, T. and Z. Virani (2011), Three linked risks for development in the Pacific Islands: Climate change disasters and conflict, Climate and Development, 3, pp.193-208.

Yamamoto, L. and M. Esteban (2010), Vanishing island states and sovereignty, Ocean and Coastal Management, 53, pp.1-9.

RIGHT TO HEALTH FOR POVERTY REDUCTION AND ECOLOGICAL JUSTICE

LARISA SKURATOVSKAYA* (RUSSIA),
LALA CORNELIUS (USA),
GUILLERMO KERBER
(URUGUAY/SWITZERLAND)
AND ELIAS C. ABRAMIDES (ARGENTINA)

*Corresponding author from Russian Academy of the Medical Sciences
E-mails: laskur@hotmail.com; lalacornelius@gmail.com; gkm@wcc-coe.org; eliasabramides@gmail.com

Abstract

In this article authors from various countries discuss the Right to Health and Women's Rights. Gender and Equity are defined within the confines of this paper. This article also addresses the necessity of understanding that gender equitable responses within and beyond the health sector are needed, and calls for a gender equity sensitivity in adaptation programs.

The last fifteen years have shown the relationship between climate, weather and health. Climate scientists were able to pinpoint human activities as being at least partly responsible for the global warming that was observed during the last century.

It has also become clear that even with the most drastic measures to curb the emission of greenhouse gases, the climate will continue to change for many decades on account of past emissions. The Third and Fourth Assessment Reports of the Intergovernmental Panel on Climate Change (2001, 2007) state through different scenarios that climate change will have an impact on food security, freshwater supply, rural and urban settlement and infrastructure. It has been agreed that in this situation, adaptation measures to respond to the climate challenges become a necessity (TAR3 2001, FAR 2007),

The United National Development Program is currently developing an Adaptation Police Framework that focuses on the specific needs and constraints for adaptation, including assessments of the expected health impacts of climate change.

In the last few years, researchers found sex-based differences in the biology of males and females, and this gender dimension of climate change was tangentially broached on the occasion of the session of the Conference of the Parties (COP) 8, held in New Delhi in 2002, and in a more formal way from COP-9 in Milan up until the COP18.

Policymakers and health professionals have taken steps to adapt to the impacts of climate change with the aim of developing gender sensitivity approaches with regards to adaptation projects and national regimes. This is of crucial importance with regards to the development of international Human Rights, in particular the Right to Health, which is specifically protected under the International Covenant on Economic, Social and Cultural Rights (ICESCR) (Skuratovskaya L. at al., 2005).

The Kyoto Protocol, adopted by the Parties in 1997, and coming into force in 2005, was a significant step forward in the international response to the climate change issue. For the first time, international binding targets and timetables have been agreed for the period 2008-2012. However, we do not have a specific reference to the gender aspect in this international treaty. The progress of development of equal rights for men and women is very slow, but it is very valuable to include gender aspects of adaptation and mitigation in negotiations between countries in the post-Kyoto period after 2012.

Firstly, let us have a working definition of "gender". Gender is used to describe those characteristics of women and men which are socially constructed, while sex refers to those that are biologically determined. People are born female or male but learn to be girls and boys who grow into women and men. This learnt behavior makes up gender identity and determines gender role (WHO, 2003). The way in which we experience our gender identities and roles are complex, vary in different contexts and can change through time. This is because gender is socially constructed and affected by a whole host of other variables such as poverty, age, (dis)abilities, occupational status, marital status, etc. For example, poor women in particular are disadvantaged in their ability to access quality health care.

Equity is a much debated and complex concept in the context of health, which we will not discuss in depth here. One widely cited definition of inequity in health is "differences in health which ... are considered unfair and unjust" (in WHO/EURO documents), which therefore state that "equity requires reducing unfair health disparities." Most medical professionals argue that an equity focus may be concerned with reducing inequalities in health outcomes or access to quality healthcare between different socio-economic groups, or maintaining a minimum level of health for all, regardless of their socio-economic status or birthplace, according to the principle of social justice (Gender&Climate Forum of the WCC-3, 2009).

Gender equity, thus, refers to fairness and justice in the distribution of benefits and responsibilities between women and men. The concept recognizes that women and men have different needs and power, and that these differences should be identified and addressed in order to reduce disparities in health experiences and outcomes. There are many challenges in ensuring gender equity in disease control programs. One particular challenge lies in the need to recognize that neither women nor men constitute homogeneous groups. In practice, for example, this may mean addressing the particular barriers faced by poor, urban, unemployed men in adhering to treatment and evaluation strategies and related research; training and resources to develop, implement and evaluate ways to promote resilience to communicable diseases; and ability to access quality preventive and curative care for different groups of women and men (WHO, 2011).

The collation and analysis of data by age, gender and socio-economic status in health information systems can provide an invaluable resource in understanding how these variables affect access to health care and health outcomes. Addressing gender bias and stereotypes and raising awareness of gender inequities and specific needs in any given context amongst health care providers and within disease control programs are also important (UN Population Fund, 2009). Gender equitable responses may involve new partnerships within and beyond the health sector.
The most often studied differences are in the reproductive system. However, over the past 20 years, and most rapidly in the last 10 years,

scientists have accumulated data differences between the sexes at many levels, from the cellular to the behavioural, from the clinical to the pharmaceutical. For example, health educators have been trying for years to tell the public about sex differences in heart disease using the general media: on average, men experience heart attacks 10 years earlier than women, and have a better rate of survival after one year. Symptoms vary by sex: women experience shortness of breath, fatigue and chest pain; most male heart attacks come on as a sudden, striking pain in the chest (Mr. Iris Tetford, 1998).

Women and men also metabolize some drugs differently. Seldan, an antihistamine, was taken off the market for this reason because it was causing arrhythmia, and in more women than men. In 2002 the total USA budget for heart diseases research was $566.2 million, with $80.7 million allocated for women and $56.9 million for men and $428.6 million for both. In comparison the total budget for heart diseases research in 1993 was $36.4 million, with $27.7 million for men. The big picture is that we believe that there's a major benefit to medicine and society by understanding the natural differences between men and women (Women & Gender Research , 2009).

Dr. Legato, director of the Partnership for Women's health at Columbia University, wrote "we look specifically at the difference biological sex makes and find differences in every system of the body, including how drugs are absorbed, metabolized, exert their effect, and the excretion from the body". Now we have a recommendation to the Journal to start asking authors to report differences between sexes in their articles, and while biomedical journals are not jumping on the bandwagon to include sex-differences drug companies are "getting on board". FDA has widely disseminated studies to those who review basic and clinical drug data to educate these scientists about the underlying scientific validity sex differences have in drug metabolism (Skuratovskaya L, Pestova N., 2012). According to recent studies there are various kinds of investments in the field of climate-change adaptation measures and initiatives to be considered. These include, for instance, projects in agriculture and forestry, and civil engineering interventions to shore up dykes and sea walls. All these sectors are characterized by a considerable involvement by women. It appears that clearly the introduction of a gender sensitive approach could assure real benefits both in efficiency and equity terms. In this context the role of education and its implications in terms of gender is very important. In reality "education increases women's ability to access relevant information and, importantly, to get their voices through to decision-makers" (NESS report, 2005).

The strategy envisaged for this action is based on the combination of adaptation and mitigation efforts. In actuality, only recently has adaptation entered the policy debate. Recent studies state that adaptation refers to changes in "process or structures to moderate or offset potential dangers or to take advantage of opportunities associated with changes in climate". Adaptation involves adjustments to decrease the vulnerability of communities and regions to climate change and variability. The adaptive capacity of a system is its ability or potential to adapt to the effects of climate. Adaptation can be spontaneous or planned, and can be carried out in response to, or in anticipation of, changes in conditions. The concept of adaptation has generally been used to focus attention on the impacts associated with climate change mechanisms and, in particular, how ecosystems and human societies can cope with them (Protecting Health from Climate Change, connecting Science, Policy and People , 2009).

The inclusion of gender equity in adaptation programmes would help to make them more effective in responding to specific challenges posed by climate change to men and women.

P.S. World Council of Churches is a very active partner with different NGOs (for the latest publication, see Theodor Rathgeber, 2012), which since the beginning has been a very clear articulate advocate of Climate Justice and Human Rights as a main goal for sustainable development at the local level and has presented these goals in Statements to all the UN Conferences of the Parties for Climate Change. When this article was ready for publication, the Intergovernmental Panel on Climate Change (IPCC) announced the Synopsis of "The Fifth Assessment Report of IPCC", which will be presented in 2013 (September) to 2014 (March), and we note two special Chapters related to the topic of our Article:

Chapter 11: Human Health.
Health inequalities, gender and marginalized population
Chapter 13: Livelihoods and Poverty.
Inequalities, gender and marginalized population.

It demonstrates that the global activity of NGOs (gender-cc network, WEDO, WCC etc.) aimed to secure special attention for gender inequalities in different aspects of life, including “Health”, is now receiving very special attention.

References

1. Cleland, J., S.Benstein, A. Faundes, A. Glazier and J. Innis, “Family planning: the unfinished Agenda”. The Lancet, 368: 1810-27, 2006.
2. “Gender&Climate Forum-WCC-3”, Geneva , WCC, 2009.
3. Gender, Climate Change and Health, WHO, 2012.
4. Mr. Iris Tetford, “Violence Against Women: A Public Health Priority” in International Conference on preventing violence, caring for survivors: role of Health professional and services in violence, CEGAT, Mumbai, India (1998), WHO.
5. NESS Report-2005, Sure Start Local Programmes.
6. Protecting Health from Climate Change, connecting Science, Policy and People”, WHO, 2009.
7. Скуратовская Л., Робра М., Репин В. «Новые знания - новые опасности.» Человек. 2005. No 1, 116 - 24
8. Skuratovskaya L., N. Pestova “Gender’s Aspects of Climate Changes” in “ Actual topic of Woman Health” by European Medical Woman International Organization, 2012.
9. Theodor Rathgeber “Climate Justice, Human Rights and the role of Churches”, 2012, WCC.
10. UN Population Fund (UNPF), State of World Population. Facing a changing World: Women, Population and Climate, NY, 2009
11. For all “Assessment Reports” of the Intergovernmental Panel for Climate Change, see http://IPCC.ch

POVERTY, FINANCIAL CRISIS AND HUMAN RIGHTS

Poverty in Antioch at the End of the 4th Century and Poverty Today: Some Reflections based on St. John Chrysostom`s Writings and Other Sources

V. Rev. Lecturer Dr. Daniel Buda

Sibiu/Geneva, World Council of Churches, Geneva, Switzerland
E-Mail: Daniel.Buda@wcc-coe.org

Abstract *It is a great honour for me to be part of this conference and to present some of my preoccupations and reflections in the field of patristics that are related with the themes we are discussing here in the Orthodox Academy of Crete. We have heard and learned a great deal about the present situation on poverty in the world; strategies for poverty reduction have been presented and perspectives for the future have been predicted. However, I invite you all to take a look into the past and to reflect on the way in which poverty was perceived by a certain personality of the Ancient Church, who, like each human being, lived and acted in a certain context and in a certain period of time. Whenever we are confronted with a challenge, we have the tendency, which I would define as natural, normal and logical, to look into the past, i. e. into history, and to see how a particular problem was perceived and addressed by our predecessors. One of the roles of historical studies, since Antiquity, is to bring into the present the experience of the past. In this sense, history has the role of being magistra vitae*[1] *(life teacher). It is important to underline that we might learn from the past how to address problems and how to avoid mistakes, tensions and complications, but we shall not copy solutions simply because the context of our times might be similar with previous contexts but never the same.*

[1] Cicero, *De Oratore,* II, 36.

The Orthodox Church pays special attention to the writings and experience of the Church Fathers and considers them as a source of authority and inspiration for addressing and solving the problems of the present. This applies not only to the doctrinal, disciplinary or liturgical challenges of the Church, but also for the ethical, social or environmental challenges of our times. Nowadays, careful research of the writings of the Church Fathers can bring us inspiring ideas on how to approach social issues, how to develop a sustainable theology of the protection of creation in order to give a Christian dimension to ecology, how to address the issue of poverty, and so on.

Today I will speak about Saint John Chrysostom and the way in which he addressed the issue of poverty in Antioch, a big city of his time, where he was active as a priest for many years. St. John Chrysostom was born around 349 in Antioch[2] into a quite wealthy Christian noble family. He studied rhetoric with the most famous Antiochian Professor of that time named Libanius, who was a pagan. He was introduced into Christian theology by Diodor of Tars, a famous representative of the so-called (Christian) "Antiochian school."[3] St. John became in 386 priest of the greatest Christian church of Antioch. In 397 he was elected as bishop of Constantinople, which was the capital of the eastern part of the Roman Empire. He served in this capacity until 407 when he passed away on the way to exile. He was condemned to exile by a synod of bishops gathered near Constantinople, mainly due to a conflict with the royal family of the East Roman Empire. St. John received in the 6th Century the surname "Chrysostom" which in Ancient Greek means "Gold Mouth" because of his remarkable talent of rhetoric – he preached as nobody else in the history of the Christian Church before him and probably also after him. He was also the most prolific Greek Church Father. His literary heritage includes 14 treatises, 820 authentic homilies and 240 letters. The highest reputation of St. John Chrysostom however is above all reflected in the fact that until today the most important Liturgy of the Eastern Orthodox Church is attributed to him. [4]

---

[2] Robert Carter, *The Chronology of St. Chrysostom's Early Life,* in *Traditio*, nr. 18 (1962), p. 357-364; Baur Chrysostomus, *Wann ist Chrysostomus geboren?*, in *Zeitschrift für Katholische Theologie* nr. 52 (1928), p. 401-406;

[3] R. V. Sellers, *Two Ancient Christologies,* London, 1940.

[4] For more details regarding Saint John Chrysostom`s life, work and doctrine see Johannes Quasten, *Patrology,* vol. III *The Golden Age of Greek Patristic Literature From the Council of Nicaea to the Council of Chalcedon,* Christian classics, Notre Dame, IN, pp. 424-482, with recommended bibliography.

It has already been mentioned that St. John was active as a priest in Antioch on the Orontes (today Antakya, a small town in the Province of Hatay, Turkey), which was in his time the third biggest city of the Roman Empire and very probably of the world, after Rome and Alexandria, having in his time around 300,000 inhabitants.[5] If we want to compare the city of Antioch from Saint John Chrysostom`s time with the cities of our time, we can say that Antioch was for the second half of 4th Century, according to some statistics, for our times the city of Cairo, Egypt, namely the third largest city in the world in terms of population. Many people compare the European Union with the Roman Empire and I agree that there are certain similarities as there are, of course, a lot of differences. Antioch was for the Roman Empire the third largest city, as Madrid (in Spain) is for the European Union. Finally, the Roman Empire was the world power in St. John Chrysostom`s time as it is obviously today the United States of America. Antioch was for the Roman Empire what Chicago, Illinois is for the United States of America, namely the third biggest city. These comparisons give us, I hope, a better image of the importance of Antioch as a city in the time of St. John Chrysostom.

Antioch was similar to the big cities of our time not only in terms of size, but also in terms of the diversity of its population. It was founded near the end of the 4th Century BC by Seleucus I Nicator, one of emperor Alexander the Great`s generals, as a Greek-Hellenistic city. Very soon, many other ethnic groups were attracted by the beauty and opportunities offered by the big city of Antioch: Syrians – who were considered as the indigenous population of the region; Jews, Armenians, Persians, Phoenicians, Arabs, Egyptians, Indians, and, after the Roman conquest by Pompeius in 64 BC, Latins etc. The cosmopolitan character of Antioch was described by authors like John Malalas[6] and the great orator Libanios[7], whom I mentioned before as being St. John Chrysostom`s professor for rhetoric. In particular, Libanios describes his native city of

---

[5] For more details about Antioch and its ancient history see Benjamin Drewery, *Antiochien* in *Theologische Realenzyklopädie* III, pp. 103-113, with recommended bibliography. The best monograph about Ancient Antioch remains Glanville Downey, *Ancient Antioch,* Princeton, Princeton University Press, 1963.

[6] Elisabeth Jeffreys, Michael Jeffreys, Roger Scott, *The Chronicle of John Malalas: A Translation*, Byzantina Australiensa 4, Melbourne, 1986, 201, 18-202,6; 346, 21-347,7.

[7] Libanios, *Antiochikos (or. XI). Zur heidnischen Renaissance in der Spätantike,* translation and commentaries by Georgios Fatouros and Tilman Krischer, Verlag Turia & Kant, Wien, Berlin, 1992

Antioch as being an irresistible attraction for all those who want to get rid of poverty.[8] The population of Antioch was not only ethnically and linguistically diverse, but also religiously. We know from other sources that in John Chrysostom`s time, the majority of the Antiochian population was Christian; however a significant part was Jewish,[9] and the pagan population was decreasing but continued to play an important and influential role especially in Antiochian intellectual life.[10]

Saint John Chrysostom is the first Christian author that offers us in his writings detailed descriptions of the daily life of Antioch. Some of these descriptions are related with the social situation of the Antiochian population. We are told about the luxury in which some citizens, both Christians and pagans, are living, but also about the cruel poverty of other citizens. Members of Nobel families were carried on the streets in their litter (*lectica*) on the shoulders of slaves surrounded by many servants and body guards. They lived in luxurious houses based around Antioch, far away from the dust and crowded streets of the city. At the same time, St. John Chrysostom describes the poverty he saw on the streets of Antioch. Beggars called loudly and with impressive words the mercy of those who walked on the streets, exposing their disabilities or cut arms and legs. Some other poor people, including children, tried to impress the walkers by offering disgusting and dangerous performances like swallowing fire or snakes, by lying in water in the winter time until their body were surrounded by ice etc. Sometimes the baggers even entered church

---

[8] Libanius (see footnote 71), Or. XI, 164, p. 43.

[9] There is enough evidence to affirm that a Jewish community existed in Antioch since the city was founded (see C. H. Kraeling, *The Jewish Community at Antioch, Journal of Biblical Literature* nr. 51 (1932), p. 131). The Jewish Antiochian community grew very fast, becoming the largest of this kind in Syria (M. Stern, *The Jewish Diaspora* in *The Jewish People in the first Century* (ed. S. Safrai şi M. Stern), vol. I, Fortherst Press, Philadelphia, 1974, p. 117-183). It has been estimated that in the first Century AD in Antioch there were 45,000 (C. H. Kraeling, as quoted above, p. 136) or even 65,000 Jews (R. N. Longenecker, *Galatians* (WBC 41), Dallas, Texas, 1990, p. 68), which meant 1/7 of the total population of Antioch. Other authors consider this estimate as being exaggerated and estimate that in Antioch there were at that time no more than 22,000 Jews (W. A. Meeks & R. L. Wilken, *Jews and Christians in Antioch in the First Four Centuries of the Common Era,* Missoula, 1978, p. 8).

[10] After the Age of Constantine the Great, the pagan population of the big cities of the Roman Empire became an intellectual elite (see Peter Brown, *Die letzten Heiden. Eine kleine Geschichte der Spätantike,* Frankfurt a. M., 1995).

buildings and disturbed the believers while attending the Holy Liturgy.[11] Antioch, as described by Chrysostom, was certainly a city of contrasts as is every big metropolis of our times. Extreme poverty and extreme wealth lived together; forming a terrible disharmony which could not remained unaddressed by a careful observer like Saint John Chrysostom.

Saint John Chrysostom`s writings contain strong condemnations of wealth which was obtained in unjust or illegal ways. One of the first sermon delivered by John Chrysostom as a priest in Antioch is entitled *About the word of the Prophet David and about hospitality*."[12] It is basically an exegesis on Psalm 49: 17. "For when he dieth he shall carry nothing away: his glory shall not descent after him". The main idea is that there is no reason to grieve the wealth of rich people. This is because wealth is often obtained, in Saint John Chrysostom opinion, on the exploitation of the poor.

> "These magnificent residential palaces are irreconcilable prosecutors, which, even if their owner is since long time dead, still raise their accusatory voice. ... Every passer who sees the height and the wide proportions of the building will say the following to his neighbour: How many tears coasted the construction of this house? How many orphans were plundered here? How many widows were treated here with injustice? How many workers were cheated here of their full pay?"[13]

It would be certainly wrong to affirm that Saint John Chrysostom considers wealth *per se* as evil. His exegesis on the story of "wealthy young man" makes clear the difference between wealth that was obtained with unjust means and by oppressing other people and legitimately obtained wealth, which is also used according to God`s commandments.[14] As a "man of the Bible"[15] Saint John Chrysostom was fully aware that

---

[11] Adolf Martin Ritter, *Situationsgerechtes kirchliches Handeln in der Spätantike und heute am Beispiel des Johannes Chrysostomos* in *Studia Chrysostomica. Aufsätze zu Weg, Werk und Wirkung des Johannes Chrysostomus (ca. 349-407),* p. 206.

[12] Homily 2 in Ps 48:17 *In Illud: Ne timueritis hom. 2* (PG 55, 511-518).

[13] Ebd. 3 (PG 55, 516 1). For similar condemnations of exploitative practices see homily 61 (62), 3 in Mt. (PG 58, 591-592); Laz 1, 12 (PG 48, 980-982).

[14] Homily 66 (67) 3, 4 in Mt (PG 58, 629-631.

[15] The German theologian Adolf von Harnack used this expression for Saint John Chrysostom, underlining in this way his excellent biblical knowledge (Adolf Martin Ritter, *John Chrysostom – a portrait in brief* in *Studia Chrysostomica ... ,* p.129-130)

biblical figures like Abraham, Jacob or Job were wealthy and at the same time loved by God.

Saint John Chrysostom does not limit his action by observing and describing poverty and wealth in Antioch, but he proposes also concrete solutions on how poverty could be limited, or even abolished in his native city. Some of his proposals sound realistic and could theoretically be implemented by those who are really committed to follow the Gospel of our Lord Jesus Christ, while some are idealistic or even utopic and caused extensive debates among modern scholars on the nature of social thinking of Saint John Chrysostom. I am going to present in the following pages of this paper a few principals of Saint John`s reflections on overcoming poverty.

1. *The principle of ìsonomìa.* In Saint John Chrysostom's writings one finds articulated the principle of *ìsonomìa* i. e. "equality of rights" which is "a central notion of platonic social utopia."[16] In its classical form, the principle of *ìsonomìa* contains the aspect of "equitable distribution" of earthy goods, as well as of the nature: sky and earth, sun, moon and stars, air and sea, fire and water; life and death are equally accessible to all human being independently of their race, religion or social status. Saint John extends the platonic notion of *ìsonomìa* to the "spiritual things" which are the sharing of the table of the Lord, Baptism as washing of the sins and regeneration of the human nature and promise of the Kingdom of God, equal access to salvation, as well as to the unspeakable eschatological goods that "eye has not seen and ear has not heard" (I Cor. 2:9)[17] This extension of the principals of *ìsonomìa* seems to be the main goal for Saint John Chrysostom social thinking. Equality of rights has for him not only natural roots, but also spiritual. Because the reality of Antioch showed huge differences between its citizens, Saint John Chrysostom appeals to his believers to overcome or at least to decrease the differences between rich and poor primarily based on the spiritual principals of *ìsonomìa.*

2. *Saint John Chrysostom urged his Christians to see Jesus Christ Himself in the poor.* In his period as priest in Antioch, Saint John Chrysostom delivered a series of homilies on the Gospel of Mathews. It is the first

---

[16] Adolf Martin Ritter, *Zwischen "Gottesherrschaft" und "einfachem Leben": Dio Chrysostomus, Johannes Chrysostomus und das Problem einer Humanisierung der Gesselschaft* in *Studia Chrysostomica* ... p. 42-43.

[17] Hom 2 in Ps. 48, 17, 4 (PG 48, 517-518).

complete commentary on the Gospel of Mathews in the history of Christianity. In Homily 50 (51) he explains the verse of Mathews 14, 23-36 where it says that our Lord was touched by the people living in the land of Gennesaret. It seems that in the opinion of the Antiochian Christians, the contemporaries of our Lord enjoyed a special and unique privilege, which was to live in the same period with our Lord Jesus Christ and to have the chance to touch His Body. Saint John Chrysostom explains to his audience that the Christians today have actually the same privilege as the contemporaries of Jesus: they can enjoy the real presence of Jesus Christ in the Eucharist. Along with this "Eucharistic explanation" that was to be expected from a theologian like Saint John Chrysostom one finds an additional one: the Eucharist is not the only one way in which we can touch our Lord: all Christians, from all places and all ages may touch him, because He is present also in the poor. This identification of our Lord with the poor is so serious that he continues by saying:

> "For what is the profit, when His table indeed is full of golden cups, but He perishes with hunger? First fill Him, being a hungered, and then abundantly deck out His table also. Dost thou make Him a cup of gold, while thou givest Him not a cup of cold water? And what is the profit? Dost thou furnish His table with cloths bespangled with gold, while to Himself thou affordest not even the necessary covering? And what good comes of it? For tell me, should you see one at a loss for necessary food, and omit appearing his hunger, while you first overlaid his table with silver, would he indeed thank thee, and not rather be indignant? Let this them be thy thought with regard to Christ also, when He is going about a wanderer, and a stranger, needing a roof to cover Him; and thou, neglecting to receive Him, deckest out a pavement, and walls, and capitals of columns, and hangest up silver chains by means of lamps. 5. And these things I say, not forbidding munificence in these matters, but admonishing you to do those other works together with these, or rather before these."[18]

In this text, St. John Chrysostom describes a reality which was obviously present in his own community in Antioch. Rich Christians were ready to spend a lot of gold and other precious metals to decorate the church building. It was considered as a special way to honour the Lord by offering a golden cup or other objects and ornaments to be used in the Liturgy. St. John shows to his flock another and, in his opinion, better way to honour our Lord, which is to help the poor among them. But at the same time St. John Chrysostom is realistic. He does not propose to Antiochians to stop

[18] *Homily on the Gospel of Mathews,* 50, 2-5.

offering precious objects for decorating the church building.[19] He knew that such a tradition was deeply rooted in the mentality of his believers and was in many ways part of their spirituality. But he invited his rich flock to prioritize their generosity: firstly to help the poor, and only then to take care of the decoration of church buildings.

3. *Each society is based on reciprocity.* St. John Chrysostom was strongly convinced that not only do the poor need the support of wealthy people, but also the poor are useful for wealthy people and can support them. In *Sermon 34 on 1st Corinthians 13, 8-13*, after explaining the famous hymn of love, St. John adds, as usual in his homilies, a "practical application" to his homiletic interpretation. He affirms that we as human beings are linked with one another by a thousand ties, and that each needs the other. Thus, he continues, the poor need the rich and vice versa. He illustrates this with the image of two cities: the city of the rich and the city of the poor:

> "For it is not, we see, the poor only who need the rich, but the rich also the poor; and these require those more than the others them. And that thou mayest see it more clearly, let us suppose, if it seem good, two cities, the one of rich only, but the other of poor; and neither in that of the rich let there be any poor man, nor in that of the poor any rich; but let us purge out both thoroughly, and see which will be the more able to support itself. For if we find that of the poor able, it is evident that the rich will more stand in need of them.
>
> Now then, in that city of the affluent there will be no manufacturer, no builder, no carpenter, no shoe-maker, no baker, no husbandman , no brazier, no rope-maker, nor any other such trade. For who among the rich would ever choose to follow these crafts, seeing that the very men who take them in hand, when they become rich, endure no longer the discomfort caused by these works? How then shall this our city stand? "The rich," it is replied, "giving money, will buy these things of the poor." Well then, they will not be sufficient for themselves, their needing the others proves that. But how will they build houses? Will they purchase this too? But the nature of things cannot admit this. Therefore they must needs invite the artificers thither, and destroy the law, which we made at first when we were founding the city. For you remember, that we said, "let there be no poor man within it." But, lo, necessity, even against our will, hath invited and brought them in. Whence it is evident that it is impossible without poor for a city to subsist: since if the city were to continue refusing

[19] Donation was a widespread practice in Antiquity, both in pagan as well as in Christian tradition. Wealthy people used to donate goods to temples or to build public edifices at their own expense. Donors enjoyed popularity and were treated with high respect.

to admit any of these, it will be no longer a city but will perish. Plainly then it will not support itself, unless it shall collect the poor as a kind of preservers, to be within itself.

But let us look also upon the city of the poor, whether this too will be in a like needy condition, on being deprived of the rich. And first let us in our discourse thoroughly clear the nature of riches, and point them out plainly. What then may riches be? Gold, and silver, and precious stones, and garments silken, purple, and embroidered with gold. Now then that we have seen what riches are, let us drive them away from our city of the poor: and if we are to make it purely a city of poor persons, let not any gold appear there, no not in a dream, nor garments of such quality; and if you will, neither silver, nor vessels of silver. What then? Because of this will that city and its concerns live in want, tell me? Not at all. For suppose first there should be need to build; one does not want gold and silver and pearls, but skill, and hands, and hands not of any kind, but such as are become callous, and fingers hardened, and great strength, and wood, and stones: suppose again one would weave a garment, neither here have we need of gold and silver, but, as before, of hands and skill, and women to work. And what if one require husbandry, and digging the ground? Is it rich men who are wanted, or poor? It is evident to every one, poor. And when iron too is to be wrought, or any such thing to be done, this is the race of men whereof we most stand in need. What respect then remains wherein we may stand in need of the rich? except the thing required be, to pull down this city. For should that sort of people make an entrance, and these philosophers (for I call them philosophers, who seek after nothing superfluous) should fall to desiring gold and jewels, giving themselves up to idleness and luxury; they will ruin everything from that day forward."[20].

From the context, we understand that the poor are here identified with the workers and ordinary people, effectively with those who earned their daily bread working "with their own hands", while the rich are those who possess enough goods to be able to hire other people for everything they needed. However, the imagine of "two cities" created by St. John`s imagination illustrates clearly that he pledged for a strong social cohesion which is absolutely necessary for a functional society.

4. *A statistic aimed to serve overcoming poverty in Antioch.* St. John Chrysostom even developed a statistic on the situation of the Antiochian population, with which he connects a concrete proposal on how poverty could be overcome in his city. According to Saint John, one-tenth of the

---

[20] St. John Chrysostom, *Homilies on the Epistles of Paul to the Corinthians*, Nicene and Post-Nicene Fathers 1-12, Edinburgh, 2002, *Homily 34,8 on I Corinthians*, pp. 282-283.

Antiochian population were wealthy, one-tenth were poor, which means that they were without possession of any kind, while the rest belonged to the "middle position." Much more important than the statistic is the solution proposed for overcoming poverty in Antioch. The church, whose wealth is also evaluated by St. John – the church possesses the income of one of the very rich and one of those of moderate means – is given as an example of deep social engagement. After describing it, the conclusion is a matter of simple calculation: "if only ten of the wealthy were willing to spend as much as the church, poverty would be banished from Antioch:"[21]

> "For ... let us inquire, if it seem good, which are more numerous in the city, poor or rich, and which they, who are neither poor nor rich, but have a middle place. As, for instance, a tenth part is of rich, and a tenth of the poor that have nothing at all, and the rest of middle sort. Let us distribute then amongst the poor the whole multitude of the city, and ye will see the disgrace how great it is. For the very rich indeed are but few, but those that come next to them are many; again, the poor are much fewer than these. Nevertheless, although there are so many of those who are able to feed the hungry, many go to sleep in their hunger, not because those that have are not able with ease to succor them, but because of their great barbarity and inhumanity. For if both the wealthy, and those next to them, were to distribute amongst themselves those who are in need of bread ... the church is possessed of a revenue of one of the lowest among the wealthy, and not of the very rich, consider how many widows it succors every day, how many virgins; for indeed the list of them hath already reached unto the number of three thousand. Together with these, she succors them that dwell in the prison, the sick ..., those that are absent from their home, those that are maimed in their bodies, those that wait upon the altar. So that if ten only were thus willing to spread, there would be no poor."[22]

5. *Wealth needs to be shared with the poor.* I already mentioned above a general opinion of St. John Chrysostom on wealth, and that he does not consider it as being a necessary evil. We can find in his writing extended debates on wealth and its nature. One of them is part of his exegesis on 1 Timothy. The exegesis on 1 Tim. 4: 1-10 gives to St. John the opportunity to make an extensive evaluation of wealth. He addresses the question as to what extent wealth is good as "everything created by God is good" and "nothing is to be rejected if it is received with thanksgiving" (1 Tim. 4:4).

---

[21] Adolf Martin Ritter, *John Chrysostom as an Interpreter of Pauline Social Ethics* in *Studia Chrysostomica ...*, p. 75.

[22] *Homily on the Gospel of Matthew* 66, 3.

"For if worldly goods, which are uncertain and perishable, are never gained by men without labor and pains, much less are spiritual. Well, saith one, but some inherit them. Yet even when inherited they are not guarded and preserved without labor, and care, and trouble, no less than those have that have gained them. And I need not say that many who have toiled and endured hardships have been disappointed at the very entrance of the harbor, and an adverse wind has caused the wreck of their hopes, when they were upon the point of possession. But with us there is nothing like this. For it is God who promised, and that "hope maketh not ashamed." (Rom. v. 5.) Ye who are conversant with worldly affairs, know ye not how many men, after infinite toils, have not enjoyed the fruit of their labors, either being previously cut off by death, or overtaken by misfortune, or assailed by disease, or ruined by false accusers, or some other cause, which amidst the variety of human casualties, has forced them to go with empty hands?

But do you not see the lucky men, says one, who with little labor acquire the good things of life? What good things? Money, houses, so many acres of land, trains of servants, heaps of gold and silver? Can you call these good things, and not hide your head for shame? A man called to the pursuit of heavenly wisdom, and gaping after worldly things, and calling them "goods," which are of no value! If these things are good, then the possessors of them must be called good. For is not he good, who is the possessor of what is good? But when the possessors of these things are guilty of fraud and rapine, shall we call them good? For if wealth is a good, but is increased by grasping, the more it is increased, the more will its possessor be considered to be good. Is the grasping man then good? But if wealth is good, and increases by grasping, the more a man grasps, the better he must be. Is not this plainly a contradiction? But suppose the wealth is not gained wrongfully. And how is this possible? So destructive a passion is avarice, that to grow rich without injustice is impossible. This Christ declared, saying, "Make to yourselves friends of the Mammon of unrighteousness." (Luke xvi. 19.) But what if he succeeded to his father's inheritance? Then he received what had been gathered by injustice. For it was not from Adam that his ancestor inherited riches, but, of the many that were before him, some one must probably have unjustly taken and enjoyed the goods of others. What then? he says, did Abraham hold unrighteous wealth; and Job, that blameless, righteous, and faithful man, who "feared God and eschewed evil"? Theirs was a wealth that consisted not in gold and silver, nor in houses, but in cattle. Besides this, he was enriched by God. And the author of that book, relating what happened to that blessed man, mentions the loss of his camels, his mares and asses, but does not speak of treasures of gold or silver being taken away. The riches of Abraham too were his domestics. What then? Did he not buy them? No, for to this very point the Scripture says, that the three hundred and eighteen were born in his house. (Gen. xix. 14.) He had also sheep and oxen.

Whence then did he send gold to Rebekah? (Gen. xxiv. 22; xii. 16.) From the gifts which he received from Egypt without violence or wrong.

Tell me, then, whence art thou rich? From whom didst thou receive it, and from whom he who transmitted it to thee? From his father and his grandfather. But canst thou, ascending through many generations, show the acquisition just? It cannot be. The root and origin of it must have been injustice. Why? Because God in the beginning made not one man rich, and another poor. Nor did He afterwards take and show to one treasures of gold, and deny to the other the right of searching for it: but He left the earth free to all alike. Why then, if it is common, have you so many acres of land, while your neighbor has not a portion of it? It was transmitted to me by my father. And by whom to him? By his forefathers. But you must go back and find the original owner. Jacob had wealth, but it was earned as the hire of his labors.

But I will not urge this argument too closely. Let your riches be justly gained, and without rapine. For you are not responsible for the covetous acts of your father. Your wealth may be derived from rapine; but you were not the plunderer. Or granting that he did not obtain it by robbery, that his gold was cast up somewhere out of the earth. What then? Is wealth therefore good? By no means. At the same time it is not bad, he says, if its possessor be not covetous; it is not bad, if it be distributed to the poor; otherwise it is bad, it is ensnaring. "But if he does not evil, though he does no good, it is not bad," he argues. True. But is not this an evil, that you alone should have the Lord's property, that you alone should enjoy what is common? Is not "the earth God's, and the fullness thereof"? If then our possessions belong to one common Lord, they belong also to our fellow-servants. The possessions of one Lord are all common."[23]

The main criterion for an objective evaluation of wealth is not its origin but also, and above all, how is it used. St. John is strongly convinced that wealth belongs to God, and a human being who is entitled to administer the wealth temporarily "should not try to keep it for oneself."[24]

6. *A call for a humble and Christian life style.* St. John called his flock to a humble and simple life-style. While those who are poor have anyway a humble life-style because they have no other choice, his call for a humble life is addressed to those who are wealthy. In order to convince his audience to such a simple life, St. John brings Biblical examples. For

---

[23] *The Homilies of St. John Chrysostom on the Epistles of St. Paul The Apostle to I Timothy, NPNF 1-13,* Homily 12, 4, pp. 597-599.

[24] Adolf Martin Ritter, *John Chrysostom as an Interpreter of Pauline Ethics* in *Studia Chrysostomica ...*, p. 76.

instance, St. Paul worked with his own hands and lived very simply and this brought him great blessings:

> "Did Paul then suffer reproach, and art thou impatient? Did Paul labor, and wouldest thou live luxuriously? But had he lived luxuriously, he would never have attained such great blessings."[25]

Wealthy people must help those who are starving and are homeless. This is one of the main ideas of St. John's ethics. That is why in his writings one can find very sharp criticisms addressed to those women who use exaggerated jewelry instead of feeding the poor. His criticism goes so far that he treated with anathema those who were living in luxury. There is no evidence that his threats of this kind became practice, but they are evidence of how committed was Saint John to the commandments of the Gospel.

7. *Two idealistic or utopian solutions?* There are two solutions for overcoming poverty proposed by St. John Chrysostom which could be labelled as "idealistic" or even "utopian."

A. The first one refers to heritage and how to deal with the goods that normally are transmitted by the parents to their children. We saw before that a wealth obtained through heritage is an acceptable reality for St. John. However in one on his sermons on the Gospel of Matthew, St. John Chrysostom advises his flock not to simply transmit their goods to their children, but give at least part of them to the poor as the best way for using them for the salvation of the soul:

> "And what, it will be said, are our children to inherit? The principal remains, and the income again is become more abundant, the goods being stored up for them in Heaven. But are ye not willing to do this? At least do it by the half, at least by the third part, at least by the fourth part, at least by the tenth. For owing to God's favour, it were possible for our city to nourish the poor of ten cities."[26]

St. John is very serious in his proposal to give one's wealth to the poor and not to transmit it to one's children. As the end of the quotation from above shows, this can be for him the solution for overcoming poverty in Antioch. However, St. John is realistic and urges his audience to try to dedicate to the poor at least part of their wealth: half of it, if not the third part, the

[25] Homily on I Timothy, 12, 4, (see footnote 23), p. 597.
[26] *Homily on the Gospel of Matthew* 66, 4.

fourth part or at least the tenth part. We know from his biography that Saint John shared with the poor some of the wealth he inherited. In this way, he showed to all of us that his proposal can be followed by those who really live according to and for the Gospel.

B. A second proposal that could easily be called idealistic or utopian is formulated by St. John in Homily 11, 3 on the book of the Acts of the Apostles, which was delivered in Constantinople. While explaining the episode of Sapphira and Ananias from Acts, chapter 5, and the way in which the first Christian community living in Jerusalem was sharing all its goods, St. John expresses his conviction that the model of the Jerusalemite community, as described in the book of Acts and interpreted by him, is capable of being implemented in the capital of the Roman Empire:

> "For this is the foundation of all that is good, this of which he now for the second time makes mention, exhorting all men to the contempt of riches: "Neither said any of them that aught of the things he possessed was his own," "but they had all things common." For that this was in consequence not merely of the miraculous signs, but of their own purpose, is manifest by the case of Sapphira and Ananias. "And with great power gave the Apostles witness," etc. (v. 33.) Not in word, but with power the Apostles exhibited their testimony of the Resurrection …And great grace," it says, "was upon them all; for neither was there any among them that lacked." (v. 34.) This is why the grace (was upon them all,) for that "there was none that lacked:" that is, from the exceeding ardor of the givers, none was in want. For they did not give in part, and in part reserve: nor yet in giving all, give it as their own. And they lived moreover in great abundance: they removed all inequality from among them, and made a goodly order. "For as many as were possessors," etc. And with great respect they did this: for they did not presume to give into their hands, nor did they ostentatiously present, but brought to the Apostles' feet. To them they left it to be the dispensers, made them the owners, that thenceforth all should be defrayed as from common, not from private, property. This was also a help to them against vain-glory. If this were done now, we should live more pleasant lives, both rich and poor, nor would it be more pleasant to the poor than to the rich themselves. And if you please, let us now for awhile suppose all here, men and women, to empty out their whole property, lands, possessions, houses,—for I will not speak of slaves, since at that time there was no such thing, but doubtless such as were slaves they set at liberty,—perhaps ten hundred thousand pounds weight of gold would be the amount collected: nay, twice or thrice as much. For consider; at what number of "*juga*" (yokes) is our city rated? How many (of the population) shall we say are Christians? shall we say an hundred thousand, and the rest Greeks and Jews? Then what thousands (of pounds) of gold would be collected! And what is the number of poor? I do not think more than fifty thousand.

> Then to feed that number daily, what abundance there would be! And yet if the food were received in common, all taking their meals together, it would require no such great outlay after all. But, you will ask, what should we do after the money was spent? And do you think it ever could be spent? Would not the grace of God be ten thousand fold greater? Would not the grace of God be indeed richly poured out? Nay, should we not make it a heaven upon earth? If, where the numbers were three thousand and five thousand, the doing of this thing had such splendid success, and none of them complained of poverty, how much more glorious would this be in so vast a multitude? And even of those that are without, who would not contribute?—But, to show that it is the living separately that is expensive and causes poverty, let there be a house in which are ten children: and the wife and the man, let the one work at her wool, the other bring his earnings from his outdoor occupation: now tell me, in which way would these spend most? by taking their meals together and occupying one house, or by living separately? Of course, by living separately. For if the ten children must live apart, they would need ten several rooms, ten tables, ten attendants, and the income otherwise in proportion. Is it not for this very reason, that where there is a great number of servants, they have all one table, that the expense may not be so great? For so it is, division always makes diminution, concord and agreement make increase. The dwellers in the monasteries live just as the faithful did then: now did ever any of these die of hunger? was ever any of them not provided for with plenty of everything? Now, it seems, people are more afraid of this than of falling into a boundless and bottomless deep. But if we had made actual trial of this then indeed we should boldly venture upon this plan. What grace too, think you, would there not be! For if at that time, when there was no believer but only the three thousand and the five thousand: when all, throughout the world, were enemies, when they could nowhere look for comfort, they yet boldly entered upon this plan with such success; how much more would this be the case now, when by the grace of God there are believers everywhere throughout the world? What Gentile would be left? For my part, I think there would not be one: we should so attract all, and draw them to us. But yet if we do but make fair progress, I trust in God that even this shall be realized. Only do as I say, and let us successfully achieve things in their regular order; if God grant life, I trust that we shall soon bring you over to this way of life."[27]

I have allowed myself to give here such a long quotation in order to demonstrate that St. John was really convinced that sharing property between Christians was capable of being implemented in Constantinople in his time, even more than in Jerusalem at the beginning of Christianity.

---

[27] St. John Chrysostom, *Homilies on the Acts of the Apostles 11,3,* Nicene and Post Nicene Fathers, 1-11, Edinburgh, 2002, pp. 108-110.

In this way poverty could be overcome, the grace of God would be abundant and many non-Christians would convert to Christianity. How was it possible for St. John to propose such a solution that has rightly been called by modern scholars a social utopia[28] or proto-socialism?[29] A possible explanation is that for St. John the Jerusalemite model of sharing goods was practised in his time in monasteries and by big families. Particularly the monastic model played an important role for St. John, and served as evidence that sharing goods was a reality of his time for those who were ready fully to follow the Gospel. For St. John monasteries were the so called *societas perfecta* (perfect society) where the Gospel was authentically lived out and served as the ideal model for the entire Church.[30]

***

After this presentation in which I have attempted to make St. John Chrysostom speak to us, I hope it is clear why he was characterized as being, probably more than other Church Fathers, a "lover of the poor."[31] His love for the poor as well as his constant and tireless appeal to the rich and middle class citizens of Antioch to have mercy on them is based primarily on the Biblical message.[32] He lived the Gospel and tried to implement its principles in the life of his flock. His call for overcoming poverty was accompanied by concrete solutions even if some of them were idealistic and/or utopian. However, one must be clear: that St. John never tried to impose his solutions by force, just as no Church Father did anything of the kind. The historical impact (*Wirkungsgeschichte*) of his

---

[28] Wolf Dieter Hausschield, *Christentum und Eigentum. Zum Problem eines altchristlichen Sozialismus* in ZEE 16, 1972, p. 48.; Robert M. Grant, *Christen als Bürger im Römischen Reich,* Göttingen, 1981, p. 135.

[29] Robert von Pöhlmann, *Geschichte der sozialen Frage und des Sozialismus in der antiken Welt, Munich 1925,* vol. 2, p. 476-477 and 484-486.

[30] Adolf Martin Ritter, (see footnote 75)…, p. 197.

[31] Mayer, *John Chrysostom on Poverty* in Pauline Allen (ed.), *Preaching Poverty in Late Antiquity,* Leipzig, 2009, p. 110.

[32] Mayer identifies several philosophical ideas of different provenances (Stoic, Cynic, Platonic, Aristotelian, Epicurean) that might have influenced St. John Chrysostom`s thinking on social issues and poverty (Mayer, *John Chrysostom on Poverty* ...., p. 81. However, this hypothesis cannot lead to Mayer`s conclusion that there is no major difference between Chrysostom`s opinion on poverty and that of the pagan Libanios (Mayer, *John Chrysostom on Poverty* ...., pp. 110-111). The same disagreement with Mayor`s opinion was formulated by Adolf Martin Ritter (see footnote 80), p. 65.

social thought confirms this affirmation.[33] In regard to the actuality of his social thought for today's struggle against poverty, I will let you decide if St. John`s ideas are still meaningful today, because for me the response can be only a positive one.

---

[33] Ritter, *Interpreter of Pauline Social Ethics* … (see footnote 75), p. 90.

REMEMBER—NEVER FORGET: THE WOMAN WITH THE ALABASTER JAR

LINDA ELAINE VOGT TURNER

MA Simon Fraser University, Vancouver Canada
Chair: Renew it. Do it. Green the Scene! 2012 Inter-Ecothee Symposium.
Bethany-Newton United Church, Metro Vancouver
Email: *l.e.turner@shaw.ca*

Abstract

Ecological Justice and Poverty are sustainability issues. Shockingly, water, oil and wealth are used and controlled by a small powerful elite. But without access to water and fuel, the poorest persons in the world suffer horrific deaths and as more and more water and fuel are consumed by a dwindling elite few, a growing number of poor people in obscure villages or garbage strewn ghettos, will suffer and die without any hope or power of sustaining their populations beyond the present age.

This paper examines the Gospel story of the Woman with the Alabaster Jar to ask, "Has the lust after heavenly treasure contributed to the ecological crisis today?" Have Christian theologians and disciples forgotten the choice given to the Woman with the Alabaster Jar: She could sell her treasure and give all the money she received to the poor and gain "more" treasure in heaven. Or she could anoint the Teacher she believed was her Messianic partner with her priceless treasure. She chose to anoint the Teacher. The Cross that ensued caused her to be separated from him without any certainty as to the likelihood of her being reunited with him. Witnessing the Crucifixion and suffering the separation and the physical pain the whole ordeal brought her is an integral part of the Gospel Story. She and the Teacher she loved accepted this Crucifixion—as did their beloved Martha, the Lady of Bethany—believing it and the Resurrection would usher in a new era where God's Love would bring victory to the people of God. We Christians need to remember what she did, so that everywhere the Gospel is proclaimed people will experience God's Resurrection. Divine Love brings the truth out into the open and puts an end to unjust laws and policies that exploit the resources of women and marginal people.

1. Introduction

In the story of the Woman with the Alabaster Jar, Jesus the Teacher says. "You will always have the poor with you and you can help them whenever you wish; but you will not always have me. She has done what she could; she has anointed my body beforehand for its burial. Truly I tell you, wherever the good news is proclaimed in the whole world, what she had done will be told in remembrance of her" (Mark 14: 7-9). Jesus says this in defense of the Woman because the disciples have scolded her for what they see as a waste of her rich resources. They think she should have sold her Alabaster Jar and its rich fragrant contents and given the money to the poor. Had she done that, the Teacher's body would not have been anointed. So perhaps the betrayal, crucifixion and burial that follows would not have happened. And then of course there would have been no need for a resurrection or a new Gospel era. Today their disciples are presented with a similar dilemma. They are called to witness this anointing and reflect upon its significance, and heed the Teacher's words. "Wherever the [Gospel] is proclaimed…what she has done will be told in remembrance of her."

2. Remember—Never Forget: The Woman with the Alabaster Jar

2.1 God's Love

As many people would agree, money is not what the poor need. They need resources and they need the freedom to use their resources to sustain themselves and their communities. They also need skills and the loving encouragement of their neighbours. Failing that they need the love and inspiration that comes from God. In my mind, as long as God's Love lives in the hearts and minds of the poor and the rich…poverty reduction is possible. Should God's Love die, be dismissed and rejected as something of little significance, the poor will remain dependent on the help of people who are kind.

So how does God show love? Isn't God's Love shown through neighbourly kindness? What does one do when one's neighbour and/or family have turned their back on them? What happens when one is suddenly shunned by one's neighbours and nobody is able or willing to talk with them or help them develop their resources? Now if one is a megalomaniac, one probably deserves to be shunned. But surely if one is a sincere lover of God, God will help them find a way to overcome their bleak situation. How do we know this? The Bible is filled with rich and poignant examples. The Woman with the Alabaster Jar is one.

2.2 The Woman with the Alabaster Jar

The story of the Woman with the Alabaster Jar appears in all four Gospels. Until 1969, Roman Catholic tradition taught that this Woman was Mary Magdalene. Since then, feminists and Protestants have refuted this claim, that Mary Magdalene was the Woman with the Alabaster Jar. They claim there is no explicit evidence to warrant such a claim (Vogt Turner 2011). However whoever this woman was historically, this woman was a faithful steward of her own resources. She had rich resources that she had control of and did not sell. Rather she used them to anoint Jesus the Teacher in a very dramatic way to declare her love and faith in him as the Christ. Yet instead of rallying round her, the disciples scolded her. In all but one of the stories, the anointing takes place in Bethany as does the Ascension and the Commission to proclaim the good news everywhere (Luke 24:50-53; Mark 16:19-20).

Without explicit evidence, people today insist that Mary Magdalene is best remembered as the one named at the foot of the Cross and the first recorded witness to the Resurrection (Vogt Turner 2011; Mark 15:40; Matthew 27: 56; John 19:25; Luke 23:49). However, I would like to point out that the Woman Jesus asks the disciples to remember everywhere they proclaim the Good News of the Resurrection is the Woman with the Alabaster Jar.

Personally, I think the Woman with the Alabaster Jar is the Magdalene, simply because Jesus asks the disciples to remember her and what she has done every time they proclaim the Good News of the Resurrection. A person needs a name or a title to be truly remembered for what one has done. More importantly, outside the tomb, Jesus speaks "face to face" with the Woman people identified as Mary Magdalene. Therefore it is a safe assumption. The Woman who anointed the Teacher in Bethany and the Woman coming to the tomb in the name of Mary Magdalene are the same person.

For me, the Alabaster Jar is a powerful symbol. It symbolizes the Woman's body and her most treasured resource. She does not sell them, lusting after more treasure in heaven. She uses them according to God's purpose (Romans 8:28).

I brought an Alabaster Jar to this conference for a very important reason. I wanted to remind this ecumenical body that God's Love and the faithful stewardship of our resources are more important than money. We need to continue to show our love as one ecumenical family and walk the way of the Cross with our neighbour. This is what we are about. In more secular language, we need to continue to show our love and solidarity with our neighbour and work toward a just sustainable ecological world that loves God with our whole heart, mind, body and soul and our neighbour as our self. For me, this is the cornerstone of Christianity. This is what the Story of Bethany and the Alabaster Jar represents.

2.3 Loving God Holds us Responsible to our Neighbour and the Truth

We who profess to be the followers of Jesus Christ are the living body of Jesus Christ and this means we have a responsibility to our neighbour. We have been blessed with the Good News and the promise of Eternal Life so that all the peoples of the Earth may have life and have it in abundance. If we sell all that we possess, hoping for more treasure in heaven, we will indeed get what we hope for, treasure.

Trusting in God's Eternal Life, we need to realize that Love for God, Neighbour and Self is the cornerstone of our faith. This cornerstone is our priceless treasure. Without this cornerstone we call Jesus Christ, we will not achieve Eternal Life. This is what I believe the Woman from Bethany had in her heart and what I truly believe the Alabaster Jar represents. Without a container for our souls, the sweet perfume of who we are, we cannot inherit Eternal Life. Sure our spirits may continue to live in space without a body. But without a body and without a heart, it will make little difference whether we have treasure or not in heaven.

To be the living temple of God, the dwelling place for God, we need Love and to love we need a heart, a mind and a body. In addition, we need earthly resources and access to those earthly resources to sustain our earthly home and our bodies. To preserve our earthly home and our bodies, we need to preserve our heart. We need to remember what the

Heart, the Mind, and the Gracious beloved body of believers have done every time we tell the Good News of the Resurrection.

Money and kindness matter. Giving money to the poor and finding ways to reduce poverty are kind and good things to do. Yet, loving our neighbour is more powerful. Loving our neighbour means being in relationship with our neighbour and working with our neighbour to find ways to have a just relationship. As we do this, we may bring monetary prosperity to the whole community and reduce poverty. But we might not. The important thing to remember is that Love is patient and Love is not satisfied with injustice. Love finds a way to lift up the Truth and bring it out into the open to expose injustice.

2.4 Remembering What the Woman of Bethany did is Foundational

Remembering what the Woman of Bethany did is foundational to Christianity. You will recall: in Mark's Gospel, Jesus says. "For you always have the poor with you, and you can show kindness to them whenever you wish; but you will not always have me. She has done what she could; she has anointed my body beforehand for its burial. Truly I tell you, wherever the good news is proclaimed in the whole world, what she has done will be told in remembrance of her."

What did she do? She did not hide or cover up the Truth. She professed her faith and love for the Teacher openly, knowing that it would bring death to their relationship. In my view, without her, the Cornerstone of our faith is an unfulfilled promise. Remembering and vindicating both of them is the first step to realizing lasting Peace with Justice on Earth and in Heaven—the Gospel.

She anointed the Teacher she believed was her Messianic partner with her priceless treasure—her faith and love. Betrayal, denial, and torture followed.

To remember what she has done is important to my understanding of the Gospel. I hope it is important to yours as well. Being a witness to the Truth means walking the way of the Cross with one's neighbour.

I'm convinced that the Crucifixion was extremely painful and hard to endure for both the Woman and the Teacher she loved. And I am convinced that they both accepted this Crucifixion—as did their beloved Martha, the Lady of the House—believing that the forgiving power of God's Love would reunite the three of them in a right relationship and restore all Creation in the process.

2.5 Concluding Thoughts

We Christians have been kind, and in our strong desire to be kind we have contributed to the Ecological Crisis. We have kept quiet. We have not been willing or able to tell the Truth of the Gospel. Sure, we have hinted at it or told the truth in such a way as to let people figure it out for themselves. But we have not told it boldly. We have kept it hidden in patriarchal language and cultural traditions that protect the powerful elite—the Teacher and his beloved Martha—at the expense of the Woman with the Alabaster Jar. It is time to do more. It is time to stand up for the Truth and recognize that the Woman with the Alabaster Jar is an integral part our faith story. Without her there is no Resurrection.

From the pages of scripture she comes alive again and again to testify to the Truth and to beseech Christians everywhere to testify to the Resurrection Truth. Poverty reduction and ecological justice in every generation depend upon holding ourselves and our leaders accountable to the Truth and being in right relationship with God, Neighbour, Self and all Creation. From a position of being in "right relationship" with each other the stone guarding the tomb will roll away. Thereupon we will find a way to be forgiving, kind and just—if we Remember what Love has done everywhere the Gospel is preached. So as we remember, let us work together to find ways that all God's people can have access to and control of the resources we need to live and Renew all Creation as we live together forever.

References

The Holy Bible (1990), New Revised Standard Version. Nashville: Thomas Nelson Publishers.

Vogt Turner, Linda Elaine (2011), Mary Magdalene: Her Image and Relationship to Jesus. Vancouver: Simon Fraser University. https://theses.lib.sfu.ca/thesis/etd6959

HUMAN ACTIVITY AND THE NATURAL ENVIRONMENT: A PHILOSOPHICAL APPROACH

K. KALACHANIS,* I. MATIATOS, C. KOTAMPASI, P.STAMATIS AND P.NICOLOPOULOU-STAMATI

University of Athens, Medical School, M.Sc. "Environment and Health, capacity building for decision making"

*To whom correspondence should be send at kalahanis@hotmail.com

Abstract

Human activity is directly related to the physical environment, while it also discloses human attitudes towards nature. In Greek antiquity humanity considered itself an integral part of nature, a perception that originates from both philosophical thought and religious tradition. By contrast, modern civilization treats nature just as a resource to be exploited. As a result many health problems have occurred, while many countries have been condemned to poverty.

1. Introduction

The relationship that humanity develops with the natural environment is not simply related to survival through the use of natural resources, but is a relationship aiming at economic development and improved quality of life. It is clear that people have not just been content to survive in a hostile environment, but have also attempted to improve their life conditions, by using natural resources. This objective has determined their activity and its impacts on the natural environment. In this paper we attempt to examine how the human way of thinking and acting has affected the natural world.

Specifically, we will study human activity from antiquity up to the industrial revolution and the modern era.

2. Man and Environment in Antiquity

Since ancient times, people have thought of themselves as an integral part of the natural environment, having observed that their lives depended on the plant and animal kingdom. Greeks in antiquity understood this relationship between humanity and mother Earth, which was indeed worshipped as a supreme Goddess, who was honoured with festivals, praising the rebirth of nature in Spring. Such a festival was the Anthesteria, which used to take place in the end of the month of Anthesteriona (end of February to the first days of March). The name of the festival comes from the ancient Greek word 'anthos', which means flower. It is beyond doubt that the name of the festival was intended to praise the rebirth of nature, something which created a very cheerful and pleasant atmosphere for the participants.

The perception of the connectedness between humanity and nature was accepted by ancient Greek philosophers, particularly those of the Presocratic period (6th-5th centuries B.C.). The main characteristic of Presocratic thought was the attempt to find naturalistic explanations of the world, by attributing the origin of the Universe and of all beings to an initial first principle. Philosophers like Thales of Miletus, Anaximenes and Empedocles, having observed the life-cycle of all beings, concluded that not only the nature we understand, but also the entire universe is a single entity. The philosophers from the city of Miletus (in Asia Minor) used elements from their everyday life to symbolize this entity. Specifically, Thales regarded 'hydor' (water) as the first principle of all beings, while Anaximenes held that the first cause was 'aer' (wind). Also Anaximander intended to show the transcendental nature of the first cause, by naming it 'apeiron' (without borders) [Diogenes Laertius (1964)]. Thus, scholars from Miletus introduced a philosophical conception known as 'monism', according to which the fundamental character of the universe is unity.
Apart from Thales, Anaximander and Anaximenes, Heraclitus, a scholar from Ephesus (a Greek city in Asia Minor), disagreed with the philosophers from Miletus, since he claimed that the first element of the Universe was fire [Diogenes Laertius (1964)]. However, the basic idea of his thought was the origin of the Universe from an initial first cause.
In ways such as these, the ancient Greeks established a holistic approach, which was taken forward later through Christianity. St. Gregory, Bishop of

Nyssa, a great Byzantine philosopher and theologian of the 4th century, upheld the view that humanity is a microcosm of the cosmos, being rightly praised with this name [St. Gregory of Nyssa (1972)].

It is manifest that this view of the Universe had a very significant impact on human activity, since their buildings were in total harmony with the environment. In particular, temples of the god of medicine, Asclepius, were usually built in places which the location, the lush vegetation, the climate, the availability of natural sources of clean water and natural beauty made suitable not only for treatment, but also for worship [Stavropoulos (2000)].

Hippocrates, the famous doctor from the island of Kos, had already understood that since humanity interacts with the environment, there must a direct impact on human health, if this relation is disturbed. That is the reason. This is why he considered that human activity, especially when a city is being built, should take into account the impact suffered by human health. Therefore he considers it essential for the maintenance of human health to study the climatic factors that affect it, such as the quality of wind and water, which it is important to take into account when a city is built. So, the Hippocratic vision of man as integrated in the cosmic system and the environment should be a guideline for how human activity should take place, namely with respect for nature and health [Kalachanis (2011)].

However, is this unity of man and Universe a fact? All the chemical components of the human body (carbon, nitrogen, oxygen etc.), are supernovae remnants, formatted in star nuclei. It is know from the field of astrophysics that the explosion of a star of great mass expels much of its material at a velocity of up to 30,000 km/s (approximately 10% of the *speed of light*), driving a *shock wave* into the surrounding *interstellar medium*. This shock wave sweeps up an expanding shell of gas and dust, called a *supernova remnant*. This gas consists of hydrogen, carbon, oxygen and helium. When these gases collapse through gravity, new stars are created, and also solar systems and of course life. It follows that every single molecule of the human body consists of chemical bonds created in deep space. The stellar origin of matter composing the human body shows our connection with the universe [Danezis, Theodosiou 1999].

Summarizing, it is apparent that according to ancient Greek thought all human activity should take into account the fact that human beings are parts of a whole, which is the environment. Therefore, any harmful intervention in this system may also have harmful impacts on human health.

3. Human Activity and the Environment in the Modern Era

The concept of the harmonious co-existence of humanity with our physical environment was gradually changed due to significant changes in technology, which have in turn triggered changes in economic structures. While the economy from ancient times was based on craft, agricultural production and trade (marine and land), the rapid development of technology changed everything.

A significant relevant landmark is the Industrial Revolution (1750-1850), which initially took place in Britain. During this period the development of the means of production necessitated the use of natural resources. Massive changes in agriculture, manufacturing, transportation and technology had a profound effect on the social, economic and cultural conditions of the times. Since then, the production of goods has greatly increased, and nowadays appeals to wider groups of people; so it becomes imperative for industry to find readily available source materials.

The transition to this new era of industry undoubtedly contributed to improving the quality of life, but also has contributed to the alteration of human attitudes towards the environment. Nature is no longer the "home" of humanity, but rather resources for exploitation. Humankind, instead of considering itself part of nature embedded within this harmony, has placed itself at the top of the environmental system. Beyond the undeniable deterioration of the physical environment, human activity has also resulted in the occurrence of health problems. Impacts of environmental contamination have been observed in many systems of the human body, such as the reproductive, respiratory and endocrine systems, and also a significant increase in cancer rates has been observed, especially in the developed industrial world [Nikolopoulou-Stamati (2007)].

Besides health problems, environmental degradation due to uncontrolled human activity has resulted in economic disaster in many countries of the Third World. Millions of people are condemned to poverty for the sake of economic profit. Therefore humanity needs to examine the priorities of its civilization. Man should not only consume, but care for our home –Earth – because our entire activity has serious impacts on our planet. Also a form

of economic growth is needed not only aiming at profit, but also at the protection of the natural environment. Such a strategy, if adopted by decision makers, would have as its basic parameter the assumption that man is an integral part of the environment requiring mutual respect.
This attitude will combine economic growth with protection of the natural environment and will eliminate poverty. In this respect the primary objective of harmonious living with a natural environment will be fulfilled.

4. Summarizing

It is apparent that from ancient times to the modern era, human attitudes towards the environment have changed dramatically. While the natural environment was treated as the "home" of man, the development of industry and the overconsumption of natural resources has not only caused environmental deterioration, but also condemned millions of people to poverty. Therefore a sustainable development is required which will combine both the protection of the natural environment and the quality of human life.

References

Danezis, E. and Theodosiou, E. (1999): The Universe that I loved: Introduction to Astrophysics, Diavlos Publications, Athens [in Greek].

Diogenes Laertius, (1964): Vitae Philosophorum, ed. H.S. Long, Clarendon Press, Oxford, books 1-2 and 9.

Kalachanis, K. (2011): Environmental impact on Health in the work of Hippocrates, Final Thesis, Post-Graduate Program "Environment and Health, Capacity-building for Decision-making", University of Athens, Medical School [in Greek], 22-30.

Nicolopoulou-Stamati, P. (2007): "Introduction, concepts in the relationship of congenital diseases with the environment" in: Congenital diseases and the environment, Springer pp. 1-14.

St. Gregory of Nyssa. (1972): De creatione Hominis sermo alter: ed. H. Horner, Leiden, Brill, 65,13-14.

Stavropoulos, S. (2000): Asclepeia of Peloponnesus: Aeolos editions, Athens, 20-21 [in Greek].

THE UN HUMAN RIGHTS: A TOOL FOR THE PEOPLE'S AGENDA ON SUSTAINABILITY

THEODOR RATHGEBER[1]

United Evangelical Mission
University of Kassel (Germany), Department on Social Sciences (FB 05).
E-mail: TRathgeber@gmx.net

Abstract

The term 'sustainability' has become prominent since 1987 when the report of the so-called Brundtland Commission was released. The report addressed three issues of major relevance: the needs of the poor, the environment, and indicators of sustainable development. Accordingly, the text deals with issues such as re-organising society beyond merely improving technical knowledge and considers what kind of orientation can be given in favour of a fundamental rethinking. In the text it is argued that human rights comprise such a normative framework for orientation towards a better rule of law, good governance and democratic participation, and that in particular they enable people on the ground to make their choice and to be part of decision-making. The text relates to the United Nations' human rights system, which provides norms, instruments and mechanisms that are relevant to addressing governance effectively in view of ecological crisis, climate change, sustainability, eco-justice and social mobilisation. Within this context, the churches form a genuine

[1] Theodor Rathgeber, Political Scientist (PhD), freelance consultant on human rights, international law and development policies, dealing with human and minority rights. Since 1987 lecturer at the University of Kassel (Germany), Department on Social Sciences (FB 05). Since 2003, official observer to the UN Commission on Human Rights, since 2006 to the UN Human Rights Council assigned by the German NGO network "Forum Human Rights" (Forum Menschenrechte). Since then cooperation with the World Council of Churches on issues such as migration as well as climate change and its adverse impacts on human rights including various publications..

platform for dealing with such topics in terms of human dignity, through strengthening the perspective of the victims and the poor. The option for sustainable development is fundamentally related to a human rights-based approach.

1. Introduction

The term 'sustainability' has become prominent since 1987 when the report of the so-called Brundtland Commission (i.e. *United Nations World Commission on Environment and Development*) was released. The report "*Our Common Future*"[2] declared that development should predominantly meet the needs of the poor while not increasing environmental problems and not hampering the future of coming generations. The report addressed three issues which are relevant for the purpose of this text: the needs of the poor, the environment, and indicators of development in order to meet such ends. This chapter concerns how society should be re-organised beyond improving technical know-how, but also argues in favour of a fundamental rethinking and setting a pertinent normative framework.

In the debates at the academic as well as at activists' level, such re-structuring of society is widely based on key concepts such as rule of law, legal framework, good governance, participation, democratic selection, choice and decision making, indices of a good life or happiness instead of mere growth, the right to say 'No', and re-cognising economy as an ultimately valued-based structure.[3] Out of the multiplicity of fields for discussion, this text concentrates on norms, institutions and procedures suitable for making policies and politics effectively oriented towards sustainable development, providing justice for the poor and, in particular, helping the underprivileged to articulate their own voice.

History shows that intellectual evidence is not automatically translated into politics. Even setting mere incentives towards sustainable development requires more than simply better information and insight but also needs to address power and power structures. Thus, we need to talk about social agents of change too. Overall, this text relates to norms,

---

[2] UN document A/42/427 plus Annex.

[3] See Thomas Sedlacek (2011): *Economics of Good and Evil: The Quest for Economic Meaning from Gilgamesh to Wall Street.* Oxford University Press; see the early involvement of the WCC in the debates on human rights and the ecological crisis at http://fore.research.yale.edu/religion/christianity/projects/wcc_jpc.html.

instruments and mechanisms which are fundamentally relevant to properly addressing governance in view of ecological crisis, climate change, sustainability, eco-justice and social mobilisation. It will be argued that human rights and the United Nations' human rights system form a very pertinent regime in order to give particular stress to the people's agenda dealing with sustainability and political decision- making. Within this context, the churches are a genuine platform to deal with starting from its own background in terms of human dignity, and in particular strengthening the perspective of victims, the poor, and the underprivileged respectively.

Thus, I briefly explains in the following section key aspects of this essay focussing on the relationship between key terms and their normative context in the human rights system. The following two sections then illustrate international institutions and procedures as well as normative standards, embedding the people's agenda related to sustainability. The following section describes the genuine relevance of church institutions, followed by concluding remarks which wrap up what has been presented in the previous sections and relate the issue of sustainability to a rights-based approach.

2. Key Aspects in Brief

> *"A quarter of a century ago, the Brundtland report introduced the concept of sustainable development to the international community as a new paradigm for economic growth, social equality and environmental sustainability. It argued that sustainable development could be achieved by an integrated policy framework embracing all three of those pillars. The Brundtland report was right then, and it remains right today."*[4]

The report of the UN Secretary-General's High-level Panel on Global Sustainability proceeds to argue that sustainable development remains a generally agreed concept, rather than a day-to-day practical reality on the ground. Sustainability and sustainable development need to be understood as a dynamic process. The concept comprises two key aspects to be considered when the present as well as the future of human generations is at stake: conserve resources, promote the most efficient use, replace non-renewable with renewable resources wherever possible, and thus meet the

[4] United Nations Secretary-General's High-level Panel on Global Sustainability (2012): *Resilient People, Resilient Planet: A future worth choosing.* New York: United Nations, p. 11.

needs of the present without compromising the ability of future generations to meet their own needs.[5] In this way, a number of issues for decision-making are addressed such as economy, investments, environment, ecology, patterns of consumption, social and global justice, political discourse and decision-making. The challenge relates to public and private sectors as well as developing a principle of solidarity, strengthening of institutional capacity and empowering the people in order to take their own informed decisions.[6] The report of the UN High-level Panel concludes that despite the dimensions of the challenges, there is no need to feel helpless, or to feel like a victim of determinist forces of history. People have the chance to choose their future.In short, there is one critical condition: democratic governance and full respect for human rights are a must in order to enable people to make their own informed choices.[7]

Sustainability as such is not codified in the UN human rights regime, although in the discussions on post-MDGs (Millennium Development Goals)[8], the UN Conference on Sustainable Development Rio+20[9], and Sustainable Development Goals (SDGs)[10] there is a strong call to make human rights a legally binding reference for any future goals, the process as such and its indicators. The final document of the Rio+20 conference in 2012 in Brazil (*The Future We Want)* [11] generally refers to human rights, rule of law, gender equity and democracy (para. 8), and in subsequent paragraphs in particular to the right to food (para. 108), the right to water and sanitation (para. 121), the right to health (para. 138), and the right to education (para. 229). The UN High Commissioner for Human Rights, Navi Pillay, had pronounced at the eve of the Rio+20 conference that any debate on future development necessarily needs to take human rights into consideration, including principles such as participation, accountability,

---

[5] Op. cit., pp. 9, 15, 21.

[6] See Markus Vogt (2009): *Prinzip Nachhaltigkeit. Ein Entwurf aus theologisch-ethischer Perspektive.* Hochschulschriften zur Nachhaltigkeit 39, München.

[7] United Nations Secretary-General's High-level Panel on Global Sustainability 2012, op.cit., p. 10.

[8] Millennium Development Goals; for details and further links see http://www.un.org/millenniumgoals.

[9] UN Conference on Sustainable Development as follow up to the conference in Rio de Janeiro in 1992; see details and further links at http://www.uncsd2012.org.

[10] For an overview and further links see http://sustainabledevelopment.un.org/index.php?menu=1300.

[11] See UN document A/CONF.216/L.1; accessible via https://rio20.un.org/rio20/documents.

monitoring, non-discrimination and focussing on vulnerable groups.[12] This may sound self-evident, even though informal government groupings such as *G-8*[13] or *G-20*[14] would hesitate to really foster such a statement.

Next to sustainability, the concepts of environment and environmental degradation are critical to identifying a safe livelihood conducive to people's living in dignity today as well as for future generations. Although there is at present no internationally recognised human right to a safe environment – which varies under different regional and national laws[15] – conceptually, environment is a precondition to the enjoyment of human rights, and human rights are conversely tools to address environmental issues. At both levels, procedurally and substantively, both are needed in order to achieve appropriate conditions and levels for human life responding to the very basic requirements for a 'good life' (or *buen vivir*).[16] Thus, the concept of environment includes also access to information, public participation and access to justice beyond topics such as natural resources or public health.

The UN Conference on the Human Environment in 1972 was the first United Nations conference which dealt with the relationship between human rights and environment.[17] In August 1989, the UN Sub-Commission

---

[12] Navi Pillay (2012): *Wanted: Human Rights in Rio.* Speech delivered to a panel discussion on 2 May 2012; accessible via http://www.ohchr.org/EN/NewsEvents/Pages/WantedHumanRightsinRio.aspx; see also Open Letter by Navi Pillay on 30 March 2012, accessible via http://www.ohchr.org/Documents/Issues/Development/OpenLetterHC.pdf.

[13] Group of 8: Canada, France, Germany, Italy, Japan, Russian Federation, United Kingdom, USA.

[14] Group of 20: Argentina, Australia, Brazil, Canada, China, France, Germany, India, Indonesia, Italy, Japan, Mexico, Russian Federation, Saudi Arabia, South Africa, South Korea, Turkey, United Kingdom, USA, European Union.

[15] In 1994, the so-called *Ksentini Report* on human rights and environment found that over 60 countries had established constitutional provisions on environmental protection (para. 241). This number of constitutions including references to environmental rights and responsibilities has increased in 2010 up to 140.

[16] Emerged as concept in the international debate predominantly in the context of indigenous peoples; see e.g. the basics of the concept at http://www.cicaregional.org/leer.php/4084261.

[17] See the Declaration of the United Nations Conference on the Human Environment (Stockholm Declaration), accessible via http://www.unep.org/Documents.Multilingual/Default.asp?documentid=97&articleid=1503.

on Prevention of Discrimination and Protection of Minorities (later called the UN Sub-Commission on the Promotion and Protection of Human Rights), a subsidiary body of the then UN Commission on Human Rights, mandated the preparation of a study of environmental degradation and its relation to human rights. The final report was presented in 1994 and offered for the first time the possibility of comprehensively considering environmental problems with a focus on human rights.[18] Finally, in 1995, the Commission on Human Rights appointed the Special Rapporteur *on the adverse effects of the illicit movement and dumping of toxic and dangerous products and wastes on the enjoyment of human rights*.[19] In 1997, the Sub-Commission discussed a working paper on the right of access of everyone to drinking water supply and sanitation services[20], which in 2008 turned into a mandate for a Special Rapporteur on this subject.[21]

In more recent years, climate change and its adverse effects for the future of the earth in general and for human rights in particular have emerged as a specific and rather urgent issue to be dealt with. The discussion about a normative platform within the context of the United Nations Framework Convention on Climate Change (UNFCCC) [22] started only recently. The Conference of the Parties (COP) No. 13 on the UNFCCC met in December 2007 in Bali, and identified four pillars of a future climate change settlement: mitigation, adaptation, finance, and technology transfer. Since then, the technical management of the impacts of climate change has remained at the heart of the COP meetings. Only COP 16 in Cancún (Mexico, 2010) made a reference to human rights, emphasising that human rights should be fully protected in all climate change-related actions.[23] The Parties to UNFCCC agreed that climate change implicates a

---

[18] See UN document E/CN.4/Sub.2/1994/9.

[19] Today the mandate relates to *the implications for human rights of the environmentally sound management and disposal of hazardous substances and wastes;* for details of the mandate, see http://www.ohchr.org/EN/Issues/Environment/ToxicWastes/Pages/SRToxicWastesIndex.aspx.

[20] See UN document E/CN.4/Sub.2/1998/7.

[21] See Human Rights Council, resolution A/HRC/RES/7/22; for details of the mandate see http://www.ohchr.org/EN/Issues/WaterAndSanitation/SRWater/Pages/SRWaterIndex.aspx.

[22] Adopted in 1992; see details at http://unfccc.int/2860.php.

[23] See UNFCCC (2011): *Report of the Conference of the Parties on its sixteenth session, held in Cancun from 29 November to 10 December 2010*, Addendum Part

bundle of direct and indirect impacts for the full and effective enjoyment of human rights. COP 17 in Durban (South Africa, 2011) did not elaborate further on this subject and thus human rights are still not part of the UNFCCC process as a normative reference.[24]

The people's agenda for sustainable alternatives necessarily comprises also the issue of overcoming poverty. Without going too much into details, the concept of poverty in terms of human rights is understood – in analogy to Armatya Sen – as denial of the comprehensive development of the capacities of a human being, i.e. among others social, political and jurisdictional exclusion. As such, the term 'poverty' does not only refer to economic, social and cultural rights but also to the right to freedom of expression, freedom of assembly and to legal protection as well.[25] Overcoming poverty addresses not only social security but also the State's commitment to generate a public order which fosters human rights and the corresponding articulation of them by the people.

In overcoming poverty, the concept of justice needs to be considered too. Justice is a central principle and stands for the fundamental right to a dignified existence and the right to have access to the pertinent resources that enable such an existence. Although many different aspects can be emphasised, within our context it means that populations and countries affected by poverty as well as by social and political marginalisation should be treated preferentially and given additional support in order to overcome their situations. A normative structure is required which aims to redirect the macro-level and its sources of wealth towards the primary goal of guaranteeing human existence as such, human dignity, and lastly a sustainable form of being. It is within the logic of the subject to refer to human rights as a key element, in order to bring underprivileged people

---

Two: Action taken by the Conference of the Parties at its sixteenth session. Document FCCC/CP/2010/7/Add.1, Operative Paragraph 8; see also Preamble Paragraph 6.

[24] For an overview, see Theodor Rathgeber (2012): *Climate Justice, Human Rights and the Role of Churches*. WCC Publications, Geneva.

[25] Office of the High Commissioner on Human Rights (2004): *Human Rights and Poverty Reduction: A Conceptual Framework.* Document HR/PUB/04/1, Geneva; see further UN Special Procedures (2006); *States must address poverty with utmost urgency, say UN Independent Experts on the occasion of Human Rights Day.* Press Statement, Geneva, 8 December; as well as Magdalena Sepúlveda Carmona (2012): *Final draft of the guiding principles on extreme poverty and human rights, submitted by the Special Rapporteur on extreme poverty and human rights, Magdalena Sepúlveda Carmona*. Document A/HRC/21/39, Geneva.

into a position where they gain access to the minimum standard of living, to participate in the discourse in order to define what justice means for them, what priorities should be set, and to make the victims' needs prominent in the paradigms of development. The UN Declaration on the Right to Development provides, for instance, such a conceptual structure on equity and justice in combining international cooperation and core human rights norms.[26]

3. Institutions and Procedures

The UN High-level Panel on Global Sustainability concludes that sustainable development requires a kind of international governance which itself must be developed and strengthened. Such a process would at best make use of existing relevant institutions, seeking their engagement and dynamic involvement. Against the asymmetric distribution of access for ordinary people to public articulation and the decision-making process, it is of critical importance for them to be able to make reference to a normative structure, to established institutions and to measures which are pertinent to make their concerns and interests articulated. The human rights approach is not just about law, but also stresses the importance of relevant institutions in order to be implemented. Such institutions can be found within the United Nations system and agencies, but also within regional bodies. Among the UN institutions I have selected in this chapter ones which are authorized to assess and monitor the issues mentioned, in conformity with normative standards, and taking into account the viewpoint of victims. Within the UN human rights regime, we will therefore focus on the Human Rights Council, the UN Special Procedures and the UN Treaty Bodies. The Office of the High Commissioner for Human Rights (OHCHR) also forms part of this system.

UN Human Rights Council

The UN Human Rights Council (HRC) has been dealing with subjects like environment, climate change, development and, most recently, with sustainability. On environmental issues, the main threats in terms of

---

[26] Declaration on the Right to Development, UN General Assembly A/RES/41/128 (1986), accessible via http://www.un.org/documents/ga/res/41/a41r128.htm; for the current discussion on the Right to Development see Stephen P. Marks (2011); *The Politics of the Possible. The Way Ahead for the Right to Development.* FES, International Policy Analysis, Geneva.

human rights violations were identified as atmospheric-related environmental impacts such as e.g. air pollution, climate change, ozone-layer depletion, land-based environmental threats such as land degradation, deforestation and desertification, degradation in water quality, freshwater scarcity, threats to oceans such as collapse of fisheries, hazardous waste, chemical contamination and pollution, loss of bio-diversity, and the increase of natural disasters in general.

Since 1995, there has been a mandate of the Special Procedures on the implications for human rights of hazardous substances and wastes[27], and since 2012 a mandate on human rights and the enjoyment of a safe, clean, healthy and sustainable environment. Thus in March 2012, the resolution A/HRC/RES/19/10 established the mandate of an Independent Expert on the issue of "*Human rights obligations relating to the enjoyment of a safe, clean, healthy and sustainable environment*". The mandate holder is entitled to conduct studies, to identify best practices, to make recommendations, to apply a gender perspective, and to coordinate with other Special Procedures.[28] Although the current Independent Expert is a highly recognised person because of his expertise, commitment and sincerity, concerns remain about the mandate, because it is also understood as an instrument to split the demand for a mandate on climate change and thus spreading the dynamics of the topics 'climate' and 'environment'. Since 2010, a controversial debate had emerged about which term – 'environment' or 'climate change' – might be more appropriate, and to address the causes of human rights violations in the context of e.g. hurricanes and floods. Considering the number of people who have been displaced or have migrated or fled because of climate change impacts, there is an urgency but definitively also a political sensitivity about this subject.[29]

---

[27] Op.cit.; see http://www.ohchr.org/EN/Issues/Environment/ToxicWastes/Pages/SRToxicWastesIndex.aspx.

[28] For details see http://www.ohchr.org/EN/Issues/Environment/IEEnvironment/Pages/IEenvironmentIndex.aspx.

[29] For an overview see Bread for the World/ Oikumene Pacific/ World Council of Churches (2010): *Conference on Protection and Reparations for „Climate Refugees": Imperatives & Options.* Compendium of a conference in May 2010 in Chavannes-de-Bogis, Switzerland.

The issue of climate change had become prominent since 2007. The Maldives, supported by other Pacific and Caribbean island nations that will be threatened by rising sea levels, launched in March 2008 the resolution with the document number A/HRC/RES/7/23 on *Human rights and climate change*, which requested the OHCHR to carry out an analytical study focusing on the relationship between human rights and climate change. This report in 2009[30], as well as a later report in December 2011[31], concluded that climate change is interfering with a wide range of human rights, including the rights to life and health, food, water, housing and shelter, and the preservation of specific cultural characteristics in cases of resettlement and migration, but also touches political and civil rights such as freedom of information and opinion, the right to dissent or dispute the environmental or climate policy taken, access to a legal review of decisions, participation in the political opinion-forming and decision-making process, and self-determination as well. The report of 2008 took up further the issue of international cooperation and referred to the disputed principle of '*extraterritorial state obligations*' according to Art. 2.1 of the UN Covenant on Economic, Social, and Cultural Rights. HRC resolution A/HRC/RES/10/4 (2009) invited Special Procedures to consider such issues in their future reports, and launched a panel discussion in June 2009 on the relationship between climate change and human rights.

These initiatives were followed by an HRC Social Forum session on the adverse effects of climate change (resolution 13/17 in 2010). In its recommendations, the Social Forum proposes that HRC establish a new mechanism under the Special Procedures on climate change and human rights. In 2011, the HRC further requested the OHCHR to convene a seminar in order to address the adverse impacts of climate change on the full enjoyment of human rights. The seminar took place prior to the nineteenth session of HRC in March 2012, and made a strong call to forge cooperation between the human rights and climate change affected communities, as well as supporting a special mechanism on the issue of climate change and human rights, monitoring the impacts of climate

---

[30] Office of the High Commissioner for Human Rights – OHCHR (2009): *Report of the Office of the United Nations High Commissioner for Human Rights on the relationship between climate change and human rights.* Document A/HRC/10/61, Geneva.

[31] Office of the High Commissioner for Human Rights – OHCHR (2011): *Analytical study on the relationship between human rights and the environment.* Report of the United Nations High Commissioner for Human Rights. Document A/HRC/19/34, Geneva.

change and corresponding measures on the realisation of human rights. Despite all these efforts, till today no such mandate has been established, and this may reflect the hesitance (or reluctance) among predominantly industrialised countries to deal with climate change in a normative context. Nevertheless, human rights have emerged as a platform within the UN system to allow peoples and individuals affected by climate change to raise their voices.

UN Special Procedures

Within the institutional setting of the UN human rights system, and in particular related to the Human Rights Council, there are so-called 'Special Procedures', or mandates of independent experts who are entitled to undertake in-depth investigations and to present reports on country situations as well as on thematic issues, sometimes to the UN General Assembly too.[32] As mentioned earlier, two mandates of the Special Procedures explicitly deal with the issue of environment. In addition, the Special Rapporteur on the Right to Food has been actively involved in the debates on environmental and climate change effects on food security. He devoted a special report to agro-ecology, and emphasised the recycling of nutrients and energy and the diversity of species. These jointly enhance the sustainability of food systems and their resilience to climate change. Agro-ecology would therefore contribute to a development approach which integrates the local environment and cares for biological tenets and cycles. The Special Rapporteur on the Right to Safe Drinking Water and Sanitation elaborated a position paper specifying benchmarks on the same subject.[33]

With regard to climate change and its adverse effects on human rights, there are several Special Rapporteurs within the thematic mandates involved, such as on the Rights of Indigenous Peoples, the Right to Safe Drinking Water and Sanitation, the Right to Food, the Right to the Enjoyment of the Highest Attainable Standard of Physical and Mental Health, the Right to Adequate Housing, and on Internally Displaced

---

[32] For general information see http://www.ohchr.org/EN/HRBodies/SP/Pages/Welcomepage.aspx.

[33] Olivier de Schutter (2010); *Report submitted by the Special Rapporteur on the right to food.* Document A/HRC/16/49, Geneva; Catarina de Albuquerque (2010); *Climate Change and the Human Rights to Water and Sanitation.* Position Paper, Geneva.

Persons.[34] Already in 2008, the then Special Rapporteur on the Right to Food documented examples of how disasters resulting from climate change threatened people's food security.[35] Also in 2008, the Special Rapporteur on the Right to Health warned the UN General Assembly that insufficient measures to counter climate change would further endanger the lives of millions of people already at risk. According to this report, the extent to which education and public health facilities were expanded would determine the severity of the effects of climate change.[36] All mandate holders of Special Procedures are entitled to address concerns held by the people of the places affected, and function therefore as a complaint mechanism too.

The Representative of the UN Secretary-General on the human rights of internally displaced persons has looked in particular into issues such as displacement that can be attributed to climate change.[37] He has identified five such situations: increased hydro-meteorological disasters (hurricanes, flooding); environmental degradation and slow onset disasters (i.e. desertification, sinking of coastal zones, or increased salinisation of groundwater and soil); sinking of small island States; relocation of people from high-risk zones; and violence and armed conflict due to scarcity of resources such as water or inhabitable land.[38] The then Representative is meanwhile Envoy of the Chairmanship of the so called Nansen-Initiative, a government based initiative launched by Norway and Switzerland in 2012.[39] The initiative aims to build – in a bottom-up process – consensus

---

[34] See brief descriptions of each of the mandates via http://www.ohchr.org/EN/HRBodies/SP/Pages/Themes.aspx.

[35] Jean Ziegler (2008); *Report of the Special Rapporteur on the right to food.* Document A/HRC/7/5 plus Add.1-3, Geneva

[36] Paul Hunt (2008); *Right of everyone to the enjoyment of the highest attainable standard of physical and mental health.* Report of the Special Rapporteur, Document A/62/214, New York

[37] Meanwhile turned into the category of a Special Rapporteur; see http://www.ohchr.org/EN/Issues/IDPersons/Pages/IDPersonsIndex.aspx.

[38] Walter Kälin (2009): *Report of the Representative of the Secretary-General on the Human Rights of Internally Displaced Persons, Walter Kälin. Addendum - Protection of Internally Displaced Persons in Situations of Natural Disasters.* Document A/HRC/10/13/Add.1, Geneva. The current legal framework at international level does not recognise climate or environment- induced refugees and migrants.

[39] For details to the Nansen Initiative see http://www.fmreview.org/preventing/kalin. Further initiatives dealing with displacement and migration in the context of environmental disasters or climate

among states about how best to address cross-border displacement in the context of sudden- as well as of slow-onset disasters. Among others, the Nansen Principles stress that existing norms of international law should be utilised and normative gaps addressed.

UN Treaty Bodies

The UN Treaty Bodies are organised in the form of committees of independent experts that monitor the implementation of the core international human rights treaties. Each State, party to a treaty, has the obligation to ensure that everyone in the State can enjoy the rights set out in each of the treaties respectively and is obliged periodically to inform about the implementation via a so called state or national report.[40] The system of state reports is complemented by so called parallel reports by non-state actors. The assessment is carried out in addition by public hearings and informal meetings with victims and rights holders. Most of the treaties (or conventions) dispose of a complaint mechanism which enables individuals under certain circumstances to complain of the non- or inadequate implementation of the treaty.[41]

In addition, UN Treaty Bodies contribute to standard-setting by General Comments, a kind of case law,[42] and by the so called "*Concluding Observations*" and "*Recommendations*", making reference to the stage of implementation by the State party. The reporting system, the complaint procedure, General Comments, Concluding Observations and Recommendations provide an authoritative reference system in particular for victims and rights holders, which, together with advocacy efforts, is highly conducive to attracting broad and international public attention. The system also enables the dissemination of information to people and groups on the ground, and further helps generate tools and systems, build

---

change are the policy on pastoralists by the African Union, the Kampala Convention, the Great Lakes Protocol, and the Operational Guidelines of the UN Inter-Agency Standing Committee (IASC) on the protection of persons in situations of natural disaster.

[40] For general information see http://www.ohchr.org/EN/HRBodies/Pages/TreatyBodies.aspx.

[41] For details see http://www.ohchr.org/EN/HRBodies/TBPetitions/Pages/HRTBPetitions.aspx.

[42] For details see http://www.ohchr.org/EN/HRBodies/Pages/TBGeneralComments.aspx.

capacities, and, lastly, empower ordinary people including the most vulnerable.

Several committees of the Treaty Bodies have started to take environment and climate change into their considerations, in particular the Committee on Political and Civil Rights (CCPR), the Committee on Economic, Social and Cultural Rights (CESCR), the Committee on the Elimination of Discrimination against Women (CEDAW), the Committee on the Elimination of Racial Discrimination (CERD), and the Committee on the Rights of the Child (CRC). Although there is no explicit right to a safe and healthy environment, the treaty bodies have stressed in several instances the connection between the environment and the implementation of human rights, as well as the adverse effects of climate change on human rights.[43]

The CESCR expressed, for instance, its concern in relation to Australia's State report in 2009 addressing the negative impact of climate change on the right to an adequate standard of living, including on the right to food and the right to water, which is affecting indigenous peoples in particular. The CESCR encouraged the Australian government to reduce its greenhouse gas emissions and to mitigate the adverse effects of climate change as well as to adequately consult affected Aboriginal and Torres Strait-Islander peoples.[44] The CRC urged Grenada in 2010 to evolve a programme of preparedness for natural disasters, and to take into account in its policy-making that the environment as well as cooperation with regional and international partners requires relevant states to be actively engaged within the context of climate change.[45]

Given the sophisticated and widely-based system of reports, hearings, investigation and monitoring mechanisms, complaint procedures, case law decisions, authoritative comments on the scope of human rights, conclusions and recommendations towards governments, the question

---

[43] See the overviews in e.g. OHCHR (2009); *Report of the Office of the United Nations High Commissioner for Human Rights on the relationship between climate change and human rights.* Document A/HRC/10/61, Geneva; International Council on Human Rights Policy (2008): *Climate Change and Human Rights: A Rough Guide.* Versoix, Switzerland; Theodor Rathgeber (2009): *Climate Change violates Human Rights.* Volume 6 Ecology, Heinrich-Böll-Stiftung, Berlin.

[44] CESCR (2009); *Concluding Observations of the Committee on Economic, Social and Cultural Rights: Australia.* Document E/C.12/AUS/CO/4, Geneva, para. 27.

[45] CRC (2010); *Concluding Observations of the Committee on the Rights of the Child: Grenada.* Document CRC/C/GRD/CO/2, Geneva, para. 52.

remains: is this system effective, in particular from the viewpoint of rights-holders and victims? Are there any meaningful outcomes from this viewpoint? While there are different opinions on that in general,[46] it is noteworthy that the bundle of instruments and mechanisms has meanwhile reached a high level of functionality for assessing human rights situations in accordance with pertinent normative standards.

The procedural structure of the UN human rights system provides an efficient tool for clearly identifying both wrongs and progress. The studies, their conclusions and recommendations are abundant, publicly accessible, and publicly debated. The evolution of all the tools can be interpreted not only as progress in methodology, but also in terms of statements addressing the State's obligations and the challenges to re-organise national – and sometimes international – society accordingly. Human rights provide benchmarks for clarifying State responsibilities and for a fair establishment of international cooperation. Furthermore, all stakeholders, rights holders and victims are invited and entitled to participate. Thus, the mechanisms, including the complaint procedures, help to set the priorities for essential areas of intervention and prevention from the perspective of the people on the ground.

Regional Human Rights Mechanisms

At regional level, there are also institutions and procedures which function similarly compared to the UN institutions mentioned above; they will be named in the following. The African Charter on Human and Peoples' Rights and the Additional Protocol to the American Convention on Human Rights in the Area of Economic, Social and Cultural Rights make explicit references to the environment and, thus, to the livelihood conditions of people. The African Commission on Human and Peoples' Rights has focused on the rights of indigenous and tribal peoples affected by environmental degradation resulting from extractive activities and their forceful removal from their traditional lands. The African Commission has further articulated the right to benefit from natural resources and the right to development as well as set standards for consultation in terms of free, prior and informed consent. The latter is indispensable for people's real

---

[46] See details to this debate in Theodor Rathgeber (2012): *New Prospects for Human Rights? The Human Rights Council between the Review Process and the Arab Spring.* FES International Policy Analysis, Dialogue on Globalization, FES Geneva

participation and for becoming an effective part of the decision-making process.[47]

The Inter-American Court of Human Rights recognised that indigenous and tribal peoples have a right to property over the lands and territories that they have traditionally occupied, and which may be put at risk by environmental degradation or adverse effects of climate change. To reach this conclusion, the Court interpreted the American Convention of Human Rights in the light of other relevant international human rights treaties. For example, the right to ancestral lands, territories and natural resources has been supported by the right to self-determination recognised in both the International Covenant on Civil and Political Rights and the International Covenant on Economic Social and Cultural Rights as well as by the standards set by the International Labour Organisation in particular in its Convention No. 169 (1989) concerning Indigenous and Tribal Peoples in Independent Countries. The Inter-American Court has also developed a system of safeguards that apply where the State is considering development or investment projects that may impair the enjoyment of the rights of indigenous peoples.[48]

The European Court of Human Rights has found that environmental pollution can interfere with the enjoyment of several protected rights, particularly the right to life and the right to private and family life. The Court has established that the State has therefore duties to protect individuals from environmental risks. The Court has clarified that States are responsible for addressing environmental risks once they become known, by effecting an appropriate regulation, monitoring and enforcement. There are further State duties in relation to disclosing information to the public regarding environmental risks. The European Committee of Social Rights, the monitoring body of the European Social Charter, has clarified that environmental dimensions form part of the right to health.[49]

---

[47] See examples at http://www.fmreview.org/preventing/kalin and further links there.

[48] For details see International Council on Human Rights Policy (2008) op.cit.; Theodor Rathgeber (2009): *Climate Change violates Human Rights*, op.cit.

[49] See decision of the European Commission of Social Rights on Marangopoulos Foundation for Human Rights (MFHR) vs. Greece, Complaint No. 30/2005, 6 December 2006, accessible via http://www.coe.int/t/dGHl/monitoring/Socialcharter/Complaints/CC30Merits_en.pdf.

4. People's Agenda in Terms of Human Rights

The previous section revealed institutions and procedures, while this section focuses roughly on the normative substance. The overarching reference in the context of people's primary interest, sustainability, environment and climate change is the Right to Life. This right is anchored in Article 6 of the International Covenant on Civil and Political Rights (ICCPR). In its General Comments Nos. 6 (1982)[50] and 14 (1984),[51] the Human Rights Committee (CCPR) postulated the right to life as the supreme right, which may not be infringed even in times of emergency. Similar, the Committee on the Convention on the Rights of the Child (CRC) says in its General Comment No. 7 (2005) on early childhood that a healthy environment is an inalienable standard for ensuring the survival and development of children.[52] With regard to standards of living, the Convention on the Rights of Persons with Disabilities (CRPD) provides in Articles 25 (health) and 28 (standards of living) the right to non-discriminatory access and continuous improvement of livelihood. Similar provisions are found in the Convention on the Elimination of All Forms of Discrimination Against Women in Article 14.2, as well as in the Convention on the Elimination of Racial Discrimination (CERD) in Article 5.e.

Next is the right to food, laid down in Article 11 of the International Covenant on Economic, Social and Cultural Rights (ICESCR). In its General Comment No. 12 (1999),[53] the Committee on Economic, Social and Cultural Rights (CESCR) called on the signatory governments to implement appropriate economic, social, and environmental policies designed to enable citizens to feed themselves using their own local

[50] General Comment No. 06: The right to life (art. 6): 30.04.1982, accessible via http://www.unhchr.ch/tbs/doc.nsf/%28Symbol%29/84ab9690ccd81fc7c12563ed0046fae3?Opendocument.

[51] General Comment No. 14: Nuclear weapons and the right to life (Art. 6): . 09.11.1984, accessible via http://www.unhchr.ch/tbs/doc.nsf/%28Symbol%29/9c882008fd898da7c12563ed004a3b08?Opendocument.

[52] See document CRC/C/GC/7/Rev.1, General Comment No. 7 (2005), Implementing child rights in the early childhood, accessible via http://www2.ohchr.org/english/bodies/crc/docs/AdvanceVersions/GeneralComment7Rev1.pdf.

[53] See document E/C.12/1999/5, General Comment 12 (Twentieth session, 1999), The right to adequate food (art. 11), accessible via http://daccess-dds-ny.un.org/doc/UNDOC/GEN/G99/420/12/PDF/G9942012.pdf?OpenElement.

resources. Hunger should be prevented even in cases of disaster, and particularly vulnerable populations should be better protected. Within the same approach, the Convention on the Rights of the Child emphasises in Article 24.2 the obligation of the State party to adequately combat malnutrition (and illness) among children and to heed the threat of environmental pollution.

With regard to the right to water, CESCR postulates in its General Comment No. 15 (2002)[54] the right to clean, sufficient, and accessible water and expects the State party to guarantee this right, although the Covenant (ICESCR) does not explicitly mention a right to water. The Committee considers it as being covered by the ICESCR Articles 11 (adequate standard of living) and 12 (health). Although some States like Canada are still hesitant to acknowledge this right, the overwhelming majority takes it now for valid. The Convention on the Rights of the Child refers in Article 24.2 to "*clean drinking water*" as a means of preventing illness and malnutrition. Both ICEDAW in Article 12.2 and CRPD in Article 28.2 include access to water in their definition of adequate living standards.

The right to adequate housing is covered by several human rights standards: the Universal Declaration of Human Rights (UDHR) in Article 25, ICESCR in Article 11, ICERD in Article 5.e.iii, ICEDAW in Article 14.2, ICRC in Article 27, the International Convention on the Protection of the Rights of All Migrant Workers and Members of Their Families in Article 43.1, and the Convention on the Rights of Persons with Disabilities in Articles 9.1, 28.1, and 28.2. The Committee on Economic, Social and Cultural Rights interprets the right to adequate housing in its General Comment No. 4 (1991)[55] in such a way that the right also comprises accessibility, habitability and adequate location, i.e. requiring that the law and administrative rules on housing should not allow construction on environmentally polluted sites. In its General Comment No. 12 (1999)[56]

---

[54] See document E/C.12/2002/11, General Comment No. 15 (2002), The right to water (arts. 11 and 12 of the International Covenant on Economic, Social and Cultural Rights), accessible via http://daccess-dds-ny.un.org/doc/UNDOC/GEN/G03/402/29/PDF/G0340229.pdf?OpenElement.

[55] See document The right to adequate housing (Art.11 (1)): . 13.12.1991, CESCR General comment 4. (General Comments), accessible via http://www.unhchr.ch/tbs/doc.nsf/%28Symbol%29/469f4d91a9378221c12563ed0053547e?Opendocument.

[56] See document E/C.12/1999/5, General Comment 12 (Twentieth session, 1999),

the same Committee (CESCR) postulates, e.g. in paras. 18, 21, 24 and 25, that the rights holder is entitled to a secure, peaceful, and dignified existence as well as to access to necessary services, material, and infrastructure. The CESCR's General Comment No. 7 (1997)[57] states that the right to property also encompasses protection from displacement or resettlement in dangerous areas.

The right to the highest possible standard of physical and mental health is covered by Article 12 of the ICESCR, but also addressed by ICEDAW in Articles 12 and 14.2, the Convention on the Rights of the Child in Article 24, ICERD in Articles 16.4, 22.2, and 25, as well as in the Convention on the Protection of the Rights of All Migrant Workers and Members of Their Families in Articles 43, 45, and 70. In General Comment No. 14 (2000),[58] the Committee for Economic, Social, and Cultural Rights lists a "healthy environment" as one of the preconditions for this right to be fulfilled.

A fundamental principle of international law is the right to self-determination. The two core human rights treaties – the International Covenant on Civil and Political Rights and the International Covenant on Economic, Social, and Cultural Rights – each use the same wording to articulate this right in their Article 1. It means that no one should be forced to give up his or her means of subsistence against his or her will; i.e. clearly relevant in the context of longing for sustainable development. More specifically, the social and cultural identities, for instance, of indigenous peoples are protected by Article 27 of the ICCPR and Article 15 of the ICESCR.

It may be remembered that nearly all of these human rights standards dispose of a complaints mechanism either by the text of the document or in the form of an Optional Protocol which was ratified at a later stage. Although the use of that mechanism is tied to certain criteria, it provides

---

The right to adequate food (art. 11), accessible via http://daccess-dds-ny.un.org/doc/UNDOC/GEN/G99/420/12/PDF/G9942012.pdf?OpenElement.

[57] See document The right to adequate housing (Art.11.1): forced evictions : . 20.05.1997, CESCR General comment 7. (General Comments), accessible via http://www.unhchr.ch/tbs/doc.nsf/%28Symbol%29/959f71e476284596802564c3005d8d50?Opendocument.

[58] See document E/C.12/2000/4, General Comment No. 14 (2000), The right to the highest attainable standard of health (article 12 of the International Covenant on Economic, Social and Cultural Rights), accessible via http://daccess-dds-ny.un.org/doc/UNDOC/GEN/G00/439/34/PDF/G0043934.pdf?OpenElement.

individuals a legal channel that is independent of the political and legal parameters set by national governments. In addition, the implementation of the standards by the State is monitored and assessed by the independent experts in the committees, and stakeholders are entitled to participate with the monitoring process by witnessing and providing critical data in order to verify the State's report.

Altogether, the substance of human rights and the related mechanisms are sufficient to adequately address problems in the realisation of human rights referring to sustainability, and are adequate to deepen the incidence of human rights onto the discussions and standard setting in relation to this subject. They provide to the people on the ground in particular an internationally agreed language, and methods of understanding the scope of the norms as well as the challenges, and entail a steady ethical compass for requesting the needed transformation of society, politics and economy. The legal substance and the mechanisms place these instruments into the hands of affected people on the ground by providing participation in the reporting system as well as through complaint procedures. People are therefore in a position to demand protection for and implementation of their rights, setting the priorities via the international standards. Victims are further enabled to negotiate a fair compensation based on the indicators for the minimum standards.

Altogether, the normative standards and their mechanisms allow us to determine a fairly accurate stage-set for the people's agenda concerning how the urgency, the priorities and the centrality of rights should be handled with relation to a sustainable future. Such agenda, based on agreed language, can further be transferred to the international level of cooperation, negotiation and debate. Whatever may be the outcome, concept-wise such a rights-based approach includes the element of social mobilisation and self-organisation by the rights holders. It goes beyond a purely State-related understanding of policy making and beyond dispensing charity to the poor. Human rights norms and mechanisms provide poor and vulnerable people in particular with norms and instruments in order to articulate their concerns, interests and demands through their own initiatives. This is even more relevant considering the asymmetries in access to information, opinion-building and decision-making.

5. Challenges for Churches

The call for justice and peace has a long-standing legacy within the churches and the ecumenical movement as they focus on human dignity, the common good and, in more recent times, respect for human rights. In search of a new and different global order and a new ethos, the ecumenical movement has played an important role in recent years, for instance, by its commitment and campaigns against apartheid, and by the decade to overcome violence (2001-2010).[59] In all these commitments, it has stood for justice, fair international cooperation, and empowerment of underdeveloped nations as well as underprivileged people. Conversely, the ecumenical movement has called for responsibility from wealthy nations and the establishment of practical steps on the ground.[60] Obviously, the call for a shared common standard of living in combination with human rights has been uncomfortable and disruptive of conventional thinking in its quest for selfish benefits and greed.

Churches have always been involved in different ways in mobilising ethics and norms in order to gain a different perception of reality, which is orientated to the needs of the most vulnerable. Churches and the ecumenical movement have participated in developing the call for fair conditions in order for people to be able to make a choice for a self-determined future. Churches have taken part in and contributed to a discourse on sharing, and the distinction between legitimate and illegitimate wealth, and have dealt with the impacts on people's lives and minds. The Gospel warns against the idolatry of '*mammon*': it will corrupt social relationships, exhaust and destroy natural resources, obstruct or impede the access of future generations to means for living their lives in dignity and, thus, consolidate a structure of discrimination, inequality and violence. Although the term 'sustainability' has been more prominent in political discourse, the concept and program on 'Justice, Peace and the Integrity of Creation' is just a translation of this into a more biblical language.

---

[59] For details see http://www.oikoumene.org/en/resources/documents/wcc-commissions/international-affairs/peace-and-disarmament/peace-concerns/the-decade-to-overcome-violence-dov-churches-seeking-reconciliation-and-peace.html.

[60] See Guillermo Kerber (2011): *„Peace with the Earth" in the Context of the Decade to Overcome Violence.* In: The Ecumenical Review. Volume 63, Number 1, March 2011, Geneva, pp. 35-43.

Ecumenical work has been rooted in the wholeness of creation and the biblical imperative of commitment to justice which pays special attention to the poor. It is its shared understanding that local communities should be empowered and local people given a voice in order to articulate their concerns and demands for themselves. The "*Ecumenical Declaration on Just Peace*" highlights the intimate relationship between peace and justice. In the "*Peace with the Earth*" perspective, "*Just Peace*" refers to climate justice too, since the challenges posed by climate change are also considered as a matter of justice: "*those who suffer most due to the impact of climate change are impoverished and vulnerable communities who contribute only minimally to global warming*".[61]

The "*International Ecumenical Peace Convocation*" in Jamaica, May 2011, underlined the churches' commitment to the integrity of creation and a change of daily lifestyle in order to overcome both the environmental and the ethical and spiritual crisis. The message of Jamaica affirms the churches' readiness to join global civil society in urging governments radically to reconstruct economic activities, and urgently to reduce the extensive use of fossil fuels and greenhouse gas emissions. The message of Jamaica further states: "*The ecological debt of the industrialized countries responsible for climate change must be considered when CO2 emission shares and plans for adaptation costs are negotiated.*"[62]

In the genuine context of human rights, the churches, in particular the ecumenical movement, have played a basic role during the foundation of these norms by looking for ways that tend to build a new global order and ethos. To achieve this order, the political capacity became imperative to mobilise opinion.Subsequently, programs and campaigns have been developed in order to address the major threats to life, such as on Justice, Peace and Integrity of Creation. The World Council of Churches (WCC) raised its voice during the World Summit on Social Development in Copenhagen 1995, and in the aftermath of the Fifth World Trade Organisation (WTO) Ministerial Conference in Cancún (2003) on trade,

---

[61] See Guillermo Kerber (2011), op. cit.; World Council of Churches – WCC (2006); *Climate Justice for All.* A statement from the World Council of Churches to the High-Level Ministerial Segment of the UN Climate Conference in Nairobi COP12/MOP2, 17 November 2006, Nairobi.

[62] See the message at http://www.overcomingviolence.org/en/resources-dov/wcc-resources/documents/presentations-speeches-messages/iepc-message.html; see the program of the Convocation at http://www.oikoumene.org/en/news/news-management/eng/a/article/1634/international-ecumenical-8.html.

development, food and livelihood security, against the social, political and cultural exclusion of people in the context of politics imposing structural adaptation programs, or within the debates on the MDGs. To meet the conditions for the full enjoyment of human rights was always a key element for the WCC in addressing poverty and social security.

Thus, churches and the ecumenical movement have substantially contributed to make people aware about optional choices and about their genuine rights (and obligations). The ecumenical movement has helped people to get empowered and claim their rights, and also provided tools and mechanisms for self-organisation. People and communities have learned cooperative behaviour, trust between individuals and groups, and trust in a more friendly future. Sometimes even the churches' vertical architecture has been helpful in directly and inclusively linking up the grass-roots level with political decision-making at the top. Because of the wide range of groups, congregations and parishes at the local level up to the advocacy work at the governmental and intergovernmental levels, churches and the ecumenical movement are well placed to address a comprehensive policy on the integrity of creation and a sustainable future. Within this large area of potential activities suited to make the integrity of creation and sustainability a prominent aspect, there are some specific challenges which can be genuinely adopted by churches and the ecumenical movement.

First of all, churches are asked to provide hope and spiritual accompaniment. Furthermore, the churches can enhance the participation of people(s) in order to ensure that victims become active stake-holders. They can contribute to increasing local capacity, educating households and building information-sharing networks. Conversely, the knowledge and expertise of local people in addressing environmental degradation and climate change disasters, including local but adequate knowledge on resilience, can be explored, organised and channelled. Collective action and sense of community can be established by praying, speaking and acting together. Local communities are able to identify the most vulnerable, documenting community concerns and to foster the views of the people. At local level churches are further asked to undertake urgent activities. Beyond the local level, the ecumenical movement is in a good position to develop platforms of exchange and exposure, reaching out to social movements, community-based and non-governmental organisations and to members of parliament or parliamentary committees and governments.

A comprehensive understanding of environment, climate change and adverse effects from its impacts, challenges in terms of sustainability and the scope of the new policy, starts with language, the way of identifying the problem and the methodology of fact finding. It is necessary to generate a genuine narrative, based on the views of the people on the essence of their life and the alterations experienced by them. A narrative which enables the people to talk in their own language, their stories and parables, their way of assessment, their science and understanding of the essence of their life. All our known facts and perspectives around the subjects mentioned are radically incomplete, and the churches can make a difference that emerges with such an authentic narrative from the people's view.

Giving the people a voice, their own voice, will challenge existing power relations. So far as churches are involved, it is a twofold task. Churches are by their structure in a position to make poor and underprivileged people familiar with the international normative standards which provide protection, with a specific concept of 'vulnerable people' and mechanisms for complaint and demanding. Churches are also asked to provide hope to the most affected by environmental degradation and climate change. Secondly, churches and the ecumenical movement are in a position legitimately to demand a fundamental transformation of patterns and structures guiding the economy and consumption and the corresponding policy-making.

Human rights will be an indispensable instrument for carrying out this endeavour. They primarily bring the people's view to public attention and facilitate moving their concerns forward. Human rights will bring justice to any policy dealing with environment, climate change, sustainability, justice to victims and rights holders. They entail a wide range of procedures in order to seek accountability and to petition governance structures. At the ethical as well as at the legal level, human rights provide a substantial platform for priorities from a people's view. Such activities within and by the churches and the ecumenical movement can make all the difference.

6. Conclusion

Human rights – including the normative standards of the International Labour Organisation (ILO) – provide an international agreed language in order to address the subject of sustainability and its options from the people's viewpoint. They provide a platform for understanding the scope of the problem, and entail an ethical compass in the needed transformation of society, economy and governance. Human rights norms and mechanisms give priority to the human being, and provide poor and vulnerable people in particular with a framework of understanding and with mechanisms to articulate their concerns, interests and demands on their own behalf. The human rights approach attracts attention to priorities, for instance, in terms of food security, supply of clean water or health care. As the decision-making process on questions related to a sustainable future is not only a legal matter, human rights should also be considered as a genuine platform for social mobilisation and self-organisation, although there is no automatism. Experiences conveys that the human rights approach helps people to escape from the feeling of powerlessness. The reference to human rights legitimates challenging governance, as well as stimulating people to become active participants on their own behalf, starting with testimonies, reports and documents. Thus, such tools allow the identification of interests and conditions for a life based on self-determination by the people on the ground. Conversely, institutions and procedures within the United Nations human rights system have already developed guidelines, benchmarks and case law in order to prioritise the victims' and rights holders' view on issues like resource-management, protecting the environment and establishing an infrastructure which includes the sustainability of the family household. Some regional institutions even encompass judiciary procedures for such an end. The human rights approach provides references and orientation for emerging with fair conditions for international cooperation including extraterritorial State obligations, although this concept is still disputed. The Right to Development aims to support underdeveloped countries in overcoming the asymmetric relationships in the world market and, thus, to forms of sustainable development.

Nevertheless, the human rights approach is not a '*deus ex machina*' which suddenly resolves the problems of the world. Although some of the mandate holders of the UN Special Procedures deal with unfair relationships at world market level and the obstacles to establishing a more sustainable pattern of economy – such as on foreign debt, impacts of the international financial crisis, extreme poverty – there has only been low impact on the ground to date, if any. Nevertheless, those mandates at least show the scope and potential of the human rights approach, and how far such an analysis, and the demand for implementation and action can contribute to draw up a national as well as international governance that fosters sustainable development, considering in particular the viewpoint of the people on the ground. To make such endeavours effective, social mobilisation and attracting public attention are needed. In this context, churches and the ecumenical movement are important transmitters to put such orientation on the international agenda. Against the current dynamic of the predominant economy and politics, the human rights approach provides for ethics and rights to get engaged in the debate. This is no small reason for hope.

References

Albuquerque, Catarina de (2010); *Climate Change and the Human Rights to Water and Sanitation.* Position Paper. Geneva, accessible at http://www2.ohchr.org/english/issues/water/iexpert/docs/Climate_Change_Right_Water_Sanitation.pdf

Bread for the World/ Oikumene Pacific/ World Council of Churches (eds.) [2010]: *Conference on Protection and Reparations for „Climate Refugees": Imperatives & Options.* Compendium of a conference in May 2010 in Chavannes-de-Bogis, Switzerland; accessible via http://www.oikoumene.org/fileadmin/files/wcc-main/documents/p4/climate/Compendium%20final.pdf

Hunt, Paul (2008); *Right of everyone to the enjoyment of the highest attainable standard of physical and mental health.* Report of the Special Rapporteur, Document A/62/214, New York

International Council on Human Rights Policy (2008): *Climate Change and Human Rights: A Rough Guide.* Versoix, Switzerland

Kälin, Walter (2009): *Report of the Representative of the Secretary-General on the Human Rights of Internally Displaced Persons, Walter Kälin. Addendum - Protection of Internally Displaced Persons in Situations of Natural Disasters.* Document A/HRC/10/13/Add.1, Geneva

Kälin, Walter (2010); Conceptualising Climate-Induced Displacement. In: McAdam, Jane (ed.); Climate Change and Displacement, Multidisciplinary perspectives. Oxford-Portland, pp. 81-85

Kerber, Guillermo (2011): „Peace with the Earth" in the Context of the Decade to Overcome Violence. In: The Ecumenical Review. Volume 63, Number 1, March 2011, Geneva, pp. 35-43

Ksentini, Fatma Zohra (1994): Review of Further Developments in Fields with which the Sub-Commission has been Concerned: Human Rights and the Environment. Final Report, Document E/CN.4/Sub.2/1994/9 + Corr.1, Geneva, accessible via http://www.unhchr.ch/Huridocda/Huridoca.nsf/0/eeab2b6937bccaa180 25675c005779c3

Marks, Stephen P. (2011); *The Politics of the Possible. The Way Ahead for the Right to Development.* FES, International Policy Analysis, Geneva

Office of the High Commissioner on Human Rights – OHCHR (2004): *Human Rights and Poverty Reduction: A Conceptual Framework.* Document HR/PUB/04/1, New York-Geneva, accessible via http://www2.ohchr.org/english/issues/poverty/docs/povertyE.pdf

—. (2009): *Report of the Office of the United Nations High Commissioner for Human Rights on the relationship between climate change and human rights.* Document A/HRC/10/61, Geneva, accessible via http://daccess-dds-ny.un.org/doc/UNDOC/GEN/G09/103/44/PDF/G09 10344.pdf? OpenElement

—. (2011): *Analytical study on the relationship between human rights and the environment.* Report of the United Nations High Commissioner for Human Rights. Document A/HRC/19/34, Geneva, accessible via http://www.ohchr.org/Documents/HRBodies/HRCouncil/RegularSessi on/Session19/A-HRC-19-34_en.pdf

Rathgeber, Theodor (2009); *Climate Change violates Human Rights.* Volume 6 Ecology, Heinrich-Böll-Stiftung, Berlin

—. (2012*): Climate Justice, Human Rights and the Role of Churches.* WCC Publications, Geneva, accessible via http://www.die-klima-allianz.de/wp-content/uploads/2012/11/CC-Human-Rts-and-Churches_final_18102012_1.pdf

—. (2012): *New Prospects for Human Rights? The Human Rights Council between the Review Process and the Arab Spring.* FES International Policy Analysis, Dialogue on Globalization, FES Geneva

Schutter, Olivier de (2010); *Report submitted by the Special Rapporteur on the right to food.* Document A/HRC/16/49, Geneva

Sedlacek, Thomas (2011): *Economics of Good and Evil: The Quest for Economic Meaning from Gilgamesh to Wall Street.* Oxford University Press

Sepúlveda Carmona, Magdalena (2012): *Final draft of the guiding principles on extreme poverty and human rights, submitted by the Special Rapporteur on extreme poverty and human rights, Magdalena Sepúlveda Carmona.* Document A/HRC/21/39, Geneva, accessible via http://www.ohchr.org/Documents/Issues/Poverty/A-HRC-21-39_en.pdf

UN Committee on Economic, Social and Cultural Rights – CESCR (2009); *Concluding Observations of the Committee on Economic,*

Social and Cultural Rights: Australia. Document E/C.12/AUS/CO/4, Geneva

UN Committee on the Rights of the Child - CRC (2010); *Concluding Observations of the Committee on the Rights of the Child: Grenada.* Document CRC/C/GRD/CO/2, Geneva

UNFCCC (2011): *Report of the Conference of the Parties on its sixteenth session, held in Cancun from 29 November to 10 December 2010,* Addendum Part Two: Action taken by the Conference of the Parties at its sixteenth session. Document FCCC/CP/2010/7/Add.1

UN Secretary-General's High-level Panel on Global Sustainability (2012): *Resilient People, Resilient Planet: A future worth choosing.* New York: United Nations, accessible via http://www.un.org/gsp/sites/default/files/attachments/GSP_Report_web_final.pdf

UN Special Procedures (2006); *States must address poverty with utmost urgency, say UN Independent Experts on the occasion of Human Rights Day.* Press Statement, Geneva, 8 December, accessible via http://www.unhchr.ch/huricane/huricane.nsf/0/8BAD3E92C5955773C125723E005E1E42?opendocument

Vogt, Markus (2009): *Prinzip Nachhaltigkeit. Ein Entwurf aus theologisch-ethischer Perspektive.* Hochschulschriften zur Nachhaltigkeit 39, München

World Council of Churches – WCC (2006); *Climate Justice for All.* A statement from the World Council of Churches to the High-Level Ministerial Segment of the UN Climate Conference in Nairobi COP12/MOP2, 17 November 2006, Nairobi

Ziegler, Jean (2008); *Report of the Special Rapporteur on the right to food.* Document A/HRC/7/5 plus Add.1-3, Geneva

Abbreviations

CCPR	Committee on Political and Civil Rights, also: Human Rights Committee
CEDAW	Committee on the Elimination of Discrimination against Women
CERD	Committee on the Elimination of Racial Discrimination
CESCR	Committee on Economic, Social and Cultural Rights
CRC	Committee on the Rights of the Child
CRPD	Convention on the Rights of Persons with Disabilities
COP	Conference of the Parties

G-8	Group of Eight, world's wealthiest countries: Canada, France, Germany, Italy, Japan, Russian Federation, United Kingdom, United States of America
G-20	Group of Twenty; group of finance ministers and central bank governors from 20 major economies: Argentina, Australia, Brazil, Canada, China, France, Germany, India, Indonesia, Italy, Japan, Mexico, Russian Federation, Saudi Arabia, South Africa, South Korea, Turkey, United Kingdom, United States of America plus the European Union, represented by the President of the European Council and by the European Central Bank
HRC	Human Rights Council
ICCPR	International Covenant on Civil and Political Rights
ICEDAW	International Convention on the Elimination of All Forms of Discrimination Against Women
ICERD	International Convention on the Elimination of Racial Discrimination
ICESCR	International Covenant on Economic, Social and Cultural Rights
ILO	International Labour Organisation
MDGs	Millennium Development Goals
NGO	Non-Governmental Organisation
OHCHR	Office of the High Commissioner for Human Rights
SDGs	Sustainable Development Goals
UDHR	Universal Declaration of Human Rights
UNFCCC	United Nations Framework Convention on Climate Change
WCC	World Council of Churches
WTO	World Trade Organisation

GREEN ECONOMY AND FOOD SECURITY

GREEN ECONOMY: THE CONTRIBUTION OF CORPORATE SOCIAL RESPONSIBILITY

P. NICOLOPOULOU-STAMATI,[1] N. XIANGLI,[2] I. MATIATOS,[1] C. KOTAMPASI,[1] P. STAMATIS,[1] A.J. SASCO[3] AND L. HENS[2*]

[1]Medical School, National and Kapodistrian University of Athens, Greece
[2]Vlaamse Instelling voor Technologisch Onderzoek (VITO), Mol, Belgium
[3]Université de Bordeaux, France
[*]To whom correspondence should be send at *luchens51@gmail.com*

Abstract

"The future we want", the main document summarizing the action areas advocated by the Rio+20 conference (Rio de Janeiro, Brazil, June 20-22nd), advocates "green economy" as a main instrument to eradicate poverty, while maintaining the healthy functioning of the environment. "Green economy" is a reply to global capitalism and the excesses of its elite practitioners, as they became manifest during the recent economic crisis. Its roots lie in the observation of parallels between the world's ecological crisis and the financial crisis. Both are products of a failed economic and social system. The green economy model holds the potential to bring together the needs for economic security and environmental protection, and to integrate an array of reforms in policy initiatives, regulatory actions, business standards and personal lifestyle behaviours. A classical contribution of the private business sector to sustainable development is corporate social responsibility (CSR). This concept makes room for the doctrine that a company is not only responsible for a positive economic performance, but also has to take care about the environmental, social and ethical aspects of its activities. Companies have to transparently report on these activities in their sustainability or CSR report. One of the main external advantages for CSR-committed companies

is that they are included by banks in ethical and sustainability portfolios. These funds, although originating in the US Methodist and Quaker traditions, are among the faster-growing sections of the products offered by European banks. The CSR system is criticized by developing countries and NGOs for its improper use of green economy products ("green washing"). This paper shows that a "green economy" will be a necessary part of a more sustainable society in the future. CSR is a useful instrument contributing to this evolution. Opting for CSR should be based on ethical considerations and the conviction that "going green" is important for the future of industry and services, rather than on the expectation of short-term monetary profitability. More research on assessment methods and more coherence in the practical implementation of CSR are mandatory.

1. Introduction

The concept of sustainable development (SD) [WCED (1987)] entails a symbiosis between environmental, social, and economic aspects. While in general the environmental component of SD is clear, the social and economic pillars are less specified.

The Rio+20 conference, which was held in Rio de Janeiro (Brazil, June 2012), followed up the UNCED (Rio de Janeiro, Brazil 1992) and WSSD (Johannesburg, South Africa, 2002) conferences. Like these previous conferences, the Rio+20 meeting also aimed at specifying the SD concept. On this occasion in particular the aspects related to a green economy were detailed.

Part of green economy is economic stewardship. Companies not only have an economic (profit) and societal (employment, product quality) responsibility, but also have to deal with their impact on the environment. This notion is closely associated with the idea that managers should be the instigators and guardians of a social change towards sustainable development. This is the related notion of a managerial duty towards the environment. Elements of both stewardship and managerial duty merge in the idea of corporate social responsibility (CSR).

This concept is also subject to criticisms and limitations. NGOs raise the question to what extent CSR is associated with window dressing and the improper use of green products ("green washing").
This paper addresses the issues of a green economy and the significance of CSR in this context. It specifies the content of a green economy, situating

this sustainable development pillar. It analyses the basic elements of CSR and deals with the more pragmatic question how CSR in the international banking sector contributes to sustainable development.

2. Green Economy

2.1 Definition

There is no generally accepted definition of 'green economy'. Rather there are a number of ways describing what is captured by the concept.

Table 1. Principles for a green economy [UNEP (2011)]

1. Equitable distribution of wealth: Between nations, and among rich and poor. Achieve social and economic justice. Realize a fair share of the world resources, including biodiversity.
2. Economic equity and fairness: Reduce the gap between the developed and developing countries by financial transfers and capacity building. Support environmental sustainability worldwide.
3. Intergenerational equity of environmental resources and ecosystems: Meeting the needs of future generations.
4. Precautionary approach: Scientific uncertainty on impacts shall not lead to avoidance of measures to prevent environmental degradation.
5. The right to development: Empower individuals and societies to achieve positive social and environmental outcomes.
6. Internalization of externalities: Market prices must reflect real social and environmental costs and benefits, so that the polluter bears the cost of the pollution.
7. International cooperation: The application of environmental standards within a country should take into account the possible impacts on other states.
8. Environmental measures related to trade: fair trade should support sustainable resource use.
9. International liability: Actions by countries can cause impacts beyond national jurisdictions, requiring issues of international law to be solved.
10. Information, participation, and accountability: Citizens should have access to environmental information and the opportunity to participate in decision-making processes. Accountability is necessary in this respect.
11. Sustainable consumption and production: This is based on equitable resource use.

12. Strategic, coordinated and integrated planning: These are core instruments in delivering sustainable development, a green economy and poverty alleviation.
13. Just transition: Developing countries must have access to appropriate financial and technical assistance; citizens and communities must have access to new skills and jobs to make the transition to a low-carbon, green economy.
14. Redefine well-being: GDP is an inadequate tool for measuring social well-being and environmental integrity. Well-being, quality of life, and environmental health should be the guiding objectives for economic development.
15. Gender equity: Women have a vital role in changes for environment and sustainable development. Their actions should be rewarded and their skills enhanced.
16. Safeguard biodiversity and prevent pollution: Develop a system of governance that protects the resilience of ecosystems to prevent irreversible damage.

A green economy is associated with an increase in economic prosperity, new and more decent jobs, reduction of poverty, and healthy and active communities. A green economy necessitates a proactive, environmentally informed economic transition [Dodds et al. (2011)]. A green economy is equally a reply to the excesses of capitalism and its elite practitioners in their too extensive pursuit of self-interest and pure economic efficiency. It provides an answer to market failures, which yields more environmental and social benefits. In this way it is an essential part of sustainable development.

UNEP (2011) described a green economy in 16 principles (Table 1) which closely relate the concept to the Rio Declaration (1992), which outlined the principles of sustainable development. Inter- and intra-generational equity, the precautionary approach, the polluter/user pays, common but differentiated responsibilities, and internalization of (environmental and social) externalities are rules of the thumb appearing in both documents.
Although it is useful to describe a concept in terms of principles, this is a weak approach. More impact probably attaches to the search for a definition. Elements of such a definition should include:

(-) Transition to a green economy should be pursued in an equitable way. The green transition should be fair.

(-) Green economy should create decent green jobs, among others in sectors which contribute to environmental quality and welfare.
(-) Areas which help moves to a green economy include government actions (taxes, funds, and subsidies), city initiatives (e.g. carbon neutral cities), sustainable production and consumption, transparent financial markets, new economic indices, national wealth accounting, and corporate social responsibility (CSR) and accounting.

CSR and in particular CSR for the banking sector, as part of the green economy, is the subject of this paper.

2.2 The Rio+20 context

In the aftermath of pioneering scientific work (Daly and Cobb (1989); Pearce et al. (1989)) and the calls of international organizations and events (G20 Seoul Summit Declaration, 2010), the green economy appeared on the agenda of the Rio+20 conference, following a discussion with key experts during a two-day meeting in January 2011. The focus on a green economy accompanied the observations that:

(-) Much of what has been promised at UNCED (Rio de Janeiro, Brazil, 1992) and at the WSSD (Johannesburg, South Africa, 2002) had not been delivered.
(-) A green economy might lead to more jobs and alleviate poverty, but scientific research support on the issue is meagre.
(-) Elaboration of the subject might bring more coherence and balance between the three pillars of sustainable development.

During this preparatory phase for the conference there were also concerns raised about:

(-) The green economy as a new form of protectionism.
(-) The connotation of "green washing", which refers to the improper use of green economy products.
(-) The limited success of economic instruments such as taxes and subsidies, to improve environmental quality.

A green economy in a context of sustainable development should:

(-) Promote the transition to a low-carbon, resource efficient economy.
(-) Avoid risks such as water scarcity, the effects of climate changes, and depletion of fishery stocks and virgin forests.
(-) Contribute to economic growth and poverty alleviation.

In accordingance with these considerations "The future we want" [UN (2012)] – the main document which came out of the Rio+20 Conference on Sustainable Development – considered a green economy as an important tool to implement sustainable development and poverty alleviation. The document put emphasis on the fact that a green economy should promote the three core dimensions of sustainable development: environmental, social and economic. More specifically, a green economy should not only promote sustained economic growth, but also enhance social inclusion, contribute to human welfare, and create opportunities for employment and decent work for all, while mainstreaming the healthy functioning of the earth's ecosystems.

The action areas towards a green economy which are listed in "The future we want" include:

(-) Phasing out unsustainable patterns of production and consumption. In earlier documents [UN (1992); UN (2002)] the United Nations already indicated that overall consumption in the industrialized countries should be limited to levels corresponding with the carrying capacity of the environment, while in particular in the least developed countries consumption should increase "on a par" with poverty alleviation. Addressing sustainable production and consumption should include environmental sustainability and the promotion of nature conservation and biodiversity. Ecosystems should be used to an extent such that their regeneration capacity is not hampered and natural resources are not over-used. The concept equally includes for the UN the promotion of sustained and equitable growth.
Promotion of sustainable consumption and production patterns should get rid of its negative connotation of reduction and less opportunities. On the contrary, it should contribute to progressing towards social and economic development within the carrying capacity of ecosystems through win-win solutions.

(-) Job creation which requires equipping workers with the necessary skills through education and capacity building.

(-) Health promotion which should guarantee at least a working and general environmental quality which does not damage human health.

(-) Stakeholders, their partnerships, platforms, and networks should be involved in developing green economy policies. The United Nations call for responsibility of governments and the private sector in particular.

(-) To llustrate the main points to achieve a green economy, the following list of priority action areas was launched:

- Consumer information.

- Sustainable lifestyles and information.
- Sustainable public procurement.
- Sustainable buildings and constructions.
- Sustainable tourism, including ecotourism.

Effective actions in these areas necessitate:

(-) Interventions in the financial markets: in particular the 2008 financial crisis showed how deleterious for society too much focus on self-interest on the part of banks and financial organizations can be.

(-) New economic indexes: in a context of sustainable development the (growth of) the gross national product (GDP) reveals only part of the wealth of a society. Welfare, natural resources and their depletion rate should matter in a context of a green national product.

(-) Taxation and subsidies are the instruments most frequently used by governments to correct market failures and control redistribution of money in a way which is more acceptable for society. These instruments should be handled towards more sustainability, an improved carrying capacity and a more fair distribution of wealth.

(-) Corporate social responsibility (CSR) and accountability are often seen as the most important contributions of industry to a more sustainable society. This is the subject of the next section of this paper.

3. Corporate Social Responsibility (CSR)

3.1 CSR in a context of sustainable development

Calls of people to invest in "good causes" are perennial. More recently these calls have gained momentum because civil society (including environmental ant development NGOs) has become more vocal. This increased vocal stance of civil society is not only related to a call for more democracy, but also to the regressing, if not failing role of governments, the growing interest of companies to promote values, and the high cost and limited benefit of e.g. bank operations (a combination of high transaction costs, poor information, and high delivery costs). An important segment of this civil society in a CSR context is the fair and equitable trade/responsible investment movement.

It is difficult to provide a complete list of factors which have contributed to this evolution. The following elements are however important concerns:

(-) Economies should not only be driven by the "invisible hand of the market". A responsible economy should be about "common goods" or "commons", such as core environmental resources (clean air, drinking water, good quality soil), which are considered abundant and therefore hardly valued at all in free market economies.

(-) The call for more transparency, which is based on more and better information about markets, companies, organizations, and networks.

(-) The increasing recognition that companies not only have an economic responsibility, but also social and environmental targets to pursue. Environmental and social factors should no longer be externalized, but form an intrinsic part of company management.

(-) Experience from the past in countries with a delayed water treatment policy, and also the current refusal to deal with climate change and atmospheric pollution, illustrate that a remedial environmental policy is more expensive than a preventive and proactive one [Bénabou and Tirole (2010)].

3.2 CSR definition

Responding to this discussion, business leaders now consider corporate social responsibility. The fast increasing literature on CSR contains numerous definitions of the construct. Much-cited is the definition of the European Commission (2010), which defines corporate social responsibility as "a concept whereby companies integrate social and environmental concerns in their business operations and in their interaction with their stakeholders on a voluntary basis". More recently the Commission omitted explicit reference to the environment and proposed a more generic definition which reads as "the responsibility of enterprises for their impacts on society" [EC (2011)].

These and other CSR definitions have in common that firms should (voluntarily) go beyond their legal and contractual obligations. To meet the CSR criteria enterprises should put in place a process to integrate social, environmental, ethical, human rights and consumer concerns into their business operations, and develop a related core strategy in close collaboration with their stakeholders. This process should aim at:

- maximizing the shared value for their owners/shareholders and for the other stakeholders and society at large;

- identifying, preventing and mitigating their possible adverse impacts.

3.3 Critical parameters

Critical parameters in CSR entail corporate policies on:

(-) Environmental aspects which cover mainly the areas of environmental policy and management, environmental 'performance', and "green" products. Companies should establish a vision on their environmental responsibilities, and formalize this in a policy. They should implement it using an efficient environmental management which is preferentially certified according to international standards. Among these latter ISO14001 is the most common, while in the European Union also EMAS is used. Environmental performance relates to decreasing emissions into air (e.g. greenhouse gases, SO_2, NO_x, volatile organic substances), water (BOD, COD), and soil (heavy metals, waste). It involves environmental risk management and responsibility towards biodiversity and mobility. "Green" products refer to the involvement of companies to move production patterns towards greater sustainability: examples include hybrid and other new technology cars for the automobile industry, bio-food and products of sustainable fishery for the retailing sector, and certified hard wood and non-hazardous materials in the building sector.

(-) Human resources aspects entail decent labour relations, responsible re-organizations, employee welfare, and safety. Employee welfare is about e.g. stress management, on-site health clinics, and health and pension support. Workplace safety is about limiting the health risks at the workplace and presupposes specific management schemes in accordance with international standards (ILO-OSH, OHSAS18001).

(-) Human rights concern the implementation of the rights defined by the International Labour Organization (ILO), among which freedom of organization (worker unions) and non-discrimination (gender, population, and ethnicity) are crucial.

(-) Community involvement is rooted in the philanthropic aspects of CSR. The issue at stake is to deal in a responsible way with the societal impacts of a firm's production and products. It is about the mining company which invests in a local hospital and a school for the neighbouring community where its workers with their family live, and about the energy company which helps its clients with saving energy (tariff structure, audit).

(-) Business behavior should be targeted at fair operating practices, product safety, information for consumers, and fair relationships with suppliers. Fair operation has many applications: Lipton, of which the teas are certified by the "Rainforest Alliance", Starbucks, which is selling fair trade coffee and tea, and clothing which is made from organic cotton. The Dutch "Max Havelaar" organization started pioneering work in this area already in the 1970s. More generally, three relevant groups of labels exist: fair trade labels, social labels, and environmental labels. The relationships with suppliers aim at looking for supplying companies which embody comparable CSR standards to those of the company itself. This is one of the main structural elements promoting the proliferation of the system.

(-) Corporate governance addresses transparency, company audits, reporting in accordance with international standards, remuneration schemes, and shareholder relationships. Transparency, in particular regarding the issues listed above, is central in CSR. Transparency can be promoted by specific information entailed in high quality environmental and sustainability CSR-reports, or reports to society. The information which is provided should be correct and controllable. It should contribute to the reliability of the company. Remuneration schemes have recently attracted attention because of the high wages and bonuses in the financial sector, which hardly correspond with the societal counterpart of CSR functions. Rational and motivated compensation is thus at stake in the CSR discussion.

(-) Stakeholders' management relates to activities which focus on stakeholder groups which are different from the investors. In view of the wider social responsibility of companies the concept of stakeholders has widened. Originally stakeholders were seen in close association with the company, e.g. as individuals which depend on the company for the realization of their personal gain and on whom the company depends. Socio-economics have more recently put emphasis on society which has entrusted to business large amounts of its resources to accomplish the company's mission. Conversely, business is expected to manage these resources as a wise trustee for society. This means that the "moral community" of stakeholders today is much wider than before. They are both directly and indirectly involved; they are located both inside and outside the company; they are not limited to the shareholders but involve civil society and society as a whole. This multi-stakeholder recognition has major management consequences, in particular when incompatibility between the agendas of the various stakeholder groups emerges, not least in the case of transnational companies [Hussain (1999)]. In practice, however, the group of external stakeholders is more limited. Apart from the investors, community organizations, suppliers, customers and future generations are as a rule the most important groups.

(-) Apart from the above, also a set of negative criteria apply to CSR companies. They should not be involved in the production or trade of weapons and health threatening products such as cigarettes and alcohol. They should not deal in pornographic materials, gambling or non-medical drugs. They should be concerned about animal welfare for the animals strictly needed in tests, and they should be most careful in dealing with genetically modified organisms. They should not invest in obviously non-sustainable technologies. On these issues it is not always easy to put forward clear borders. The social acceptability of e.g. nuclear electricity generation changes over time. CSR companies should therefore gradually bring down their involvement in nuclear energy and invest more in sustainable energy generation, distribution and use.

3.4 Benefits and limitations of CSR

In summary, a CSR organization is transparent, accountable, and socially responsible. It is characterized by a comprehensive management which establishes effective incentives on the above listed critical parameters. It should do so because a strategic approach to CSR is increasingly important for the competitiveness of enterprises. If well managed, it brings benefits for the internal risk management, cost savings, access to capital, customer relationships, human resources management, and innovation capacity [EC (2011)]. In short, CSR should improve the competitiveness of the company, while behaving in a more responsible way to society.

Apart from these considerations, the question arises if and to what extent CSR companies make more profit. The strategic options are not financially neutral and involve both monetary and non-monetary benefits:

(-) CSR can lead to revenue increases from higher sales and market share. This is related to an improved and clearer brand image and the developing CSR market.

(-) CSR acts on more efficient energy and material consumption. It aims at reducing waste and transport, while replacing this latter by distance communication.
(-) Environmental and social risks are more intensely monitored. This should reduce the number of calamities and infractions. It should result in a more cost efficient way in responding to legal, competitive, and other contingencies.
(-) CSR companies attract, motivate and retrain employees more intensively and better. This should not only result in an increased quality of performance, but also reduce the need for costly performance evaluation and measurement systems.
(-) As policy promotes CSR, the companies benefit in many places from subsidies, tax relief or credits. They also benefit from green procurement.
(-) CSR companies attract the attention of the media for their engagement in sustainable development.

On the other hand organizing and maintaining CSR also entails costs. CSR management involves all levels of the organization. At every level such management not only necessitates changes in the company culture, but also investments in personnel and operational costs, which may exceed the savings. CSR equally needs coordination, an efficient communication system, and company-wide training and certification [Sprinkle and Names (2010)].

On the overall financial performance of CSR companies, the literature does not report unequivocal results. There is no guarantee that new energy firms which produce solar energy facilities or windmills will make more profit than companies producing nuclear or coal-based electricity. Companies engaging in sustainable agriculture are not more profitable than the monoculture giants producing pineapples or bananas. Even worse, up to now weapon and cigarette producers often perform financially better than CSR companies which focus on healthy and environmentally sound products. Moreover it has been shown that the ("ex ante") evaluation of the economic impact and the probability of the CSR risks is often methodologically difficult [Weber (2008)].

Therefore the main motivation for enterprises to engage in CSR is not immediate monetary profit. It is rather the conviction that investing is more than just checking figures; also non-financial characteristics of firm policies and performance matter. In particular values such as efficiency, transparency, and credibility will contribute to the long-term outcome of the organization.

4. CSR for Banks

4.1 Background and motivation

There are a number of reasons for the financial sector, including banks and insurance companies, to invest in CSR:

(-) The 2008 banking crisis showed that when a bank pursues maximizing profit too unilaterally, it impairs its basic social function of financial intermediate. As a result, society's confidence in the financial sector took a serious blow. Banks are determined to win back this necessary confidence by recognizing their impact on society, by promoting sustainable growth, by being part of the real economy, and by being transparent.

(-) An increasing volume of money is circulating in the sustainability and ethical sector of banking products. In the United States, one in every eight dollars invested is subject to social or ethical screening [Social Investment Forum (2006)].

(-) In contrast to other companies, banks offer CSR investment portfolios. These funds invest in sustainable companies (or in aspects of such companies like environmental, new energy, water related, or agricultural firms). Consequently banks as organizations themselves have the role of setting an example in matters of CSR.

(-) Investors in these sustainability and ethical funds not only include individuals who do not want to invest in military or nuclear enterprises or genetically modified organisms, but also e.g. pension funds which are interested in sustainability and (religious) organizations which have ethical motivations. For them the optimal mix between investments, risk and return is not the only target. In spite of the average return of these funds, ethical investments grow fastest in many European financial markets. Next to these traditional groups which are interested in CSR investment portfolios, more and more mainstream investors recognize the importance of non-financial information about environmental, social, and governance topics with regard to stock performance.

(-) Banks are increasingly involved in financing economic activity which aims at sustainable development. An increasing number of them screens loans and lenders for their contribution to sustainable development, prior to deciding on their lending conditions.

4.2 Results on CSR performance of banks

In this context of moving towards sustainability, the question arises whether and to what extent banks deal with CSR as a reply to social responsibility concerns. To classify banks according to their CSR performance, researchers use a comprehensive assessment scheme. It entails a two-step methodology:

(1) A set of criteria which are measurable and which represent aspects of sustainability is established. Criteria for CSR banks in general entail both generic (human resources, environment, business behavior, corporate governance, community involvement, human rights) and specific parameters (long term economic policy). Each of these criteria is subdivided into measurable indicators. Thus environment (for example) might consist in environmental strategy, policy and management, development of "green" products and services, and internal environmental performance on energy use, emissions, green procurement, waste and transport. Moreover specific attention can be paid to (legal) compliance and controversies.

(2) Step 2 determines the (relative) weight of each measurable indicator. This reveals the relative social or ethical importance of the parameter (e.g. rational remuneration of the board of administrators and the top management can be considered more, less, or as important as the composition of the board). Moreover the determination of these weights allows each rating body to introduce its own accents and policy. Consequently there is no uniform method defining these weighting factors, and no consensus list of optimal weightings.
Scholtens (2009) published the ranking of 32 international stock exchange listed banks, based on their environmental and social conduct. His indicators covered four CSR domains ((1) codes of ethics and sustainability, (2) environmental management, (3) responsible financial products, and (4) social conduct), and were based on non-proprietary information (in contrast to the commercial data banks such as EIRIS or KLD). In total 29 indicators were used in the study.
Noticeable conclusions were:

(-) Almost all banks in the sample had an explicit environmental policy defined by transparent quantitative objectives. The author concluded that by the second half of the first decade of the new millennium, environmental policy and management were widely spread among banks.

(-) Most of the banks perform environmental risk analysis related to their lending policies.

(-) The banks offer increasingly green or societally responsible financial products. E.g. two thirds of the banks in the sample offer opportunities for socially responsible investment.
(-) An increasing number of banks support local communities and NGOs.
(-) All banks have a code of ethical conduct.
(-) The Dutch “ABN Amro” and “Rabobank” came out as the best performers, together with Barclays and HSBC. The Scandinavian banks in the sample attained the lowest scores.
(-) CSR performance is positively correlated with the size of the bank.

These results have only a relative importance, e.g. because they reflect the CSR status in a limited sample at a particular period of time. A more recent comprehensive analysis of 134 international banks takes into account 5 CSR domains for which in total 18 criteria were defined. These 18 criteria were subdivided into indicators which were quantified using EIRIS data. Most of the relative weight was allocated to (long-term) economic policy and corporate governance. For the financial sector, reporting about themes such as risk management, good governance, transparency and ethical sales techniques is considered essential. Environmental impact is relatively a less material theme in this assessment. For sectors such as industry, the converse applies (Febelin, 2012). This analysis showed that Australian, Canadian, and Scandinavian banks came out as CSR front-runners, while Japanese banks were the poorest performers. The South Korean Shinan Financial Group obtained the lowest rank and showed an absolute substandard performance. Barclays performed well in this analysis as in the Scholtes’ one; HSBC holdings on the contrary was ranked in the 23th position in this study. “ABN Amro” and “Rabobank” were not assessed in this sample. The ranking also showed that rather well performing banks on the economic and governance aspects of CSR, such as Europe’s oldest “Banca Monte dei Paschi de Siena” (rank 14) and “The Royal Bank of Scotland Group” were not immune from the 2008 banking crisis [Duchi (2013)].

The comparison between both studies shows that the results need to be interpreted with care. On the one hand there is hardly any doubt that mainly during the past ten years most major banks embarked on a relevant policy and achieved results in one or more CSR domains. On the other hand the lack of a consensus method for the ranking of banks, and the major differences in choosing domains, criteria, and indicators to measure CSR, in combination with the varying values of the weighting factors, make it almost impossible to point to the best performers in the sector. The problem can only be solved by more research on CSR activities and their impact on the sector and society. Moreover ranking exercises such as the ones described above might have far-reaching practical implications (e.g. in assembling CSR funds), but assessing the real contribution of banks to a sustainable society equally necessitates investigating questions on the drives behind the banks' CSR performance.

These uncertainties also raise more fundamental questions on the real added value of CSR to sustainable development. One might argue that the benefits to a sustainable society are currently still marginal and of hardly any macro-economic significance. It is currently unclear whether the CSR activities of today are window dressing, covering an attempt of business and industry to continue a "business as usual" scenario. These considerations are probably unfair, particularly in the banking sector, but can only be counteracted by more transparency, more targeted information, and more research.

5. Conclusion

This paper has reviewed basic principles of a green economy as discussed by the world community at the UN Rio+20 conference. Overall, a green economy is shown to be driven by less carbon-based energy, less pollution, less wasteful consumption of materials and less waste. The concept is equally associated with more and better jobs, more welfare, and less poverty. An analysis of the principles of a green economy showed that they make explicit the concept of sustainable development in some detail. The economic dimension of this idea is now as well established as the environmental one.

CSR is often referred to as the most important contribution of the industry and services sectors to sustainable development. While the elements which cover CSR are clear and will almost certainly contribute to a more ethical, fair and transparent economy, their implementation proves to be most vulnerable to interpretation. This results in a wide variety of answers to the question: "which companies are sustainable as based on the implementation of the CSR paradigm?" The example of international banks which is analyzed here illustrates the division present in this variation.

Over-all the review provided in this paper shows that a green economy is increasingly considered a necessary and even unavoidable element of a sustainable society. CSR is a complex and in theory useful instrument to pursue this goal of moving towards a more sustainable society with less poverty. However the available assessments not only point to the need for more research on the subject, but equally illustrate that its implementation is still pioneering work in need of being sharpened towards better defined, measurable targets.

References

Bénabou R., Tirole J. (2010), *Individual and corporate social responsibility*. Economica 77, 1-19.

Daly H., Cobb J. (1989), *For the common good: redirecting the economy towards community, the environment and a sustainable future*. Beacon Press, Boston, Ma. Pp. 315.

Dodds F., Strauss M., Strong M. (2011), *Only one earth: the long road via Rio towards sustainable development*. Earthscan, London, UK. Pp. 312.

Ducchi D. (2013), *Personal communication.*

EC - European Commission (2010), *Corporate social responsibility (CSR)*. Available online at http://ec.europa.eu/enterprise/policies/sustainable-business/corporate-social-responsibility/index_en.htm. Last consulted on December 28th, 2012.

EC – European Commission (2011), *Green paper: Promoting a European framework for Corporate Social Responsibility*. COM (2011) 366 final, 18.7.2011. Commission of the European Communities. Brussels, Belgium.

Febelin – Social responsibility report of the Belgian financial sector (2012) A road to a more sustainable and vigorous financial sector. Available on line at http://www.bankingforsociety.be/sites/default/files/flipbookEN/files/assets/common/downloads/publication.pdf. Last consulted on January 12th, 2013.

Hussain S.S. (1999), *The ethics of "going green": the corporate social responsibility debate*. Bus. Strat. Env. 8, 203-210.

Pearce D.W., Markandya A., Barbier E.B. (1989), *Blueprint for a green economy*. Earthscan, London, UK. Pp. 278.

Scholtes B. (2009), *Corporate social responsibility in the international banking industry*. J. Bus. Ethics 86, 159-175.

Social Investment Forum (2006), *2006 report on socially responsible investing funds in the United States*. SIF, Washington DC. Pp. 134.

UN – United Nations (2012), *The future we want*. A/CONF.216/L.1. United Nations, NY. Pp.53

—. (2002), *Plan of implementation of the World Summit on Sustainable Development*. A/CONF.199/L.1. United Nations, NY. Pp. 71.

—. (1992), *Agenda 21*. United Nations, NY. Pp.925.

UNEP – United Nations Environmental Program (2011), *Towards a green economy: Pathways to sustainable development and poverty eradication*. Nairobi, Kenya. Available online at http://www.unep.org/greeneconomy/GreenEconomyReport/tabid/1375/Default.aspx. Last consulted on December 30th, 2012.

WCED – World Commission on Environment and Development (1987), *Our common future*. Oxford University Press, Oxford, UK. Pp. 263.

Weber M. (2008), *The business case for corporate social responsibility: a company level measurement approach for CSR*. European Management Journal 26, 247-261.

GLOBALIZATION AND THE *HUMAN RIGHT TO FEED ONESELF*: THE IMPACT OF THE BLUE REVOLUTION ON THE FOOD SECURITY OF SMALL-SCALE FISHERPEOPLE IN TAMIL NADU, INDIA

EBERHARD WEBER

The University of the South Pacific,
Laucala Campus, Suva, Fiji Islands
E-mail: Eberhard.weber@usp.ac.fj

Abstract

On December 26, 2004 a tsunami triggered by an earthquake west of Sumatra caused serious damage in India, Indonesia, Thailand, Malaysia and Bangladesh. Almost 300,000 lives were lost and several million people put at risk of sliding into even deeper poverty. Particularly coastal fishing communities all over the region were severely affected. They suffered heavy losses in lives, and their means of production were destroyed and their sources of livelihood put at risk.

This chapter takes the tsunami in the Indian Ocean as a starting point to reflect on social and economic change in fishing communities along the Coromandel Coast of Tamil Nadu, India. It argues that processes of globalization have made small-scale fishermen more vulnerable than they had been before. Forces triggered by globalization have taken away people's ability to control their destinies and adequately respond to challenges. Today they may be richer than a hundred years ago, but at the same time they are less resilient and less able to resist and recover from events that put their lives and livelihoods at risk.

1. Introduction

Now, in the second decade of the 21st century the economic and social well-being and maybe even the survival of large parts of humankind are threatened by

- the growing social marginalization and economic oppression of the poor,
- a high and unsustainable level of consumption of non-renewable natural resources in the industrialized and to a growing extent also in the so-called developing countries.

It is not necessary to analyze these challenges again and again. They are widely recognized – and largely ignored. The governments in the industrialized and the developing countries still preach the ideology of unlimited growth, maybe even more strictly than ever before. They preach that by globalization, by liberalizing the national economies of developing countries, and by pushing the state out of the economic sphere, the countries of the South have the chance to eradicate poverty and ensure a life in dignity for all their citizens. They don't realize that economic growth also creates poverty and prevents people from living in dignity. In 1996 the World Food Summit declared that all efforts should be made to reduce by half the number of hungry and malnourished people by 2015. In 1974, when the world community gathered for the first time to discuss how to abolish malnutrition and starvation, targets were set up to eradicate hunger within a decade: no child, no woman nor man should go to bed hungry after that. The targets of 1996 have little chance of being achieved in the remaining three years up to 2015. In 2010, 925 million undernourished people were in the world, 137 million more than in 1995 the year before the world started its effort to half hunger by 2015.

Looking at fisheries development in India over the past 100 years, one can distinguish four phases: before independence marine fisheries had been at a rather low technological level. Towards the end of the 19th century the colonial power considered fisheries to fight famine. During this time much groundwork was laid in marine science and a number of trials with modern fishing equipment were conducted. A second phase started right after independence when the modernization of India's fishing sector was started. A third phase began, when the export potential of marine products became evident. During this phase, which started in the late 1960s, the idea of a fishing sector providing high quality and cheap protein to India's poor was virtually given up. Marine fishing became a tool to provide good investment opportunities for private companies to make a quick buck exporting shrimps and other marine products. This phase is characterized by a growing conflict between small-scale fishermen and modernized mechanized fishing. In the later part of this stage, starting in the late 1980s and picking up pace during the 1990s, aquaculture came up concentrating on export-oriented shrimp culture.

At present we are in the fourth stage: a stage that started on December 26, 2004. Soon after this dreadful day hope was expressed that after all the destruction of the tsunami it would be possible to rectify mistakes of earlier decades. Hopes were high that more sustainable forms of resource use were possible, forms which consider small-scale fishermen as major actors and stakeholders and where private investment and operations are restricted to deeper waters. It was also hoped that it would become possible to show fishermen "a way out of fishing", reducing pressure on marine resources that resulted from far too many large- and small-scale users.

2. The Right to Feed Oneself in an Indian Context

At the World Food Summit in 1996 it was declared that every human being has a right to food. This declaration seems to be very progressive, but it is nothing new, as in the International Covenant on Economic, Social and Cultural Rights (ICESCR) the *Right To Feed Oneself* is one of the most important among the Economic Human Rights. Even if the *Right to Feed Oneself* can be considered as a soft law whose enforcement capacities are limited, it is more than a mere moral commitment, and the Indian Government, which signed the covenant, has certain obligations under this right.

The *Human Right To Feed Oneself* – like any other Human Right – endows every human being today and in the future with a legitimate claim against his/her state. Human Rights are rights a citizen has towards the state, and Human Rights violations therefore deal with oppression by the state and its authorities or through state authorities allowing Human Rights to be violated by others. The *Human Right to Feed Oneself* can be safeguarded in various ways. First of all it is achieved when people have access to resources like land, water, forests etc. so as to grow enough food, catch enough fish or collect enough forest products that they can meet their food requirements. The *Right To Feed Oneself* is also achieved when people receive for their work enough recompense, whether in kind or in cash, such that they don't have to go hungry. The *Right To Feed Oneself* also includes the Right To Social Security, for human beings have to be protected against old age, illness and unemployment, as these sources of insecurity are closely connected with malnutrition and hunger.

The *Right To Feed Oneself* refers generally to an adequate standard of living including food, housing, clothing and health. In its core content it refers to the right to be free from hunger and malnutrition through one's earning activities. The *Right To Feed Oneself* cannot be derived from the International Covenants on Economic, Social and Cultural Rights alone, for it is also among the Directive Principles of the Indian Constitution: in particular article 39 (a) (right to adequate means of livelihood), article 41 (right to work and social security), article 43 (right to a living wage) and article 47 (duty to raise the level of nutrition) (Constitution of India, 22-23). Concerning Human Rights the state has three obligations: to respect, to protect and to fulfill these rights. The first obligation imposes a duty on states not to destroy people's livelihood, people's personal security or health, people's cultural identity etc. These obligations exist unconditionally in the sense that the state can always be expected to restrain its activities once they turn out to violate people's Human Rights. The second obligation requires the state to protect people against violation of Human Rights by a third party. In societies, groups are in conflict with each other. Often these conflicts are conflicts over access to natural resources by different user groups. Frequently the state is part of such conflicts in one way or another. In protecting economic human rights the state, for example, has to act in such a way that it protects the livelihoods of vulnerable sections in society against business interests. The third obligation imposes a duty on the state to fulfil human rights: this means to support deprived persons who are not in a position to realize their basic requirements by their own efforts. The *Right To Feed Oneself* is much more than the Right To Food, primarily because it aims to retain the ability of people to provide for their livelihood; and thus it is much more than charity.

In this paper I will look at processes in the fisheries sector of India during the last fifty years. I will show how the lives of people are influenced by events and decisions which take place outside their spheres of influence. I argue that marginalization and poverty are the consequence of external as well as internal power structures. In the development process those groups are able to prevail which can influence the process according to their interests. Groups with insufficient power are threatened with becoming marginalized. In such conflicts the state also takes a role. It backs the interests of outsiders entering the fisheries sector, and by this endangers the livelihoods of traditional fisherpeople. On the other side state authorities come forward with social welfare programs which should soften the consequences of modernization.

3. The State's Failure to Respect Fishermen's Right to Feed Themselves

In fisheries you find the situation the Indian Government aims to achieve also for its agricultural sector: to get high prices for commodities in international markets and to have unrestricted access to these markets. By exporting high value marine products India earns much foreign exchange. One would think that the aim of protecting fishermen's livelihood ought to be achieved easily as most of the 450,000-odd active fishermen working on traditional vessels should be in the position to earn quite a good income, as most of the economically important species are caught in coastal waters. They are easily within the reach of the small-scale fishermen. Changes within the Indian fisheries sector during the last five decades, however, are responsible for the fact that small-scale fishermen still belong to the poorest sections of Indian society, even though the fisheries sector has become a lucrative business. During the past 50 years quite a lot of catches have been made on boats which were unknown in India 50 years back.

The Blue Revolution, the modernization of fisheries, is more than the introduction of new technologies. It has also influenced the processing of fish and its distribution. As a result, poor regions as well as poor consumers are more and more excluded from this once cheap source of protein. In 1956 an Indian consumed about 14.5 kg fish on average. By 1982 this had decreased to 9.4 kg. At the beginning of the 1950s fish was one of the cheapest sources of protein; in 1985 it was the most expensive (Ibrahim 1992). In 1956 about 47 percent of oil sardines were marketed locally; 20 years later less than 30 per cent of this cheap fish reached local consumers (Galtung 1984). For small-scale fishermen modernization of fisheries had devastating consequences. Trawlers and purse-seiners that were introduced fish in a very unsustainable way and have caused overfishing in many places. The modern vessels depend on subsidies given by the Government. In Kerala net profits for capital invested, which were about 14 per cent in 1969, became negative in the 1980s (Kurien 1987). The *Kalavar-Committee* in the middle of the 1980s found that about 56 per cent of the trawlers operating in Kerala were economically unnecessary (Choudhury 1986).

Earning foreign exchange was not the only reason for protecting the interests of an industry which could not survive without support of the State. Small-scale fishermen can easily catch prawns in a much more sustainable way than trawlers. The reason for the expansion of the mechanized fishing sector is deeply rooted in the political economy of the country. In contrast to the business interests entering the fishing sector, small-scale fishermen do not have much power. Exporting marine products has become a good business, and powerful companies became aware of this sector. For companies, investing in fisheries is not different from investing in any other trade. They try to get as much profit as possible and when the marine resources are depleted companies shift to other sectors. That's the logic of capitalism. Small-scale fishermen follow a different kind of logic. They are not able to shift to other fields. They have to take care that the resource they depend on for their livelihood will not become devastated. Johan Galtung summarizes his experience with the modernization of the Indian fisheries sector:

> "My evaluation is that it is a scandal, and not a partial scandal but a total scandal. [...] the INP project failed in four ways: less protein became available to the population, the level of living of the fishermen decreased, partly violent conflict between the traditional and modern sectors emerged, and depletion of the raw material, particularly the shrimps, set in. Still, however, the project was a success in the sense of being a major source of foreign currency" (Galtung 1984: 259).

During the past fifty years the Indian Government and governments of coastal states have failed to respect *fishermen's right to feed themselves.* They followed policies that have threatened fishermen's livelihoods.

4. The State's Failure to Protect Fishermen's Right to Feed Themselves

Trawlers and purse-seiners are fishing technologies which were introduced into Indian fisheries during the last 50 years. These technologies were invented in industrialized countries, where they were banned long before the same nations started to promote them in India. Bottom trawling causes great damage to the marine environment. It destroys the habitat of fishes and their breeding grounds. As the mesh-size of the trawl-nets is very small, juvenile fishes are caught and the reproduction of fishes is affected in a negative way. In the long run excessive trawling leads to overfishing. This was known to the authorities long before trawlers were introduced into India. In 1908, F.A. Nicholson, director of the Department of Fisheries in Madras, warned *"[...] that Madras did not need steam trawlers and that to jump from the catamaran to the steamer was impossible and would be unwise even if it were possible, because revolutionary methods here as elsewhere were a mistake" (Anugraham 1940, 208).*

Already in 1940 Anugraham saw the threat of overfishing and competition between small-scale fishermen and industrialized fisheries. *„If steam trawling is undertaken, it will certainly revolutionise fishing in India and give rise to an intense competition between the poor fishermen and the rich capitalists. [...] The fishermen are likely to complain of a depletion of stock in the fishing grounds because of the activities of a trawler anywhere near their fishing limits" (Anugraham 1940: 210).*

To prevent conflicts Anugraham suggested creating fishing zones where only small-scale fishermen should fish, and from which trawlers would be banned entirely. It took more than 40 years before state governments introduced Marine Fisheries Regulation Acts which banned trawlers and purse-seiners from coastal waters. Still today the implementation of these laws to protect small-scale fishermen is very poor. Restrictions imposed on the operation of modern vessels don't work, as about 80 - 90 per cent of the high value marine products live in coastal waters. If trawlers and purse-seiners were really to fish in areas assigned to them, most would be bankrupt soon. It seems that state Governments enacted laws they never meant to enforce. There is a contradiction in promoting modern fishing techniques like trawling by providing subsidies and restricting the operations of trawlers to areas where few prawns are available. It becomes obvious that the State is not only not able to protect the *fishermen's right to feed themselves*, but it is also not willing.

5. The State's Failure to Fulfil Fishermen's Right to Feed Themselves

The third obligation of the state under the *Human Right to Feed Oneself* is to fulfill this right when people are not able to do so by own efforts. Here Social Security plays an important role. Did the state at least establish effective social security measure for the needy, when it was not able to respect and protect their rights?

Social security refers to protecting people against risks like illness, unemployment, old age, accidents, maternity, disability etc. which make people unable to secure their livelihoods. To achieve social security the state can use *promotional measures* aiming at averting insecure conditions and contingencies such as low incomes/unemployment/sickness through special programs or *protective measures* which are to (re)establish social security whenever the *promotional measures* have failed or are not sufficient (Burges and Stern 1991; Osmani 1991). As mentioned above the Indian Constitution mentions important aspects of social security in articles. 36 – 51 *(Constitution of India 2007:21-25)*, particularly in article 39 which says:

> *"The State shall, in particular, direct its policy towards securing*
> *(a) that the citizen, men and women equally, have the right to an adequate means of livelihood;*
> *(b) that the ownership and control of the material resources of the community are so distributed as best to subserve the common good;*
> *(c) that the operation of the economic system does not result in the concentration of wealth and means of production to the common detriment;*
> *(d) that there is equal pay for equal work for both men and women;*
> *(e) that the health and strength of workers, men and women, and the tender age of children are not abused and that citizens are not forced by economic necessity to enter avocations unsuited to their age or strength;*
> *(f) that children are given opportunities and facilities to develop in a healthy manner and in conditions of freedom and dignity and that childhood and youth are protected against exploitation and against moral and material abandonment" (Constitution of India 2007:21-22).*

Establishing social security falls under the obligation of the State as well as Central Government. The system for the formal sector of the economy in India is quite modern. Since independence many laws have been introduced which protect a small minority of Indian citizens against all sorts of risks (e.g. *Workmen's Compensation Act (1923), Bombay Maternity Act (1929), Maternity Benefit Act (1961), Employees' State Insurance Act (1948), Provident Fund Act (1925), Employees' Provident Fund Act (1952), Employees' Family Pension Scheme(1971).*

This social security system is highly subsidized as the State gives contributions to most of the different schemes and also allows deductions from the income tax for private contributions to provident funds, life insurance policies, loans from housing societies, etc. For the informal sector protection is far less (Table 1).

Table 1: Social Security Initiatives for the informal sector in Tamil Nadu

Program	year of intro-duction	target group	benefit	before 1979	1979	1982	1989	1990	1992	1995	1996	1998	2000	2006
Old Age Pensions														
(1) general	1962	for needy persons above 65 years	monthly pension	20	25	35	50		75	100		150	200	400
(2) for destitute agricult. Labourers	1981	above 60 years				35	50		75	100		150	200	400
(3) for destitute widows	1975	above 40 years		20	25	35	50		75	100		150	200	400
(4) for destitute, deserted women	1984	above 30 years					50		75	100		150	200	400
(5) for handicapped	1974	above 45 years		20	25	35	50		75	100		150	200	400
For surviving dependents	1989	death of the Head of Household agricul. Labourer, marginal farmer	lump sum	-----	-----	-----	2000	3000	5000	5000	7500	7500	10000	10000
		HH-income less than		-----	-----	-----	5000	5000	6400		6400	6400	7200	7200
Survivor's pension for handloom weavers	1991	deceased has to be member of the co-operative society	monthly for 10 yrs							250				350
Maternity Benefit	1989	HH-income below 5,000 Rs p.a. given only for the first two children and wife's age at amrriage more than 18 years	2 months before / after delivery							4*75			4*125	
Marriage grants for females														
(a) general	1989	HH-income below 12,000 Rs p.a. wife's marriage age above 18 years passed eigth standard	lump sum							5000				10000
(b) for female orphans	?	HH-income below 12,000 Rs p.a. wife's marriage age above 18 years	lump sum							2000				10000
(c) Widow Remarriage	?	HH-income below 12,000 Rs p.a. wife's marriage age between 20 - 30 years	lump sum											10000
Accident insurance for certain professions	1977	fishermen, scavengers, pesticides-sprayers, agricultural / construction coolies, Rickshaw-drivers, etc.	lump sum (in case of death)							21000			35000	50000
Saving Scheme	1982	fishermen	yearly			90			720					600

In the informal sector there is no comprehensive system of social security. People who are most vulnerable against a number of uncertainties are protected worst. They belong to the poorer sections of Indian society; they often have to live under unhygienic conditions and are more vulnerable to diseases. Many of them cannot afford a balanced diet and the risk of becoming ill due to under- and/or malnutrition is high. Unskilled labourers without fixed employment are more vulnerable to economic exploitation than those working in the organized sector of the economy, where trade unions take care that labour laws are respected. For a huge section of India's society, at least 70 per cent of the country's population, there is no system of social security. There are different schemes which try to protect people in the informal sector against some contingencies, but they usually are far too weak to achieve this.

In Tamil Nadu for example there is an old-age pension scheme which entitles different groups to a monthly pension of 75 Rs. In the early 1990s about 515,000 persons received such a pension, about 12 per cent of the elderly poor. As the Tamil Nadu Government didn't provide sufficient funds those who missed out were put on waiting-lists (Guhan 1992). The program is hardly more than charity as nobody has a legal right to get a pension. The schemes are prone to corruption as it is more or less left to the authorities to whom they give a pension and to whom they don't.

For small-scale fishermen there are two schemes: a group insurance scheme and a saving scheme. The insurance scheme was introduced in 1984. In the early 1990s dependants of a fisherman who died at sea received a compensation of 21,000 Rs. The premium of 20 Rs annually was shared by the Tamil Nadu and the Central Government. Since 1982 a saving scheme for fishermen had been introduced to balance seasonal income fluctuations. First fishermen contributed 10 Rs per month for a period of 9 months. The Tamil Nadu Government gave the same amount (90 Rs), In October, November and December fishermen then received 60 Rs monthly. In 1992 the program was changed. Now fishermen contribute a monthly installment of 45 Rs, and the State as well as the Central Government each added the same amount. 1080 Rs came together in 8 months and each fisherman now gets 270 Rs during the four months of the monsoon season (Weber 1995).
There are a few more social security initiatives for the informal sector in Tamil Nadu. However most of them are insufficient to provide meaningful protection. The monthly old-age pensions are far too low to enable pensioners to cross the poverty line. The disparities in comparison with the formal sector are striking. An annual pension in the informal sector is much lower than the monthly pension of an employee of the Tamil Nadu Government. Just to

protect State pensioners against inflation their monthly pensions were increased by 270 Rs on average between November 1991 and May 1993. In comparison a monthly pension of 75 Rs given to a few in the informal sector looks quite modest.

The biggest constraint of these programs is that people do not have a ***right*** to benefit from a scheme. Among the people of a fishing-village in Madras not a single one was found who received a pension.

To sum up: the State also failed to fulfil *fishermen's right to feed themselves* by failing to establish a comprehensive social security system. It is much more difficult to establish such a system for the informal sector as many in this sector are not able to contribute to such schemes, as their income is too low. Many in the informal sector are self-employed. Contributions by an employer (as is the case in the formal sector) are missing. This makes financing such schemes difficult. On the other hand the state is subsidizing social security highly for the formal sector. Why not cut or reduce subsidies for employees whose salaries allow them to take care of their social security needs by their own efforts? The money then could be used to establish systems of social security which at least give a minimum protection to members of the informal sector.

6. The Impact of Mechanization on Small-Scale Fishermen

Mechanization of fisheries has had a negative impact on the socio-economic conditions of small-scale fishermen. A study in a fishing settlement of Madras city indicates that small-scale fishermen's vulnerability has increased a lot as a result of modernization and conflicts over fishing resources, particularly prawns, the Pink Gold. A substantial part of fishermen's income is earned in a very short period of the year, the prawn season. Then even small-scale fishermen can earn a lot.

Figure 1 shows that there is a distinct seasonality in the incomes from fishing. A big share of the income is earned in only one month, for example, during December 1991. If we have a look on the composition of the catches (Figure 2) it becomes obvious that this seasonality of incomes is due to prawn catches.

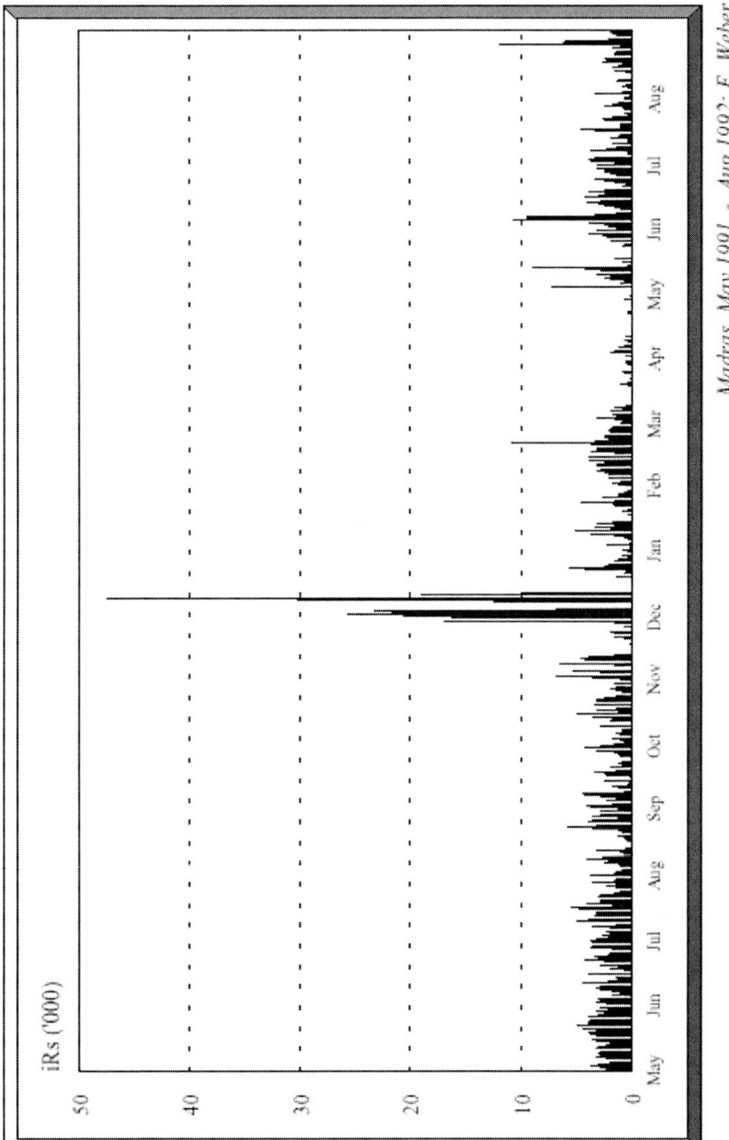

Figure 1: Value of marketable catches of 37 fishing-teams

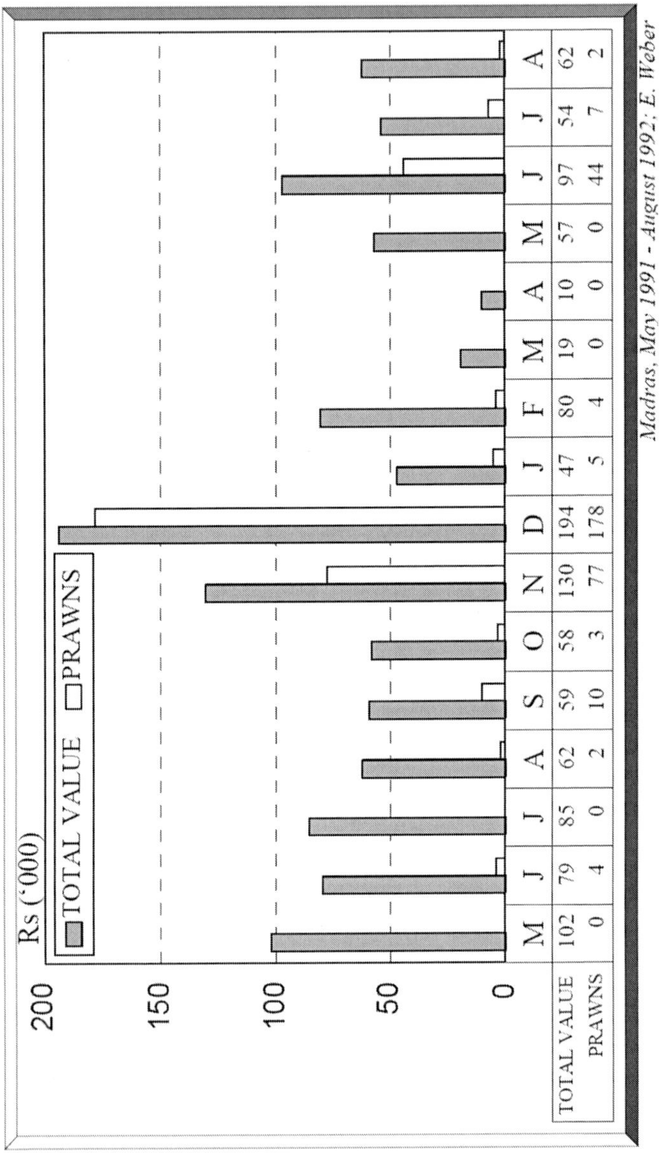

Figure 2: Composition of Fish landings

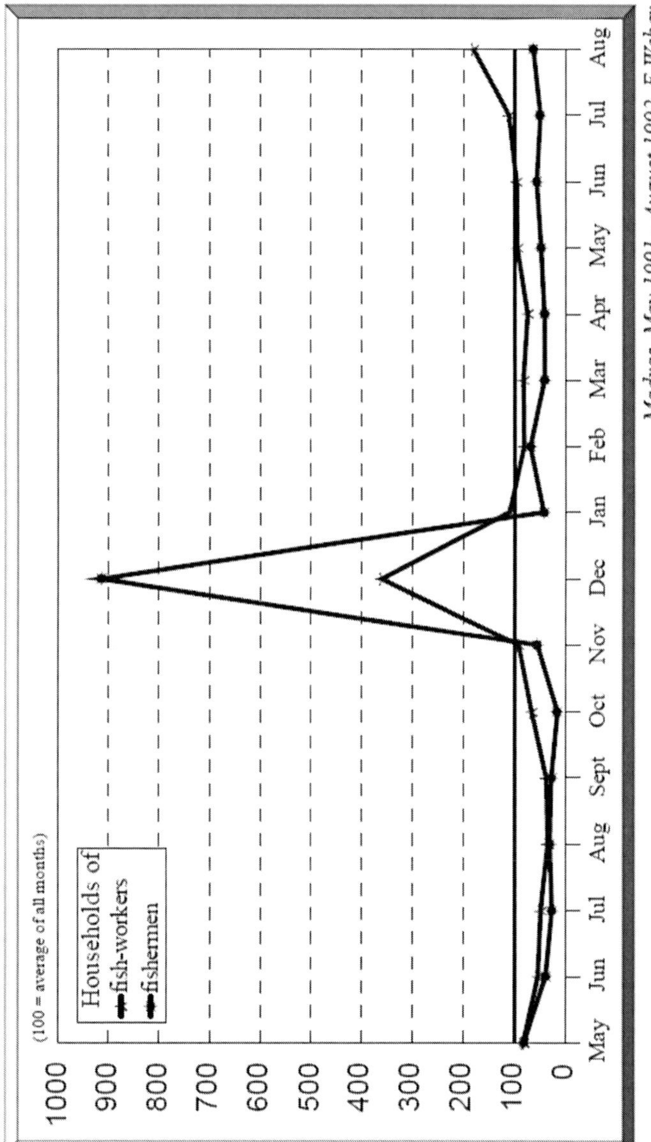

Figure 3: Repayment of loans

In Tamil Nadu November and particularly December are by far the most important months for catching prawns. During the 16 months of investigation, prawns were the source of about 30 percent of the income of 37 fishing-teams. But there are only a few days in the year when prawns can be caught by fishermen. During these days fishermen earn a big part of their income. During 12 fishing-days in December 1991 more than 20 percent of the income from 16 months was earned. Some boats earned even more than 30 per cent of their income of 16 months (Weber 1995). One can easily imagine what would happen if this income cannot be realized. During the period when fishermen can earn a fortune, weather conditions are worst. Mechanized boats do not depend so much on favourable weather conditions. They can go on fishing when small-scale fishermen have to stay on the shore. The risk that mechanized boats catch most of the prawns during those days is very high. If the income from prawn fisheries decreases, the economic and social basis of the small-scale fishermen is at risk. Most of the fishermen are in danger of being caught in a debt-trap. As Figure 3 indicates, the fishermen repay the loans which have been accumulated throughout the year during the prawn-season. If the income from prawn catches decreases, the fishermen might also become unable to raise the necessary capital for investments. Thus, it is no exaggeration to say that their socio-economic status is at risk.

7. Conclusion

Despite high prices for prawns in the international market and few entry restrictions to the markets of industrialized countries, the vast majority of small-scale fishermen in India are still living in poverty. Because of the highly profitable business of prawn exports, big business has entered the marine sector, shrinking income opportunities of small-scale fishermen considerably. The state actively promoted the mechanization of fisheries and thereby played a crucial role in the marginalization of small-scale fishermen. Protecting the small-scale fishermen against the mechanized trawlers however is not enough to help to improve the socio-economic status of the fishermen. What is needed is a policy that creates employment outside fisheries, so that the fishermen can shift to attractive alternatives. However in the present economic scenario this seems to be impossible, as much emphasis is given to capital intensive investment and only slight emphasis to employment generation.

When on December 26, 2004 the Indian Ocean tsunami fell as another blow on fisher-people in Tamil Nadu, the worst damage had already been done: marginalized and deprived of their sources of livelihood, the fishermen and their families had little remaining resilience against the damage caused by this natural hazard. The disaster however consisted not in the waves of the tsunami, but in the marginalization that happened in the decades beforehand.

Annex

Obligations of States under the International Covenant on Economic, Social and Cultural Rights

	National obligations								
	Respect	Protect			Fulfill				
		Enact	Monitor	Enforce	Macro Policy	Hierarchized System of state support	Monitoring and Analysis	Identification of especially disadvantaged groups in society	Specific Programs
		Laws, acts and regulations							
Access to Legally protected interest	No dispossession / resettlement without adequate compensation	Granting of land titles, user rights, security of tenure and tenancy, respect of traditional land rights and user rights	land register, arbitration board to solve land conflicts, land surveying	independent legal procedure in case of land conflicts	Landreform	Landreform, Employment programs, continuing education, rural extension services, social security systems (unemployment benefit, direct income transfers), Support in times of harvest failure emergency relief	regular studies and surveys to get relevant information on regional levels of a country considering the situation of women, Identification of particular vulnerable and disadvantaged groups. Tendencies regarding fulfillment of Human Rights, Obstacles and Challenges, which oppose the realization of Human Rights	e.g. Landless groups small tenants indigenous groups women	Multiple supporting activities, depending on specific needs of target groups and basic conditions
					Price Incentives for agricultural products. Infrastructure development			informal Sector	
	No redundancy without social plans	Fixation of minimum-wage to safeguard existence of employee and family (or other measures that are suitable to safeguard mere subsistence, labor laws and Occupational Health and Safety Regulations.	Monitoring the employers regarding compliance of minimum wage and safety regulations	Industrial tribunals and sanctions against companies which violate employee's rights and labor laws	Full employment policy			e.g. children old people sick people unemployed unskilled laborers	
	Respect self-help potentials when drafting development programs and resettlement, etc.	Fixation of minimum quantities and establishment of quality criteria in public food supply systems.			Establishment of social security systems, tax reduction for low income groups.			e.g. Orphans, widows pavement dwellers old sick	
sufficient in a. Quality and b. Quantity	No taxation of subsistence level				Policies that favor self-reliance in food sector. Agricultural Price policy that supports food production.			malnutrition resp. Childrens; rural and urban poverty	
		consumers laws. Occupational health regulations in companies. environmental standards.	Food safety monitoring, monitoring companies reg. Environmental and health standards.	Sanctions against companies that violate Occupational health standards, environmental standards and consumer protection laws.				Malnutritions of children and other vulnerable groups	
culturally accepted	Consideration of cultural / religious food habits in case of public food programs.							Religious and other minorities, indigenous populations	
sustainable	Consideration of importance of traditional farming systems and social structures when drafting agricultural / rural development programs.	Environmental standards to protect agricultural land, fiv, forests, ...	Monitoring companies' compliance with environmental obligations	Sanctions against companies that violate environmental obligations.	Ecologically sustainable agriculture, forestry and fisheries.			Migrants and seasonal laborers, conditions is ecologically sensitive regions.	

References

Anugraham, D.D. (1940), The Fisherfolk of Madras: an economic survey, University of Madras, unpublished M.A.-Thesis).

Burges, R. and Stern, N. (1991), Social Security in Developing Countries: What, Why, Who and How?, in: Social Security in Developing Countries. Ahmand, E., Dreze, J., Hills, J., and Sen, A. (eds), Oxford University Press, Oxford, 41-80.

Choudhury, A. (1986), Traditional Fisheries, in: Debacle, 2(4), 18-22.

Galtung, J. (1984), The Indo-Norwegian Project in Kerala: A "Development" Project Revisited, in: Internationales Asienforum, 15, 3/4, 253-274

Guhan, S. (1992), Social Security Initiatives in Tamil Nadu 1989, in: Themes in Development Economic, ed by S. Subramanian, Delhi, 242–283.

The Constitution of India as modified up to the 1st December, 2007, Government of India, Ministry of Law and Justice, New Delhi.

Kurien, J. (1978), Entry of big business into fishing - its impact on fish economy, in: EPW, 9.9.1978, pp. 1557-1565

Kurien, J. (1987), The Impact of New Technology Introduction into Fishing in Kerala, in: Programme for Community Organization, Fisheries Crisis and Policy Approach in Kerala, 69-98.

Osmani, S.R. (1991), Social Security in South Asia, in: Social Security in Developing Countries. Ahmand, E., Dreze, J., Hills, J., and Sen, A. (eds). Oxford University Press, Oxford, 305-355.

Weber, E. (1995), Globalisierung und Politische Ökonomie der Armut in Indien, Chennai/Limbach.

Household and Commercial Food Waste Reduction in Times of Increased Food Insecurity and Unsustainable Natural Resources

Dimitrios Dimitrakopoulos

Medical School, National and Kapodistrian University of Athens, Greece
Environment and Health. Capacity Building for Decision making.
E-mail: ddimitrakop@gmail.com

Abstract

Recent studies regarding further income inequality and an increase in food production to meet the future demands of a developing, increasing and more affluent world, while depriving the population of food security through food price increases at times of austerity, falling income and labour costs, guide us to two important consequences. The first one is that the prosperous section of the population overconsumes and wastes thousands of tons of food ending in landfills, while households below the poverty line experience malnutrition. The second consequence is that bad management of food overproduction ends up not only in increased compostable waste but in depletion of the rest of the resources related to this process with an impact on the soil, the water, the air and the climate. In the developed G20 countries up to 60% of all food produced is wasted in times where the food security of the majority of people in these same societies has a downward trend due to the economic recession. The management of these events must be collective, implementing fundamentals from historical, cultural, economic and belief system examples that provide solutions for resource depleters, political regulators, food producers and resellers and mainly the household end-users that are most affected by the current financial crisis.

1. Global food waste generation

Each year the European Union produces 89 million tonnes of food waste from all economic activity [Eurostat, Waste Statisics, (2010)]. This amounts to 179 kg of food for every man, woman and child, while 79 million citizens in the EU-27 live beneath the poverty line and 16 million depend on food aid from charitable institutions each year (EU Parliament, Food wastage, 2012). In the U.S. the average American throws away 115 kg of edible food annually [Gustavsson et al. (2011)]. If 15% less food were wasted in the U.S. it would be enough to feed 25 million Americans [NRDC (2012)]. With food insecurity rising due to the recession, the European Parliament has called for urgent measures to halve food wasted in the EU by 2025 and to improve access to food for needy EU citizens [EU Parliament, Agriculture, (2012)].

Food losses relate to developing countries and systems that have deficiencies in their infrastructure, such as irrigation, pests, harvest storage and processing, but also occur due to climate change [Venkatt (2011)]. A 2011 study by the *Swedish Institute for Food and Biotechnology* (SIK) on behalf of *Food and Agriculture Organization of the United Nations* (FAO), *Global Food Losses and Food Waste*, distinguishes between "food loss" and "food waste":

- Food loss/spoilage, refers to the decrease of edible food quantity or quality occurring in early stages of the FSC, that is, production, postharvest and processing, which makes it unfit for human consumption [Grolleaud (2002)]
- Food waste is applied in the retail and consumption stages of the FSC and generally relates to behavioural issues [Parfitt, *et al.* (2010)] in developed countries – that is, the throwing away of food.

Given the fact that developing countries are adopting the consumer behaviour of developed countries, we will refer to both food losses and food waste as 'food waste'.

As European society has grown wealthier, it has created more and more rubbish while every percentage growth in the GDP means an increase in waste. For every EU citizen of a country with at least a BBB+ credit rating (S&P) there is a corresponding amount of 520 kg of municipal waste. The responsibility for food waste according to the *Preparatory study on Food waste across EU27* corresponds to:

- households: 42% (60% of which is avoidable)

- manufacturers: 39%
- retailers: 5%
- catering sector: 14%

Initiatives to reduce food waste must start from the consumer in households, focusing on behavioural, educational and ethical issues. Ethical issues expand in unexpected areas, and waste reduction could have an economic impact on saving household income or providing food and reducing poverty, and an environmental impact by saving ecosystems and land from deforestation due to industrial cultivation, or saving land from landfilling. Landfilling not only takes up more and more valuable land-space; it also causes air, water and soil pollution, discharging carbon dioxide (CO_2) and methane (CH_4) into the atmosphere, and chemicals and pesticides into the earth and groundwater. This, in turn, is harmful to human health, as well as to plants and animals [EC, Waste (2012)]. The most affected people are the rising numbers of the recently homeless and malnourished that have to relocate due to property foreclosures, and often move to environmentally degraded areas. The link between environmental quality and human health is recognised in the majority of Poverty reduction Strategic Papers, but is seldom elaborated [Jean Hugé *et al.*, (2007)]. The newly poor can be affected by known diseases of poverty that tend to be largely *infectious diseases*, or the result of poor living conditions. These include tuberculosis, asthma, and intestinal diseases [Singh *et al.,* (2008)].

2. Austerity measures, the newly poor and food insecurity

In Seattle, Washington, the Cruz family up to 2008 belonged to the middle class of thc country. Today, it has to say "no" to its children's requests for a $1 ice cream. Cutting back became a necessity for many American families, but what the Cruz family and a growing number of other once-thriving middle-class families didn't expect was to find themselves qualifying for — and needing — the support of federally funded food assistance programs [Garton (2011)].

Up to 2007 13 million people in the U.S. or about 11% of all households were considered "food insecure". Events escalated in 2008 with job losses, increased mortgage defaults, property foreclosures and the inability to provide debt payments. In addition there has been an increase in needed support among groups that were historically less vulnerable to hunger, according to the USDA's *Household Food Security in the United States* annual report. The report estimates that 44 million or one in seven Americans (half of whom are children) are currently enrolled in the government's largest nutritional safety-net program — the *Supplemental Nutrition Assistance Program* or SNAP [Harden (2012)]. In 2002 food security was 89%, and in 2008 with the continuing recession the percentage dropped to 85% with a forecasted decrease.

Food stamp recipients alone jumped from 17 million in 2000 to 45 million in 2011 [Cafferty (2012)]. Mark Nord, a sociologist at the U.S. Department of Agriculture's Economic Research Service, says "Not everyone who is food insecure is literally going hungry". "Some are able to head off hunger by reducing the quality and variety of their diets. But if food insecurity is severe or prolonged, it is likely to result in hunger" [Nord et al. (2009)]. Across the Atlantic, if British consumers managed better the food they waste every year, they would save 12.6 billion euros or 560 euros per household (WRAP 2008). This amount, estimated at $2 per day, is significant when for many families in the United States the aid they receive from the federal SNAP programs is not enough to cover the month. The average supplement is $130 per month, or less than $1.5 per person per meal, an insufficient amount for proper nutrition (U.S. Department of Agriculture, 2011).

Further south in Spain the report 'Exclusion and Social Development 2012', that was published by Caritas Spain, estimates that already 22% of Spanish households are living under the poverty line with a further 30% facing serious difficulties in making ends meet at the end of the month. Caritas secretary Sebastian Mora stated that "There are more poor people than last year, and they are poorer. After four years of financial hardship poverty is more widespread, more intense and it is creating a polarized society in which the difference between rich and poor is growing". "Around 580,000 Spaniards, nearly 3.3% of the population, are not receiving a salary," said the coordinator of studies of Caritas' team Francisco Lorenzo [Caritas (2012)]. Among them, 400,000 Spaniards are in the need of food.

In the Netherlands, with a credit rating of "AAA", there are now 135 food banks bailing out people trying to survive on less than 180 euros per month (Pals *et al.* 2012). Food waste during harvesting, processing, storage and marketing is estimated at 2 billion euros. That is 14 million euros for every food bank, enough to provide 3,000,000 meals, and also to feed all the needy in Spain.

In Greece many of the people that have been affected by the financial crisis find their basic food supplies from the unsold food that is left behind by producers of open air traditional markets. This already serves as a reduction in food and landfill waste. These markets provide fruit and vegetables to the general population from both middlemen that buy from farmers and from farmers who sell their goods directly to increase their profit. Other markets influenced by the "potato movement", that sell directly from farmers without middlemen, offer olive oil, honey and dry beans, and are preferred even by people in a better economic level. The Minister of Rural Development and Food Athanasios Tsaftaris announced that the government will be evaluating bids to purchase food for free distribution to 800,000 families. The budget is 23 million euros, and will include pasta, olive oil, Greek feta cheese, Cretan graviera cheese and rice.

In 2010, approximately 115.5 million people in the EU-27 (23%) were at risk of poverty or social exclusion (AROPE). With a rate of 26.9 % in EU-27, children were at greater risk of poverty or social exclusion in 2010 than the rest of the population in 21 of the 25 Member States for which data are available. The only exceptions were Denmark, Slovenia, Finland and Sweden [Antuofermo *et al.*(2012)].

3. Global world nutrition and food waste from production to consumption

In 2009 the policy document for the first Group of Eight ministerial meeting on agriculture warned that global food production needs to double by 2050 to feed a surging population and prevent international instability, caused by a permanent food crisis related to "pronounced climate changes" and higher food costs [Blas (2009)].

A former NASA climatologist, DeWayne Cecil, expressed his concern that "Earth can only sustain a half-billion people in a "Western" consumption oriented lifestyle". He also stressed that population control and fresh water resource management are "top global priorities". What is not certain is if a "Western" oriented lifestyle is sustainable to preserve the human species or will it became extinct from affluent society illnesses?

The above sentiments have been foundations for philosophers that approach the relationship of population control and resource savings from their different perspectives. In 500 B.C. Confucious cautioned that "excessive growth may reduce output per worker, repress levels of living for the masses and engender strife." Confucius also observed that, "mortality increases when food supply is insufficient ... that war checks population growth" [Neurath (1994)]. Church author Tertullian (A.D. 200) describes the relation of overpopulation and disasters, including the impact of an over-consuming population that is demanding well-being (e.g. Roman "festivities"). He wrote: "The strongest witness is the vast population of the earth to which we are a burden and she scarcely can provide for our needs; as our demands grow greater, our complaints against Nature's inadequacy are heard by all. The scourges of pestilence, famine, wars and earthquakes have come to be regarded as a blessing to overcrowded nations, since they serve to prune away the luxuriant growth of the human race" [Tertullian A.D. 200)].

Niccolò Machiavelli (1469-1527), an Italian Renaissance political philosopher wrote on the very same issue, "When every province of the world so teems with inhabitants that they can neither subsist where they are nor remove themselves elsewhere ... the world will purge itself in one or another of these three ways," listing floods, plague and famine [Neurath (1994)].

In the 20th century, population control proponents have drawn from the insights of Thomas Malthus, a British clergyman and economist who published *An Essay on the Principle of Population* in 1798. Malthus argued that human beings have the potential to increase faster than the food needed to feed them. He claimed that the problem arose from the fecklessness of the labouring classes (Monbiot, 2008). He outlined the idea of "positive checks" and "preventative checks." "Positive checks," such as diseases, war, disaster and famine, are factors that Malthus considered to increase the death rate for population control [Neurath (1994)].

In a more moderate approach, other economists have argued that poverty and famine are caused by bad government and bad economic policies, not by overpopulation. Julian Lincoln Simon claimed that, "when considering a list of countries ranked in order by population density, there is no correlation between population density and poverty and starvation. Instead, if a list of countries is considered according to corruption within their respective governments, there is a significant correlation between government corruption, poverty and famine" (Simon 1998). Therefore food insecurity that is increasing in the western world is rather attributed to an artificial manipulation of financial institutions whose traders have collapsed the housing market, plundered retirement accounts and taken advantage of inside information [Kelton (2012)]. Hunger is caused by poverty and inequality, not scarcity of resources [Gimenez (2012)]. “Poverty is not caused by the lack of material resources. It is the immediate result of our exploitation and waste” [Chryssavgis, (2008)].

By contrast, global food production has not been affected by climate change to the degree that it cannot retain sustainability. The report of the Food and Agriculture Organization of the United Nations (FAO), the International Water Management Institute (IWMI) and the Stockholm International Water Institute (SIWI) provides evidence that there is enough food production to feed the world population, and calls for actions to reduce losses and wastage by half by 2025 [Lundqvist *et al.*, (2008)].

According to Morgan and Robertson “more than 1 billion people (one-sixth of the world’s population) suffer from chronic hunger. It is not a problem of supply, since the food system continues to provide enough food to feed the world. One of the reasons for this disconnect is simply waste. By many estimates, one in two food calories is currently wasted” [Smil (2000)].

The Update of the Nutrition Situation for Sub-Saharan Africa in 1994 associates the very high mortality rates with calorie availabilities of less than 1,500, and indeed in those cases where more than 2,000 kcals/*caput*/day are available, the mortality rates are on average less than 1/10,000/day, and do not reach 2/10,000/day (UN, 1994).

In comparison to these figures, the food wasted in France could feed the under-nourished in the Democratic Republic of Congo, and the amount discarded in Italy could put an end to hunger in Ethiopia, according to a study of the ethics of modern development in agricultural technologies [Ryan, (2009)].

For the past two decades, the rate of global food production has increased faster than the rate of global population growth, in contrast with what Malthus claimed. This has happened because food production is prioritized as a commodity for economic growth, and not as a population sustainer. The world already produces more than 1½ times enough food to feed everyone on the planet. That's enough to feed 10 billion people, the population peak *we expect* by 2050 [Gimenez (2012)]. In reality, the bulk of industrially-produced grain crops goes to biofuels and confined animal feedlots rather than food for the 1 billion hungry human beings. The key to fight current global poverty and famine is less food waste, and changing consumer habits and social ethics.

Food is wasted in every stage of the Food Supply Chain "from field to fork". There are losses on the farm, in storage, in processing, in transportation, in marketing and food preparation, up until the final decision by the consumer to eat or discard the food [Monier, *et al.*,(2010)]. The exact causes of food losses and waste vary in low-income, medium-income and high-income countries, and largely depend on the specific conditions and local situation in each country. In the developed countries a large percentage is lost at the household consumer level, ending up in the landfills while still edible. Purchases above the necessary, and also appearance faults of fruit and vegetables (weight, shape) contribute to food disposal.

In the US, 30 to 50% of all food, worth $48.30 billion (EUR32.5 billion), is thrown away each year [Jones, (2006)]. It is estimated that about half of the water used to produce this food also goes to waste. Losses at the farm level are probably about 15-35%, depending on the industry. The retail sector has comparatively high rates of loss of about 26%. Overall, losses amount to around US$90 billion to US$100 billion a year [Nelleman *et al.*, (2009)]. Part of the waste on the farm comes from undesired prices at the commodity market in Chicago which often mean an entire crop is left in the field to be ploughed under. Reducing food waste by half could reduce adverse environmental impacts by 25 percent through reduced landfill use, soil depletion and applications of fertilizers, pesticides and herbicides [Harrison, (2004)].

In many African countries, the post-harvest losses of food cereals are estimated at 25% of the total crop harvested. For some crops, such as fruit and vegetables, post-harvest losses can reach 50%. Depending on the crop, approximately 15 – 35% of food may be lost in the field due to poor agricultural techniques and other factors, including droughts, flooding and pests. Another 10 – 15% is lost during processing, transport and storage, due to a lack of modern facilities, trucks and access to refrigeration (Morgan *et al.,* 2011). In Kenya, each year around 95 million litres of milk, worth around $22.4 million, are lost. Cumulative losses in Tanzania amount to about 59.5 million litres of milk each year, or about 20% of total dairy production. In Uganda, approximately 27% of all milk produced is lost, equivalent to $23 million/year [Brussels Policy Briefing no. 28, (2012)].

Although extended food loss is not confined to borders, there is significant responsibility within the affluent societies, where wasting is a way of life. The impact is not confined to household budgets. It is costly to the environment where total economic activity for the production of food wasted, plus related emissions from landfills, adds millions of tonnes of greenhouse gases annually to the atmosphere. This in turn has consequences for climate change, with heat waves and floods that destroy crops. The destroyed crops raise food prices. The high food prices create actual incidences of hunger for a growing sector of the middle class that has been affected by the recession, resulting in food insecurity [Gardner *et al.,* (2004)]. Policies or legislations will contribute little to the improvement of the situation. What is needed is changing ethics and lifestyles on the part of the recipients at the end of the food supply chain, that is the human consumers involved.

4. Neoliberal Foundations for Consumption and Waste

As persons we are all unique and different, but when we identify our identity with a consumer culture we are statistically all the same. This is one of the elements of consumer market research that multinational companies rely on to accumulate capital and control of resources in a globalized economy. A fundamental of mainstream economics since the 1920s is mass production and mass consumption [Hounshell (1985)] through mass culture. The media industry, initially network radio and Hollywood cinema and later television in the 50s, grew in perfect harmony with the emerging industries, advertising mass culture.

Consumerism dictates that it is not the industries that determine what a society should consume, but rather the free-thinking consumers. The question is who affects the consumers to decide when and how much? The facts so far show the markets through consumer culture. It is not clear if the "markets" refer to consumers or stock exchange indexes.

The second associates better with the theory of the "Chicago School" (of Economics) that markets are self-regulating and self-sustaining and that the government's role in the guidance of the economy should be restricted severely (Lippert, 2008). The methods of expansion of mass culture towards a global economy often go beyond the free thinking of the consumers (Dijksterhuis, *et al.,* 2005) or the free will of the citizens, as with market-lobbied government-imposed austerity measures. Just as the Chicago school predicts, it is the credit service companies (S&P, JP Morgan etc.) that regulate the markets and warn about the collapse of an economy but whether this is due to consumer behaviour or trader behaviour, third party or inside behaviour has yet to be determined. In addition consumers may also provide their tax money to preserve the profits or to bail out the markets as with Wall Street in 2009 [Foley, (2010)]. Other possible obstacles to expanding a mass market culture, in addition to real government regulation, could be regional and local habits, ethnic traditions (food), and religious, political or ethical beliefs. For this reason globalization has been criticized for the perceived breakdown of traditional communities and values.

The primary feature of a consumer culture is that people's identities are defined by the things they possess that become status symbols. Meaning has become invested in things and not values. In 1929 theorist *Christine Frederick* observed that "the way to break the vicious deadlock of a low standard of living is to spend freely, and even waste creatively" [Dery (2009)]. Many economists have connected happiness with over-consumption. Consumers' unlimited wants become unlimited waste of finite resources [McEachern (1996)].

According to Victor Lebow in 1955 "the standard of living to which the consumer aspires is partially shaped by the pressures put upon him by manufacturers and retailers… The real goals are to *look* better, *live* better, *dress* better, *travel* better" in relation to other consumers [Lebow, (1955)].

Madeline Levine criticized what she saw as a large change in American culture – "a shift away from values of *community*, *spirituality*, and integrity, and toward competition, materialism and disconnection [Levine (2007)]. Other researchers argue that the struggle for symbols of social distinction promoted by consumer culture creates narcissistic, hostile relations between individuals, which can be criminogenic in locations where consumer products are difficult to acquire, or where individuals simply see no limit to their acquisition»(Hall, *et al.,* 2008). Consumerist societies are more prone to damage the environment, contribute to *global warming* and use up resources at a higher rate than other societies. Dr. Jorge Majfud says that "Trying to reduce environmental pollution without reducing consumerism is like combatting drug trafficking without reducing the drug addiction" [Majfud, (2009)]. The impact to the environment is that "the world's ecological or carrying capacity is simply insufficient to satisfy the ambitions of China, India, Japan, Europe and the United States as well as the aspirations of the rest of the world in a sustainable way"[Renner, (2006)]. Currently, the developed nations of the world consume at a rate of 32, while the rest of the developing worlds' 5.5 billion people consume at a rate closer to 1» [Diamond, (2008)]. A historical milestone in the economics of prosperity is the Rostow-Taafe model of E. Taafe and W. W. Rostow of the five stages of economic growth [Rostow (1960)] with the last one being that of mass consumption [Rostow (1960)].

Victor Lebow in his article «Price Competition in 1955», five years before Rostow's model, explains: "Our enormously productive economy demands a constantly expanding capacity to produce… demands that we make consumption our way of life, that we convert the buying and use of goods into rituals, that we seek our spiritual satisfactions, our ego satisfactions, in consumption. The measure of social status, of social acceptance, of prestige, is now to be found in our consumptive patterns… The greater the pressures upon the individual to conform to safe and accepted social standards, the more does he tend to express his aspirations and his individuality in terms of what he wears, drives, eats- his home, his car, *his pattern of food serving,...* We require not only "forced draft" consumption, but "expensive" consumption as well. We need things consumed, burned up, worn *out, replaced, and discarded at an ever increasing pace*. We need to have people eat, drink, dress, ride, live, with ever more complicated and, therefore, constantly more expensive consumption [Lebow (1955)].

After decades of mass culture in 2008 the Stockholm International Water Institute observes that "distribution of food is a problem – many are hungry, while at the same time many over eat. A hidden problem is that farmers have to supply food to take care of both our necessary consumption and our wasteful habits. Despite recent rises in world market prices, food is still a relative cheap commodity except for the very poor, and many consumers have little incentive to change their wasteful behavior" [Lundqvist, *et al.,* (2008)].

The consequences of neoliberal consumption patterns are impressed on health, on the environment and with the pollution in the environment there is again another direct impact on health. In the 20th century there was an increase in diseases attributed to affluent society illnesses. The increase in protein consumption, red meat, saturated fats and sugars, alcohol and processed food - symbols of prosperity- is associated with cholesterol, hypertension, *type 2 diabetes*, *coronary heart disease*, *cerebrovascular disease*, *peripheral vascular disease*, *obesity*, *cancer*, *alcoholism*, *allergies*

[Ezzati, *et al.,* (2005)]. Other diseases occur from the production and wastage of consumer goods and they come from industry byproducts, emissions, waste or are related to the chemical substances of the goods. Chemical pollutants due to industrialization have the ability to modulate the action of hormones and cause endocrine disruption responsible for infertility and other diseases [Nicolopoulou-Stamati, *et. al.,* (2001)].

Habits of the developed world of eating larger portions, eating more often and eating faster a single meal leads to the need for industrialized and processed food. Particularly the hazards of genetically engineered food and milk production in coordination with petro-chemicals and radio-nuclides may create possible carcinogenic properties [Nicolopoulou-Stamati, *et.al.,* (2004)].

They may also be considered to include *depression* and other *mental health* conditions associated with increased social isolation and lower levels of psychological well being observed in many developed countries [Luthar, (2003)]. Many of these conditions are interrelated, for example obesity is thought to be a partial cause of many other illnesses.

The reasons that these diseases arise in developing countries are the same with those that appeared in the Industrial north in the 19th century [Rodotheatos, (2009)]. In 2008, nearly 80% of deaths due to noncommunicable diseases (NCDs) — including heart disease, strokes, chronic lung diseases, cancers and diabetes — occurred in low- and middle-income countries [WHO report, NCD, (2011)]. The trend forecast is continuous immigration from developing to developed countries to improve the consumer standard of living. But when immigrants move from one country to another country which has different rates of cancer, the cancer rate of the migrants tends to be adjusted to that of the host country [Nicolopoulou-Stamati, *et.al.,* (2004)].

Extended overconsumption and food waste is not an issue only of the developed world. It has become a way of life affecting most of the developing countries. At present, 85 countries exceed their domestic bio-capacities, compensating for their lack of local material by depleting stocks elsewhere, in countries that have "surpluses" because they're not consuming as much [Coghlan, (2009)]. Steps are needed though behavioral change in order to prevent the adverse impacts on health, the environment, the economy, the social structures.

5. Belief Systems and Conditions to Reduce Consumption and Save the Environment. An Eastern Orthodox Theological Approach

In the past there have been initiatives for the reduction of consumption with either a secular or a religious foundation. 'Deep Ecology'—whose most distinguished representative is the Norwegian philosopher Arne Naess—argues for a spirituality rooted in 'biocentric equality', affirming the unity of all beings and their equal right to live [Christiansen, (1990)]. This position is based on the teachings of Mahatma Gandhi, process philosophy, Buddhism, and ancient religions of the Earth [Bordeianu, (2009)]. Gandhi having experience of the colonial exploitation of India from multinational companies said "live simple, so others can simply live"(Elgin, 1993). Although simplicity is a step to harness humanity's consumerist impact on the environment the question remains if it is enough alone to stop regression and if not proliferate progression, at least retain sustainability.

Sociologist Amitai Etzioni has identified three types of people that pursue simplicity. "Downshifters" are those who, having achieved a given level of wealth, make a conscious choice to reduce their income; they then moderate their lifestyle so they can spend more time with family and community interests. "Strong simplifiers" are those who give up highly paid, high-status jobs and accept radically simpler lifestyles. The most radical example are the "dedicated, holistic simplifiers," who embrace radical change and adjust their entire lives around an ethical vision of simplicity, sometimes motivated by spiritual or religious ideals (Etzioni, 1998). Beyond some exceptions simple living communities remain marginal. The religious basis for them does not appeal to everyone, and the secular versions seem less resistant to the incursions of consumerism.

People search for an identity that will give prominence to them and use material and consumer goods as status symbols. They use consumption practices to show their allegiance to certain social groups and to distinguish themselves from others (Gabriel, 2006). Consumption, well being, the good life, and prosperity has been tied with the identity of many and "we have been measuring our happiness with the volume of waste we produce" [Ganotis, (2010)].

In the anthropology of the Eastern Orthodox Church the identity of humans and their following relationship with the environment can be determined by the question “who am I, where do I come from, and what is my purpose”?

Saint Gregory Palamas in 1350 says: “Not the soul alone, nor the body alone is called Man, but both of them together, which, having also been made as an image of God, must be thus called” (Gregory Palamas, PG 150, 1361). According to biblical theology humans are soul and body created to the image of their creator that contain additional qualities from the rest of the creation not as division but as differentiation.

Humans differ from all other creatures because they have a soul in the image of God that interacts with the body giving unique expressions regarding freedom of will, the ability for free self determination (*αυτεξούσιον* in Gregory Nyssa, 12, PG 46, 369C), cognitive reason, rational nature, consciousness, the capacity for thorough deliberation and decision making. These elements can be utilized by humans with the choice of direction moving their mind and their will towards or away from God. St. Maximus the Confessor (580-662 A.D.) considers *motion* as an element of the essence of the being (Maximus, PG 91, 1177A). Beings evolve progressively or regressively, respectfully, by the choice of this motion. When there is a movement towards God then humans progress and assume the invitation and the role they were made for, resembling the image of God. Adam, according to both St. Gregory of Nazianzen and St. Maximus, was created as a “mixed worshipper, an overseer of the tangible world and a partaker of the intangible”(Gregory Nazianzen, Logos 39, PG 36, 324A) with “a soul in a middle position between God and the material creation being able to partake in reciprocal directions having unifying powers to unite them using the spiritual mind (νούς) towards God and using the senses towards creation” thus being capable of uniting the spiritual and material world having parts belonging in both realms (Maximus the Confessor, PG 91, 1193). For this very reason the human being was introduced last among beings as a kind of natural bond mediating between the divisions of the beings through their proper parts, and leading into unity in those things that are naturally set apart from one another by a great interval. St. Maximus the confessor confirms that, created beings are ‘not free of relationship’ they are inter-related [Stâniloae, (2005)], so that the movement of Adam determines the direction in which the rest of creation *moves* [Bordeianu, (2009)].

Once choices are made to move progressively towards God (with prayer and physically with actions or works), man assumes responsibility to look at creation spiritually, in order to discover the divine *logoi* in it, that is the purposes implanted in creation in order for it to proliferate progressively towards a union with the Creator. This work was assigned by God to man as being the priest of creation and as mediator, as being the sovereign of creation not the possessor, as being the creator- prophet having a deep insight of the divine *logoi* within the surrounding live or lifeless matter, to administer the path to fulfillment of the existing biodiversity and species and not their destruction and extinction. These works are not only assisted by consciousness, cognitive reason, freedom that are not self sustainable (as in paganism and pantheism) but as man being a mixed worshipper develop and thrive with the presence of the uncreated Grace in order to reach the "likeness of God".

The importance and the sanctity of matter in the orthodox church can be seen through the fact that man was created so that God could be incarnated and connect with creation (Maximus Confessor, 60, PG 90, 621A). In the East Roman iconography of the Cathedral of Monreale in Sicily (12th century) the Logos appears incarnated creating man and the environment according to his image (Monreale, frescoes). St John Chrysostom explains: "in the image of Christ, this is what (Genesis) means that he (man) was made in His image" [John Chrysostom, PG 62, 353)]. The true historical order is that first man was created and then the Logos incarnated. Depicting Christ (incarnated Logos) during Creation is not an anachronism because in iconographic dogmatic expression time is suspended. The past the future and the present, become one. The icon of the incarnated Logos explains that the purpose of the creation at eschatological times is to connect with the incarnated God. According to St. Maximus this incarnation has already started with creation (PG 91, 1285-1288). Man is created from the beginning (εξ αρχής) to provide the foundation for the hypostatic union of the human and divine natures (Nicholas Cabasilas, PG, 150, 560A). The cosmos constitutes an image of God, precisely because it is the manifestation and appearance of His uncreated energies.

The material environment such as climate, plants, animals resources and their balances are to be treated with respect because there is a sanctity in their purpose (*logoi* of beings) and also their utilization towards human necessities should be done with respect to man's biological existence and destination. Sustainability cannot be achieved by technocracy but by spiritual assessment because it involves all creation. Assistance to this pathway after the Fall comes with the incarnation of God where he acquires the flesh of the Church (John Chrysostom, PG 52, 429), as St.

Gregory the Theologian (Nazianzen) says "the one (body) not taken upon is also unremedied. The one connected to God, this is the one saved (healed to its purpose, to its *logoi*)" (Gregory Nazianzen, PG 37, 181C). Christ had assumed all of our irreproachable passions, such as hunger. Thirst, sleep are also irreproachable non sinful passions. Reproachable passions are precisely the corruption of those irreproachable passions. Gluttony, as compared to hunger. Mass consumption and overeating as compare to simple sustainability. The gift of pleasure when overeating, becomes a reproachable passion. When Christ provided food security for the multitudes with a few loaves and fish on a hill in Palestine, he instructed his disciples to "gather up all of the remaining fragments, so that nothing may be lost." (John 6, 12). This signifies the respect towards the sanctity of the food that is blessed and is a blessing. This instruction should also serve as a model in a time of wasteful consumption, where even the food surplus of affluent societies can nourish entire populations [Chryssavgis, (2008)]. When food is in excess it should be redirected to those who have not, and will go hungry. As the above Biblical narrative stresses the importance between providing food to the poor and spiritual well being, other excerpts signify the relation of withholding food and identity annihilation. Prophet Ezekiel on the destruction of Sodom says: "Behold, this was the guilt of your sister Sodom: she and her daughters had pride, *excess of food*, and prosperous sluggishness, but did not aid the poor and needy. They were haughty and committed acts of abomination before me. So when I saw this I annihilated them" (Ezekiel, 16, 49-50). Centuries after these events Patristic theology confirms that from the desire of gluttony emerges the desire for porn [St. Maximus, 56, PG 90, 1035]. In the parable of the rich man and Lazarus, "there was a poor man, named Lazarus, was laid at his gate, full of sores, and desiring to be fed from the crumbs that fell from the rich man's table"(Luke 16, 19-31). One of the things to be understood of this parable is that Lazarus has an identity in this life and the eternal life and the affluent has not. The name Lazarus is derived from the *Hebrew* אלעזר, (*Eleazar*) meaning "God has helped"(Barclay, 1999). The previous passages signify one of the elements of the Original Sin. Adam wanted to be god without God; he thought that all material realities were meant to satisfy his passions and that they should *move* towards him (St. Maximus), setting the foundations, the epigenetics, for medieval and modern economic feudalism, instead of him moving everything thankfully towards God.

Humans carry the responsibility to preserve the people, the world and the environment they live in, with which they exchange matter, as administrators

(Gen. 2, 15) and not as possessors or conquerors for personal consumption. The latter is man's lost identity. "The image we have of ourselves is reflected in the way we treat the creation. If we believe that we are no more than consumers, then we shall seek fulfillment in consuming the whole earth; but if we believe we are made in the image of God, we shall act with care and compassion, striving to become what we are created to be… in the image of the loving Creator, (that calls us) to love and care for all creation"[Patriarch Bartholomew, (2009)].

In this sense man is a co-creater through preservation, an administrator, a steward, a caretaker, a sustainer, a priest of the environment. Maximus considers that Adam was unfaithful to his call to be the priest of creation, and, as a result of his fall, the world became split by tensions and differences became divisions. When these identities are lost, because "when he was in honour (of this identity) did not understand" (Romans 8,19), he is compared to the senseless beasts, and is become like them (Psalms 49,12). Correctly some environmentalists, without knowing humanity's true identity, reject anthropocentrism and say that humans act like consuming pests (beasts) to the planet, which holds true for the fallen man. Nature calls humanity to assume responsibility as "creation is subjected to frustration (e.g. climate change), not by its own choice, but by the will of the one who subjected it, in hope" (Romans 8,19) and now "waits in eager expectations for the sons of God to be revealed" (Romans 8,19), that is the people or the persons who will resume responsibility and restore the environment in a Godlike manner. Noah had this capability and administered the preservation of creation which followed to the ark. It was a work of anthropocentrism, theocentrism, ecocentrism all extremes- that philosophers tend to consider autonomous- united by the "mixed worshippcr" Noah.

Finally two important things that the orthodox church possesses for the preservation and the path to the eschatological new creation (2 Corinthians, 5, 17) are the Holy Eucharist and ascetism [Zizioulas, (2002)]. The Holy Eucharist, because the initial purpose of creation to connect with God can now be fulfilled through the body of Christ [Nellas (1987)]. Ascetism, because it is a barrier to stop irreproachable into becoming reproachable needs and a sign of our will that we are receptive to our task of uniting creation with God through His uncreated energies. With these the value of the material world is experienced as a world of God.

The products of the grain and the vineyards are not turned into waste, like fossil fuel is into carbon dioxide. They are matter to become sacred as said in the divine liturgy "We offer to You these gifts from Your own gifts". When they turn into the body and blood of Christ, man becomes receptive to the One that will transform him and creation to immortality.

Ascetism begins with "catharsis" that is the cleansing of the mind and the awareness of our identity. Being an image of God, we live in an ascetic way to achieve likeness or "the mind of Christ" (1 Corinthians 2, 16). It is a communication tool with God, humanity and the environment where the least is valued and does not become waste. In the words of Ecumenical Patriarch Bartholomew of Constantinople:

> The ascetic way is primarily a way of liberation. And the ascetic is the person who is free, uncontrolled by attitudes that abuse the world; uncompelled by ways that use the world; characterized by self-control, by self-restraint, and by the ability to say "no" or "enough." Asceticism, then, aims at refinement, not detachment or destruction. Its goal is moderation, not repression. Its content is positive, not negative: it looks to service, not selfishness; to reconciliation, not renunciation or escape. Without asceticism, none of us is authentically human.

6. Secular and Theological Convergences in Reducing Food Waste

In early October of 2012, the Greek Ministry of Development announced a new law that allows the sale of expired but edible food in order to help citizens in financial need. According to article 5 of the new law "the provision of food is allowed of past dates and minimum preservation, with the exception of that determined by law as nonsustainable, with the term that this food will be sold distinctly separated from the rest of the food and with a label that will be suspended above it that has printed with capital letters the phrase "Food beyond preservation"."

Food that is valued not only by economic terms can be found in ascetic monastery life in premedieval periods. Abba Megethius (c. 300 A.D.) was asked: "If there is left over food for the next day, would you suggest to the brothers to eat it? If it has decayed, the elder answered, let it be thrown away because we cannot obligate the brothers to eat it and become ill. If though, while it is edible it is thrown away and a new meal is cooked for the pleasure of the stomach, this is bad".

In the United Kingdom «Love Food, Hate Waste» is a campaign, launched by the *Waste & Resources Action Programme* in 2007, with the aim of reducing the amount of *food waste in the United Kingdom* [Eccleston, (2007)]. The campaign claims to have already helped almost two million households reduce their *food waste*, amounting to savings of almost £300 million and stopping 137,000 tonnes of food that was going to be wasted [Edie Newsroom, (2009)}.
EC Commissioner Ms Mariann Fischer Boel stresses the need to change the way we evaluate things: "the main responsibility lies with consumers. We have to look at our own habits. Why do we cook more than we can eat? Why do we buy three chickens for the price of two, when we actually only need one?... Why have we forgotten the good housekeeping wisdom of previous generations who knew how to cook a meal from leftovers?" [Ryan, (2009)].

Keeping in mind this last comment, an example for food security management not only for good housekeeping but also for governance regulation comes from the life of Saint Savva (C. 500 A.D). St. Savva had assigned a man named James from Jerusalem to serve in the kitchen so he could host the visitors of the monastery. For the reason that James "had no experience of this profession, he frequently faulted, as one day he cooked broad beans more than needed and there were many leftovers, that he did not save to be eaten the next day, but as inexperienced he threw away as if they were peelings. As Savva noticed this, he went secretly and collected them, dried them in the sun, marinated them and cooked them diligently, and invited James to join him for dinner in the refectory. While eating Savvas said to James. "Forgive me if the meal is not good because I did not know how to make it better". James replied "May the Lord be witness that it has been long since I've had a more delicious meal". The Saint continued "Understand, son, that the broad beans, that you considered scrap and threw into the stream, is our meal. If you are not able to prepare a pot of food, how will you become a brotherhood leader to govern human souls"?

7. Conclusions

* every year in Europe a growing amount of healthy, edible food – some estimates say up to 50% is lost along the entire food supply chain, in some cases all the way up to the consumer, and becomes waste; EE (*2011/2175(INI)*);

* reducing food waste is a significant preliminary step in combating hunger in the world, responding to the increase in demand predicted by the FAO and improving people's nutritional levels, EE *(2011/2175(INI))*;
* 79 million people are still living below the poverty line in the European Union – in other words, more than 15% of EU citizens earn less than 60% of average earnings in their country of residence; whereas, of these, 16 million have received food aid from charitable institutions;
* the threat to food security is accompanied by mirror-image rich-world scourges such as obesity, cardiovascular illnesses and cancers arising from a diet overrich in fats and proteins, the result being that the world's overfed population numbers as many as the underfed and malnourished;
* in Europe and North America, in the previous decades, when food production was abundant, food waste was not a policy priority, which led to an overall increase in food waste along the food supply chain; whereas in Europe and North America food waste occurs predominantly at the retail and consumption stage, as opposed to in the developing world, where production, harvest, processing and transport are the stages where losses are most common;
* poverty and famine are caused by bad government and bad economic policies, not by overpopulation.
* food waste has not just ethical, economic, social and nutritional but also health and environmental implications, since unconsumed food mountains make a major contribution to global warming and food waste produces methane, which as a greenhouse gas is 21 times more powerful than carbon dioxide; for every kilogram of food produced, 4.5 kg of CO_2 are released into the atmosphere; whereas in Europe the approximately 89 Mt of food wasted produce 170 Mt CO2 eq./yr.
* Societies that adopted a consumption based culture have lack of happiness in social issues such as human relations. Consumption should have a supporting not a leading role connected with personal identity.
* In spiritual terms humans vision the environment as a material reality, without any trace of the divine wisdom (*logoi*) in it. Moreover, they try to obtain consumer satisfaction in something that cannot offer them stability. Human persons exhaust natural resources only to realize that they did not find the satisfaction they were looking for in them, but that they have instead caused more damage to the environment.

* Humanity's role is to connect the divided and not separate the connected. To respect creation because through creation holiness is achieved. Man is creator not to transform all creation into waste or even recycled material but to transform without interfering with existing balances. He is co-creator not by transforming the genetic code of beings to economically own living organisms and turning the rest of biodiversity into waste, or leaving the poor unfed. Christ assumed the flesh of the Church which means he assumed creation as we know it and that is how we will be invited to experience it at the end of times, with its purposes (logoi) fulfilled.

Households can follow these guidelines to reduce food waste, poverty and resource depletion:

* Preparation of less food that can be consumed up to the next day.
* Storage of food one hour after its preparation in air tight food containers and appropriate refrigeration temperature to reduce the microbial load.
* Organizing the refrigerator with food of immediate consumption on the upper shelf, following later expiration dates in the middle shelf and more durable food such as fruits and vegetables at the bottom shelf.
* Reducing food that is thrown away by preparing the surplus portions for people in need (neighborhood, church, or charity organizations) and less edible portions for own or neighbors household pets.
* Direct donation of food to food or other businesses and charity organizations that collect for distribution to the needy
* Selection of food of local or peripheral production to reduce greenhouse emissions due to transportation and support the local economy.
* Selection of the least processed food with a greater net weight and a smaller package.
* Composting non edible food scraps in local gardens.

8. Epilogue

Alan During, who wrote *How much is enough?* in 1992 pointed out that the pursuit of consumption not only will undermine the quality of life of those in the consumer society, it will diminish the ability of those outside the consumption class to meet their basic needs. Today developed countries with profitable stock exchange indexes are those that convert the planet's resources to the largest per capita waste, either for their own consumption or for international trade. Changes must include revaluation of human identities, of economy as a cultural model and philosophy, revaluation of the meaning of financial growth, of the distribution of wealth, of the generation of waste in affluent societies and food insecurity in the rest of the world. If the world was seen as a sacred gift from God which is meant to foster and promote union, communion with God and with others, then affluent societies would intervene in areas to reduce food insecurity issues. Such a 'liturgical' use of nature by human beings leads to forms of culture which are deeply respectful of the material world giving nature the possibility to sustain its inhabitants.

All modern economies can be leaner without being doomed. The request is to sacrifice less than Lazarus' crumbs… if there is a will. To understand this helpful is Unicef's account that in order to reduce world malnutrition it would only take to redirect 2% of global grain production to those in need.

References

Antuofermo Mélina, Di Meglio Emilio (2012): 23% Of EU Citizens Were At Risk Of Poverty Or Social Exclusion In 2010, Eurostat EU. http://epp.eurostat.ec.europa.eu/cache/ITY_OFFPUB/KS-SF-12-009/EN/KS-SF-12-009-EN.PDF and http://www.eurochild.org/fileadmin/ThematicPriorities/ChildPoverty/EU/Eurostat%20publication%20poverty%20in%202010.pdf.

Barclay W (1999), "The Parables of Jesus", Westminster John Knox Press.

Bartholomew, Ecumenical Patriarch (2005), Introductory Speech on "Environment and Spirituality" at the 2005 World Exposition, Aichi, Japan - International Conference Mediating for Sustainable Development. http://www.patriarchate.org/documents/environment-and-spirituality.

Bartholomew, Ecumenical Patriarch (2009), "Ecumenical Patriarch's Message for the Day of the Protection of the Environment", Prot., 862, September 1, 2009, http://www.patriarchate.org/documents/2009-protection-of-the-environment.

Bartholomew, Ecumenical Patriarch, on "Moving Beyond Consumerism to a Divine Economy", http://berkleycenter.georgetown.edu/resources/quotes/patriarch-bartholomew-on-moving-beyond-consumerism-to-a-divine-economy.

Blas Javier (2009), "G8 warns of hunger threat to global stability", Financial Times, April 7 2009, http://www.ft.com/cms/s/0/22cadecc-230a-11de-9c99-00144feabdc0.html#ixzz2HTfCUshl.

Bordeianu, Radu (2009). "Maximus and ecology: the relevance of Maximus the Confessor's theology of creation for the present ecological crisis." Downside Review 127 (2009): 103-126.

Brussels Policy Briefing no. 28 (2012), Addressing Food Waste in Times of Crisis, European Commission, Borschette Building, Room 2D, June 26, http://brusselsbriefings.net.

Cafferty Jack (2012), "Where is the U.S. headed if more than 100 million people get welfare?", CNN, August 9th, 2012, http://caffertyfile.blogs.cnn.com/2012/08/09/where-is-the-u-s-headed-if-more-than-100-million-people-get-welfare-2.

Caritas (2012), 'Exclusion and Social Development 2012', http://www.caritas-europa.org/code/EN/soci.asp?Page=1309, http://www.caritas-europa.org/module/FileLib/ResumenDatosFOESSA2012.pdf, http://www.caritas-europa.org/code/EN/soci.asp?Page=1437.

Cecil Dwayne (2012), «Population Control and Fresh Water Resources Should Be Top Global Priorities Says Former NASA Climatologist», Special Report by DeWayne Cecil, PhD, on Sharon Kleyne Hour Radio Talk Show, Grants Pass, OR (PRWEB) October 24. http://www.prweb.com/releases/2012/10/prweb10050733.html.

Chomsky N., (1999), Profit over people: Neoliberalism and Global order, Seven Stories Press, New York, NY.

Christiansen Drew (1990), 'Ecology, Justice, and Development', Theological Studies 51, no. 1:77-81.

Chryssavgis John (2008), Cosmic Grace – Humble Prayer: The Ecological Vision of the Green Patriarch Bartholomew I, Eerdmans Publishing Company, Grand Rapids MI.

Coghlan, Andy, Consumerism is 'eating the future', New Scientist, Aug (2009), http://www.newscientist.com/article/dn17569-consumerism-is-eating-the-future.html.

Dery M., (2009), «Essay - Dawn of the Dead Mall». The Design Observer Group. 11 November 2009, http://changeobserver.designobserver.com/entry.html?entry=11747.

Diamond, J., (2008): "What's Your Consumption Factor?" The New York Times. January 2.

Dijksterhuis, Ap; Smith, Pamela K.; van Baaren, Rick B.; Wigboldus, Daniel H.J. (2005). "The unconscious consumer: Effects of environment on consumer behavior", Journal of Consumer Psychology 15 (3): 193–202.

Eccleston, Paul (2007). “Help the Environment, reduce food waste”. London: The Daily Telegraph, http://www.telegraph.co.uk/earth/earthnews/3312730/Help-the-environment-reduce-food-waste.html.

Elgin Duane (1993), Voluntary Simplicity, ed. William Morrow, New York.

Etzioni Amitai (1998), «Voluntary Simplicity: Characterisation, Select Psychological Implications and Societal Consequences», Journal of Economic Psychology, October: 619–43.

Ezzati M, Vander Hoorn S, Lawes CM, et al. (2005). “Rethinking the “diseases of affluence” paradigm: global patterns of nutritional risks in relation to economic development”. PLoS Med. 2 (5): E133

FAO. (2011). Global Food Losses and Waste: Extent, Causes and Prevention.

Foley S (2010)., « Obama: «We want our money back – every single dime», The Independent, Friday 15 January.

Gabriel Yianis and Lang Tim (2006), The Unmanageable Consumer, Sage Publications Ltd., London.

Ganotis Konsantinos (2010), “Mournings for the environment”, Peiraiki ekklesia, June.

Gardner Gary, Assadourian Erik, Sarin Radhika (2004), “The State of Consumption Today”, Vol.: The Consumer Society, State of the World, The Worldwatch Institute.

Garton Christie (2011), “New breed of Americans going hungry”, USA TODAY, May 19.

Gimenez, Eric Holt (2012), «We Already Grow Enough Food For 10 Billion People and Still Can't End Hunger», February 5th.

Gregory Nazianzen, Logos 39, To the birthday of the Savior, PG 36, 324A. Migne J. P., Patrologiae cursus completus. Series Graeco - Latina, 161 volumes., Paris 1857-1866 (reprint Brepols, Turnhout 1981).

Gregory Nazianzen, Letter 101, To Cledonion, P.G. 37, 181 C, Migne J. P., Patrologiae cursus completus. Series Graeco - Latina, 161 volumes, Paris 1857-1866 (reprint Brepols, Turnhout 1981).

Gregory Nyssa, On virginity, 12, PG 46, 369C. Migne J. P., Patrologiae cursus completus. Series Graeco - Latina, 161 volumes, Paris 1857-1866 (reprint Brepols, Turnhout 1981).

Gregory Palamas, Προσωποποιοίαι (Person Making), PG 150, 1361 BC Migne J. P., Patrologiae cursus completus. Series Graeco - Latina, 161 volumes, Paris 1857-1866 (reprint Brepols, Turnhout 1981).

Grolleaud M. (2002). Post-harvest losses: discovering the full story. Overview of the phenomenon of losses during the post-harvest system. Rome, Italy: FAO, Agro Industries and Post-Harvest Management Service.

Gustavsson Jenny, Cederberg Christel, Sonersson U. (2011) Robert van Otterdijk, Alexandre Meybeck, «Global food losses and food waste», Food And Agriculture Organization (FAO) of the United Nations, Rome.

Hall, S., Winlow, S. and Ancrum, C.(2008), Criminal Identities and Consumer Culture. Routledge, London.

Harden Gill H. (2012), «Analysis of FNS' Supplemental Nutrition Assistance Program Fraud Prevention and Detection Efforts», Audit Report 27002-0011-13, September 2012, United States Department of Agriculture, Office of Inspector General, http://www.usda.gov/oig/webdocs/27002-0011-13.pdf.

Harrison J. (2004), Study: Nation Wastes Nearly Half Its Food, The University of Arizona News, November 18, 2004, http://uanews.org/story/study-nation-wastes-nearly-half-its-food.

Hounshell David A. (1985), From the American System to Mass Production 1800-1932, The John Hopkins University Press, Baltimore Maryland.

Jean Hugé and Luc Hens (2007), "Sustainability assessment of Poverty Reduction Strategy Papers", Impact Assessment and Project Appraisal, 25(4).

IPCC, Climate Change 2001: Third Assessment Report (New York: Cambridge University Press.

Jackson Tim (2008), «The challenge of sustainable lifestyles», State of the World, The Worldwatch institute.

John Chrysostom, Letter to thc Colossians, 8,2, PG 62, 353. Migne J. P., Patrologiae cursus completus. Series Graeco - Latina, 161 volumes, Paris 1857-1866 (reprint Brepols, Turnhout 1981).

Jones Timothy W.(2006), «Using Contemporary Archaeology and Applied Anthropology to Understand Food Loss in the American Food System», Bureau of Applied Research in Anthropology. http://www.ce.cmu.edu/~gdrg/readings/2006/12/19/Jones_UsingContemporaryArchaeologyAndAppliedAnthropologyToUnderstandFoodLossInAmericanFoodSystem.pdf.

Kelton, Erika (2012), "Barclays' Traders Show How Much Fun Wall Street Has Manipulating Markets", Forbes Magazine, November 20, 2012. http://www.forbes.com/sites/erikakelton/2012/11/20/barclays-traders-show-how-much-fun-wall-street-has-with-manipulating-markets.

Knudsen, Lara (2006). Reproductive Rights in a Global Context. Vanderbilt University Press.

Lebow, V., (1955) «Price competition in 1955», Journal of Retailing, Spring 1955, http://hundredgoals.files.wordpress.com/2009/05/journal-of-retailing.pdf.

Levine, Madeline (2007). «Challenging the Culture of Affluence», Independent School, 67.1: 28-36.

Lippert John, (2008), "Friedman Would Be Roiled as Chicago Disciples Rue Repudiation", Bloomberg, December 23, 2008, http://www.bloomberg.com/apps/news?sid=a3GVhIHGyWRM&pid=newsarchive.

Lundqvist J., de Fraiture C., Molden D. (2008). Saving water: from field to fork—curbing losses and wastage in the food chain. In SIWI Policy Brief. Stockholm, Sweden: SIWI 2008.

Luthar SS (2003). "The culture of affluence: psychological costs of material wealth". Child Dev 74 (6): 1581–93.

Majfud, Jorge (2009). "The Pandemic of Consumerism". UN Chronicle. http://www.un.org/wcm/content/site/chronicle/home/archive/issues2009/toprotectsucceedinggenerations/pid/21570. Retrieved 7 January 2012.

Maximus the Confessor, To Thallasios, 60, PG 90, 621A. Migne J. P., Patrologiae cursus completus. Series Graeco - Latina, 161 volumes., Paris 1857-1866 (reprint Brepols, Turnhout 1981).

Maximus the Confesssor, PG 91, 1193, Migne J. P., Patrologiae cursus completus. Series Graeco - Latina, 161 volumes, Paris 1857-1866 (reprint Brepols, Turnhout 1981).

Maximus the Confessor, Third hundredth (century) on love, 56, PG 90, 1035, Migne J. P., Patrologiae cursus completus. Series Graeco - Latina, 161 volumes., Paris 1857-1866 (reprint Brepols, Turnhout 1981).

Maximus the Confessor, Ambigua: Tilcuiri ale unor locuri cu multe si adinci intelesuri din Sfintii Dionisie Areopagitul si Grigorie Teologul, trans. Dumitru Staniloae, Parinti si scriitori bisericesti 80 (Bucuresti: EIBMBOR, 1983).

McEachern W.A.(1996), «Principles of Economics Textbooks: Coping with Scarce Resources and Unlimited Wants», In: Nahid Aslanbeigui and Michele Naples (edit.), Rethinking Economic Principles: Critical Essays on Introductory Texts, Burr Ridge, Irwin IL.

Monbiot G. (2008), "Population growth is a threat. But it pales against the greed of the rich", The Guardian, Tuesday January 29th 2008.

Monier Veronique, Mudgal S., Escalon Victoire, O' Connor Clementine, Gibon T., Anderson Gina, Montoux Hortense (2010), "Preparatory study on Food waste across EU27", October, http://www.europarl.europa.eu/news/en/pressroom/content/20120118IPR35648/html/Parliament-calls-for-urgent-measures-to-halve-food-wastage-in-the-EU.

Morgan G., and Robertson K. (2011), BSR. Waste Not, Want Not: An Overview of Food Waste. http://www.bsr.org/reports/BSR_Waste_Not_Want_Not_An_Overview_Food_Waste.pdf, p. 1.

Nellas P. (1987), Deification in Christ. Orthodox perspectives on the nature of the human person. (Contemporary Greek Theologians, 5.) (Trans, by N. Russell of Zöon Theoumenon.) Pp. 254. Crestwood, New York: St Vladimir's Seminary Press.

Nellemann C., MacDevette M., Manders T., Eickhout B., Svihus B., Prins A.G., Kaltenborn B.P. (Eds). (2009). The Environmental Food Crisis – the Environment's Role in Averting Future Food Crises, http://www.unep.org/pdf/FoodCrisis_lores.pdf

Nicholas Kabasilas, On the life in Christ, PG 150, 580-684B. Migne J. P., Patrologiae cursus completus. Series Graeco - Latina, 161 volumes., Paris 1857-1866 (reprint Brepols, Turnhout 1981).

Nicolopoulou-Stamati P., Pitsos M.A. (2001), The impact of endocrine disrupters on the female reproductive system, Human Reproduction Update, Volume 7, No. 3, pp. 323-330, Oxford University Press.

Nicolopoulou-Stamati P., Hens L., Howard Vyvyan C., Van Larebeke N.(2004), Cancer as an Environmental Disease, Kluwer Academic Publishers.

Nord M., Andrews M., Carlson S. (2009), Household Food Security in the United States, 2008, Economic Research Report No. (ERR-83), USDA, November 2009, http://www.ers.usda.gov/publications/err-economic-research-report/err83.aspx

Neurath, Paul (1994), From Malthus to the Club of Rome and Back. M.E. Sharpe.

Pals F., Benoit Angeline, Doyle Dara (2012), «Dutch With Food Aid Shows New Economic Reality Engulfing Europe», Bloomberg, May 10, http://www.bloomberg.com/news/2012-05-09/dutch-with-food-aid-shows-new-economic-reality-engulfing-europe.html.

Parfitt J., Barthel M.,and Macnaughton S. (2010), Food waste within food supply chains: quantification and potential for change to 2050,

Philosophical Transactions of the Royal Society, London, Biological Sciences, 2010 September 27; 365(1554): 3065–3081.

Renner, Michael (January 2006). «Chapter 1: China, India, and the New World Order" State of the world 2005: A Worldwatch Institute Report on progress toward a sustainable society. New York: W.W. Norton. ISBN 0-393-32666-7. OCLC 57470324. http://www.worldwatch.org/node/3992.

Rethinking Diseases of affluence. The Economic impact on chronic diseases http://www.who.int/chp/chronic_disease_report/media/Factsheet4.pdf.

Rodotheatos Gerasimos (2009), World food insecurity. A viewpoint from the developed north, UNESCO briefings: The right to nutrition: World food insecurity and the environment, Kalokathi Athens.

Rostow W. W., (1960)., The Stages of Economic Growth: A Non-Communist Manifesto, Cambridge: Cambridge University Press.

Ryan R. (2009), "Eliminating waste could end world hunger and ease CO2 emissions", The Irish Examiner, Tuesday, February 03.

Simon, Julian. L. (1998), The Ultimate resource 2, Princeton University Press.

Singh AR, Singh SA (January 2008). "Diseases of poverty and lifestyle, well-being and human development". Mens Sana Monogr 6 (1): 187–225.

Smil, V. (2000), Feeding the World: A Challenge for the Twenty-First Century, MIT Press, Cambridge, MA, USA.

SNAP (Food Stamps), Facts, Myths and Realities, http://feedingamerica.org/how-we-fight-hunger/programs-and-services/public-assistance-programs/supplemental-nutrition-assistance-program/snap-myths-realities.aspx.

Stâniloae, Dumitru (2005), Orthodox Dogmatic Theology: The Experience of God, Vol II: The World: Creation and Deification (Brookline, Massachusetts: Holy Cross Orthodox Press.

Tertullian, An. 30.4 (CChr.SL 2:827): Corpus Christianorum: Series Latina: Tertulliani Opera: Pars I: Opera Catholica: Adversus Marcionem. Turnholti, Brepols editores pontificii, 1953.

UN (1994), Administrative committee on coordination, Subcommitte on nutrition, A report compiled from information available to the ACC/SCN, November 1994. http://www.unsystem.org/SCN/archives/rwns94update/begin.htm#Contents.

U.S. Department of Agriculture, Food and Nutrition Service (November 2012). Characteristics of Supplemental Nutrition Assistance Program Households: Fiscal Year 2011. Table D.3. http://www.fns.usda.gov/ora/menu/Published/snap/SNAPPartHH.htm.

Venkat Kumar (2011), The Climate Change and Economic Impacts of Food Waste in the United States, Int. Journal on Food System Dynamics 2 (4): 431-446.

World Health Organization. New WHO report: deaths from noncommunicable diseases on the rise, with developing world hit hardest. Geneva, 27 April 2011.

WRAP (2008), The Food we waste, https://www.ns.is/ns/upload/files/pdf-skrar/matarskyrsla1.pdf)

—. (2009) Household Food and Drink Waste in the UK, United Kingdom.

Zizioulas John, Metropolitan of Pergamon, "Man and the environment. An orthodox theological approach", Sciences, Technologies and Orthodoxy, October 4-8, 2002.

GLOBAL CRISIS AND CASE STUDIES

A MISSION IMPOSSIBLE, OR THE NEEDED CRITICAL PIECE OF THE PUZZLE?

HEINER BENKING

Council on Global Issues, Toronto/Berlin
E-mail: hebenking@gmail.com

Abstract: *Despite the complexity of the web of interlocking global problems, there are recent signs of significant progress in understanding them, not least through the Structured Dialogic Design methodology of Alexander Christakis.*

This project launched by Reynaldo Treviño Cisneros and Bethania Arango Hisijara could be characterized as "mission impossible"; and the authors are aware of it, while the endeavour deserves merit. What are the alternatives for times when we have overextended our foot- and handprints?

Some argue that we just need to reduce the world population drastically; the densities and pressures will be less, and so we could muddle through as usual. In this case, they believe we can re-start from square one with a healthy sustainable civilization by just reducing "numbers" but inflicting unprecedented suffering. Others believe in swarm intelligence and simulations, which would use and capitalize on all available data, information and knowledge. Such a "Global Earth Simulator" would then create out of all the "mess" and "noise" the blue-print for policy development and crisis management. The amount of suffering involved, and the loss of individual and cultural autonomy, are mind-staggering. We would lose our stake for a democratic harmonious development. We would be forced to stumble down the delusional path of "unrealistic optimism."[1]

[1] How unrealistic optimism is maintained in the face of reality. Shrot T, Korn CW, Dolan RJ. Source: Welcome Trust Centre for Neuroimaging, University College London, Nat Neurosci. 2011 www.ncbi.nlm.nih.gov/pubmed/21983684

I have known proponents of many extremes and wonder why, for example, population-reduction fans have not studied harmonious life-styles, and sustainable civilizations through the ages, where different attitudes towards the Earth secured much larger population-densities — even when that would not be desirable. I wonder on the other hand how "trend extrapolators" can assume that the noise and rubbish of complexity run amok can be modeled and moulded by a "super-intelligent" system. Will machines generate humane, value-based policies for the best of all humankind? Discussions about the value of models, scenarios, and symptoms have raged for 40 years ever since the first Club of Rome World Models. Has no one followed these discussions? Please see the synopsis in interviews and the event with Alexander Christakis, Yiannis Laouris, Reynaldo Treviño Cisneros, and others.[2]

Treviño and Arango were walking into a terra incognita, marked by a welter of outcomes proposed by different objectives, perspectives, expertise, and degrees of synthesis. People more famous than they had foundered in the global problematique. They began with an awareness of the 49 Continuous Critical Problems (CCPs) identified by Hasan Özbekhan in 1970 and added the 15 Challenges to Humanity identified by the Millennium Project. Then they were on the way to make sense of the global mind-boggling mess using the Structured Dialogic Design (SDD) methodology formulated by Alexander Christakis and others. Congratulations for this trail-blazing effort.

We all agree that the problem spaces have "exploded" and leave us with unpredictable repercussions. We have numerous lists of problems, but the question is: How are they connected? Can we identify root-causes with leverage-points[3], places where we can intervene in systems to make a difference for the better instead of staring at all the tipping points where earth-systems dynamics get out of control?[4]

---

[2] Interviews with Alexander Christakis in Europe's World: 40 years Problematique: http://goo.gl/QqYsqY and The Predicament of the Individual, Communities, and Humankind in the 21st Century http://goo.gl/PT5ZYO

[3] Leverage Points: Places to Intervene in a System, Donella Meadows http://www.sustainer.org/pubs/Leverage_Points.pdf

[4] I recommend the International Encyclopedia of Systems and Cybernetics and the presentation by the founding editor when introducing the 2nd edition: benking.de/systems/encyclopedia/concepts-and-models.htm

We also agree that all stakeholders cannot be brought to one room at one time to uncover the deep meanings that people are typically not aware of, or to anticipate future consequences. So what to do? Make more embryonic steps like the one commended here.

What is our future as futurists? Are we to continue doing business as usual expecting newer technology to create more positive results? What of new approaches that work with broad participation, rich definitions, abductive reasoning, and structured dialogue? Can they find their rightful place in the local, national, and global dialogue? What can we do to advance this new approach?

There are numerous signs that we as the people have a heightened interest in global/local issues across scales and times, issues like climate change, hunger, population, water availability, ethics, and alienation, and there is a heightened, intense desire to find root causes for these complex and interrelated issues. Some examples:

1. The charter of the Millennium Development Goals asks us to address root causes and come to "win, win, win" situations so that solving one problem will help us solve others. The ability of ISM to map influence relations among issues and find root causes fits that requirement.

2. The Millennium Project has amassed an immense store of information and collected 15 Challenges to Humanity for open access and discussions within the community, each managed by an editor.

3. Organizations such as the ONLIFE Think Tank[5] of the Digital Futures Task Force of the European Commission have placed great focus on the need to re-imagine the concept of democracy, along with other concepts such as distributed responsibility, privacy etc.

We have numerous opportunities to provide and publicize SDD interventions, in addition to this one, that will demonstrate SDD's greater utility at a cost of minus 2 orders of magnitude.

Most of the past successes of SDD have been in the corporate sector with some notable exceptions, such as: the Third Iberoamerican Encounter

---

[5] ONLIFEec.europa.eu/digital-agenda/en/onlife-original-outcome and www.futureworlds.eu/wiki/Reinventing_democracy and CARDIAC cardiac-eu.org

about Communication Strategies (Mexico)[6]; and the work done by Christakis with the US Federal Drug Administration, and the World Health Organization[7]. Future monographs and articles will proclaim SDD successes at the conceptual, educational, and policy levels.

[6] III Encuentro Iberoamericano sobre Estrategias de Comunicación, México, 2005: www.fisecforo.org/index.php

[7] Christakis AN and Bausch KC 2006. *How People Harness Their Collective Wisdom and Power to Construct the Future in Co-Laboratories of Democracy.* Greenwich: Information Age Publishing

LADDER TO THE MIDDLE CLASS: A NEW SELF-FUNDING MODEL FOR SOCIAL HOUSING IN PARAGUAY

MATT DOOLAN, SAREE KOYSIRIPONG, AHMAD ALDAHASH, PAILIN LOMSINSAB AND CARLOS ESCRUCERIA

Prepared for Non-Governmental Social Housing Provider
'Un Techo Para Mi País'
E-mail: matt@developthis.com.au

Abstract:

Latin American Social Housing provider 'Un Techo Para Mi País' is currently providing social housing in inappropriate 'slum' locations. This is a short term curative solution to poverty which entrenches poverty and further concentrates the disadvantaged in marginalized locations. Slum areas typically lack security of tenure, appropriate access to employment, education, health services, and transport. They may also be unsafe, contaminated, flood-prone, or geologically unstable.

The root of the problem lies in the unequal distribution of wealth and lack of social mobility. The absence of a significant Middle Class population in Paraguay prevents appropriate leadership emerging (or lacks support if it does emerge) which could permanently solve this problem. A proposed solution is here presented.

Proposed Solution: The proposal is to shift the focus of Un Techo's efforts to the provision of housing on safe, quality sites via a self-funded model. The focus will be on housing for homeless or inadequately housed families who have fallen, or are in danger of falling, into the impoverished Lower Classes of society. It is argued that, with some relatively small assistance, these select families can be placed on the path to a self-sustained and stable Middle Class existence. This is a new preventative (not curative) approach.

To achieve this, developable land parcels will be purchased in the path of the growth of Paraguay's capital, Asunción. Land value increase or 'betterment' should then occur as population grows, services are upgraded, and the city spreads outward. Initially the residents will be housed on these sites in some 100 temporary buildings. After 5 to 10 years the land will be sold for development of some 1000 dwellings. The land sale profit will fund standard housing for the original 'Un Techo' residents. Surplus funds will go to the 'Un Techo' organization. Put basically, 'Un Techo' will become a land speculator, or land banker, profiting from land value increase (betterment).

By building the Paraguayan Middle Class a societal change can commence which will reduce corruption, redistribute land and resources more equitably, stabilize the economy, allow children to be appropriately educated, and improve general health. A significant Middle Class will assist Paraguayans to help themselves in obtaining a more successful democratic and economic system.

Costs

Costs are expected to be $1 million for the first land purchase. The land can be purchased in cash (raised from donations) with little competition from financed buyers due to the current poor local credit and world credit conditions persisting.

Holding costs are expected to be low. Taxes and rates may be discounted by the government due to the anticipated social benefits arising from the project.

Benefits

The main benefit to 'Un Techo' will be a more effective program that will make real and long-term-sustainable improvements to average living standards in Paraguay by building the Middle Classes.

Call to action

By implementing this program 'Un Techo' will be proving that this new self-funding system can work effectively. Once proven it can be repeated in many cities (and in rural areas if appropriately modified) around the world to improve peoples' lives.

> ***"About 33% of the population lives below the poverty line. Inequality is among the most extreme in the world.***
> ***Corruption is a fact of life.***

Paraguay has become a regional center for smuggling, money-laundering and organized crime."
(Global Property Guide 2010)

"...Aristotle argued that the expansion of the size of the middle class was the best practical solution to the dual problems of democratic political stability and economic equality."
(Ronald Glassman 1995)

1. Introduction

Paraguay is a country of pronounced inequality. This is caused and maintained by historical inequalities, culture, and institutions which have reduced the opportunities for social mobility. To change this we must seek to build the Middle Classes. It is this social Class which will produce and support the leaders who will address inequality and install appropriate policy. This will create lasting change. The curative measures currently being implemented by 'Un Techo' will not solve the source of inequality, and are therefore only stop-gap measures against the rising flood of poverty. This proposal suggests a new social housing model which includes the purchase of the land for housing projects. The model, once seed-funded, will become self-funding by the process of land betterment.[1]

Current Situation

Paraguay is a predominantly Roman Catholic country controlled primarily by a white minority (of European descent). This elite minority hold a disproportionate majority share of wealth, land and, consequently, power. The average Paraguayan earns one of the lowest incomes in the world (98th lowest out of 114 countries reviewed by Nation Master). Table 1 below shows Paraguay's ranking of 9th worst (and growing worse according to Von Badden, 2010) in the world for the most unequal distribution of wealth, using the Gini Co-efficient method of comparison.

---

[1] Betterment, making better, is a general term used particularly in connection with the increased value given to real property by causes for which a tenant or the public, but not the owner, is responsible; it is thus of the nature of unearned increment. When, for instance, some public improvement results in raising the value of a piece of private land, and the owner is thereby bettered through no merit of his own, he gains by the betterment, and many economists and politicians have sought to arrange, by taxation or otherwise, that the increased value shall come into the pocket of the public rather than into the owner's. Wikipedia 2011

Table 1: Gini Co-Efficient Rankings by Country (2010 Extract from available data)

Rank	Country	Gini Value
1	*Namibia*	74
2	*Lesotho*	63
3	*Sierra Leone*	62
4	*Central African Republic*	61
5	*Botswana*	60
6	*Bolivia*	60
7	*Haiti*	59
8	*Colombia*	58
9	***Paraguay***	**58**
10	*South Africa*	57
85	**Australia**	**35**

Source: http://www.nationmaster.com (2010)

These low ranking countries are characterized by a relatively large majority Lower Class, small minority Upper Class, and corresponding small Middle Class. This can be compared to Australia which ranks 85th (out of 114 countries measured) and which has much more evenly distributed wealth and a large Middle Class (by Australian standards), which is reported to be just over 50% of the population (Hamilton et al, 2007). As a result of inequality in Paraguay there are serious implications, including malnutrition, amongst other health issues[2]. (Refer to Appendix 1 – Statistics of Poverty, for further information)

[2] The Food and Agriculture Organisation of the United Nations says that nearly half of the population is not only in the Lower Class but poor. "Poverty data ... show 32% of population in Paraguay being poor and 17% extremely poor." And goes on to say that… "The nutritional status of children under 5 years old at the national level … reflected low levels of underweight (4%) and stunting (8%)."
www.fao.org/ag/agn/nutrition/par-e.stm

Figure 1 - Slum in Asunción Paraguay - Source: travelpod 2011

In Paraguay there is a polarization between the small Upper Class and majority Lower Class (with only a minority Middle Class). The upper or 'elite' 20 % (disproportionally white[3]) earn 62.4% of all income and have, since European colonization, controlled much of the land and businesses[4] (as shown in Table 2 below). It is important to note that some of the institutions and religious/cultural beliefs may be contributing to maintaining this inequality[5].

[3] "...in Latin American societies...the wealthier classes are more prone to identify themselves as "white" even when they do not seem to be so to others." (Page 170, Von Badden, 2010)

[4] The land policy of the Latin American governments also greatly influenced the distribution of land, which ultimately led to the lop-sided patterns of income distribution in the region. (Page 167, Von Badden, 2010)

[5] Max Weber in his work 'The Protestant Work Ethic and the Spirit of Capitalism' makes the connection between rationally based religious and other beliefs and the success of capitalism. This is in contrast to the mixture of the mystical Catholic and local beliefs which exist in the more family centered, and feudal-like, Latin American countries. It may be the case these belief systems have stifled the emergence of a significant Middle Class. The classes believe, to some extent, their social positions are divinely allocated, and/or strongly related to their lineage.

Table 2 – Paraguay Income Distribution

Population Sector	Income Share %
Lowest 10%	0.7
Lowest 20%	2.3
Second 20%	5.9
Third 20%	10.7
Fourth 20%	18.7
Highest 20%	**62.4**
Highest 10%	46.6

Note: This information refers to income shares by percentiles of the population and is ranked by per capita income. Source: 2000 World Development Indicators

Housing in Slums

Many of the poorest people do not have appropriate shelter. Without appropriate housing in appropriate locations residents cannot aspire to have better jobs or educate their children to achieve social mobility. Incidences of poverty are often focused in one area due to lack of employment among other factors. In addition, access to essential services like education and health care are also often limited in these same marginalized areas.

The focus to date of the 'Un Techo' program has mainly been on the provision of housing within the existing poorest neighbourhoods. (However, their community building programs have also been noted as highly effective). This program has been very successful in the short term. The problem associated with this is that it results in the entrenchment and concentration of poor people in areas that may be flood-prone, subject to land-slides, distant from services and jobs, contaminated, criminally unsafe, or a combination of these and other negative factors.

This proposal suggests an alternative model, which will give residents security of tenure and the ability to climb up the social ladder into the Middle Class. Typically quality land sites have not been an option for the Lower Classes. However, the proposed self-funding model seeks to change this. (Refer to Appendix 2 – Building the Middle Class, for further details).

2. Aims and Objectives

The following objectives summarize the main targets of the proposed program.

Middle Class and Leadership

- To enlarge Paraguay's Middle Class by lifting selected people from the Lower Classes.
- To cultivate Middle Class leaders from the local population who will create a more egalitarian society through non-violent means.

Residents

- To provide emergency housing for families in danger of becoming homeless.
- To assist residents in obtaining Middle Class housing as their situation improves.
- To locate poor people close to services and facilities (including employment and schools).
- To provide security of tenure to residents.
- To encourage small business activities amongst residents.
- To ensure residents are accommodated in, and participate in the design of, the final development of the project site.
- To make the housing projects environmentally sustainable.

Community

- To develop a sense of community and 'esprit de corps' in the housing project areas.
- To promote community level self-governance.
- To develop Middle Class institutions to rival Upper Class institutions and to cultivate leadership within these institutions.

Funding

- To ensure residents contribute to the financing and building of their own houses and the community area so thatf a sense of ownership and responsibility evolves.

- To extract 'betterment' profit that provides returns to fund future 'Un Techo' projects.
- To create a funding and development model that can be modified and repeated internationally.

3. Methodology

The proposal fits within a broader social change methodology detailed in the attached Appendix 3. The methodology below discusses the details of the specific social housing model proposed.

Social Housing Methodology

To achieve the stated objectives a methodology has been created to implement the new self-funding housing model. This will be done within the project hierarchy shown below.

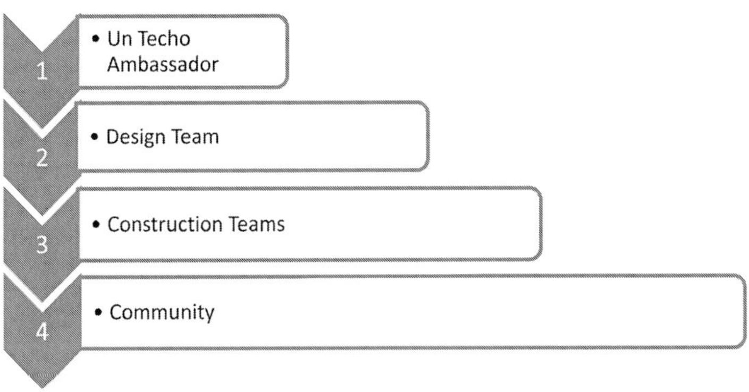

Figure 2 - Project Hierarchy

Acceptance

The first step will be to obtain agreement from 'Un Techo' to trial the new model. Once this support is obtained, then funds can be collected under their banner for the first land purchase. This will allow the remaining steps in the process (summarized below) to occur.

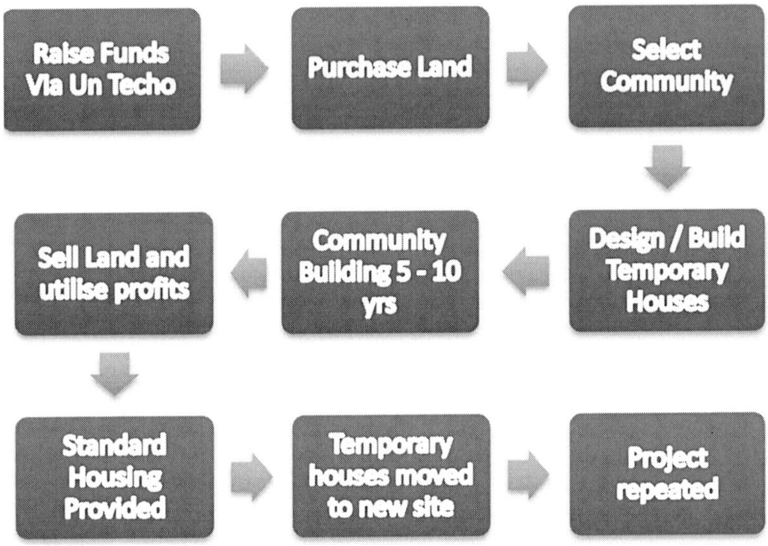

Figure 3 - Flow diagram of methodology for the implementation of social housing

Land Selection

An example site has been selected in the middle to outer suburban areas of Asunción, capital of Paraguay. This site has been selected for its position in the path of city growth which increases the probability and speed of price growth (betterment). For full details of the land selection criteria refer to Appendix 4.

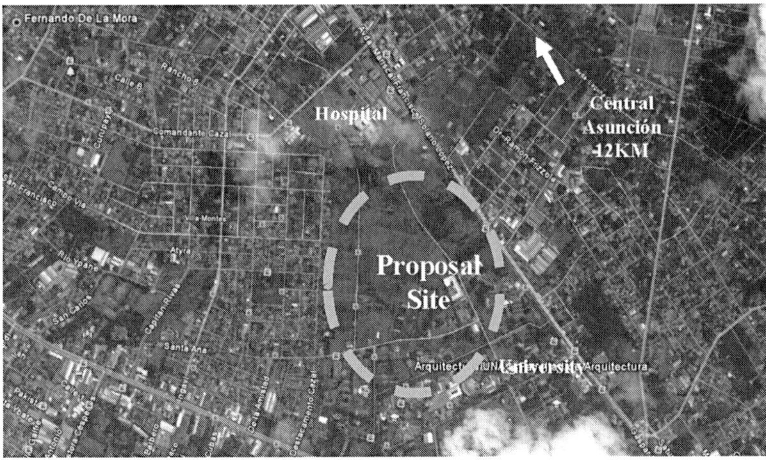

Figure 4 - Location of proposed social housing project. (Google Maps 2011)

Figure 5 - Photo taken near proposed site in Fernando de la Mora - Source: Google Earth Images 2011

Project and Design Team Selection

Once a parcel of land has been secured, a design and project team will be selected from volunteer consultants to lead and manage the project. The leadership style will be predominantly ‘participative’ to maximize resonance with, and between, team members. Dubrin’s ‘path-goal’ method will be applied to ensure outcomes are achieved using the group’s strengths appropriate to the tasks.

The design team will be overseen by the Un Techo Ambassador for Paraguay. It will be responsible for the building designs (refer to Figures below for examples), services, facilities and general town layout/planning.

The team will have a high degree of cultural intelligence in order to communicate effectively with team members and locals. As Goleman states in ‘The New Leaders’ in relation to use of the facilitative leadership style … ,

> *“In…Latin America…establishing a strong relationship is a prerequisite for doing business.”* (Page 64).

Community Selection

A community will be selected using a set criterion. The focus will be on families which show potential to rise up into the Middle Classes with minimal assistance. For details refer to Appendix 4.

Design Process

The selected community will then collaborate with the design team to prepare the plans for their new town. Some aspects of Traditional Neighbourhood Design[6] principles will be used where appropriate. This layout promotes community-based interaction, governance and management from which leaders will be cultivated and emerge.

---

[6] Traditional Neighbourhood Design or “New Urbanism is an urban design movement, which promotes walkable neighborhoods that contain a range of housing and job types. It arose in the United States in the early 1980s, and has gradually continued to reform many aspects of real estate development, urban planning, and municipal land-use strategies.” Source: Wikipedia 2011.

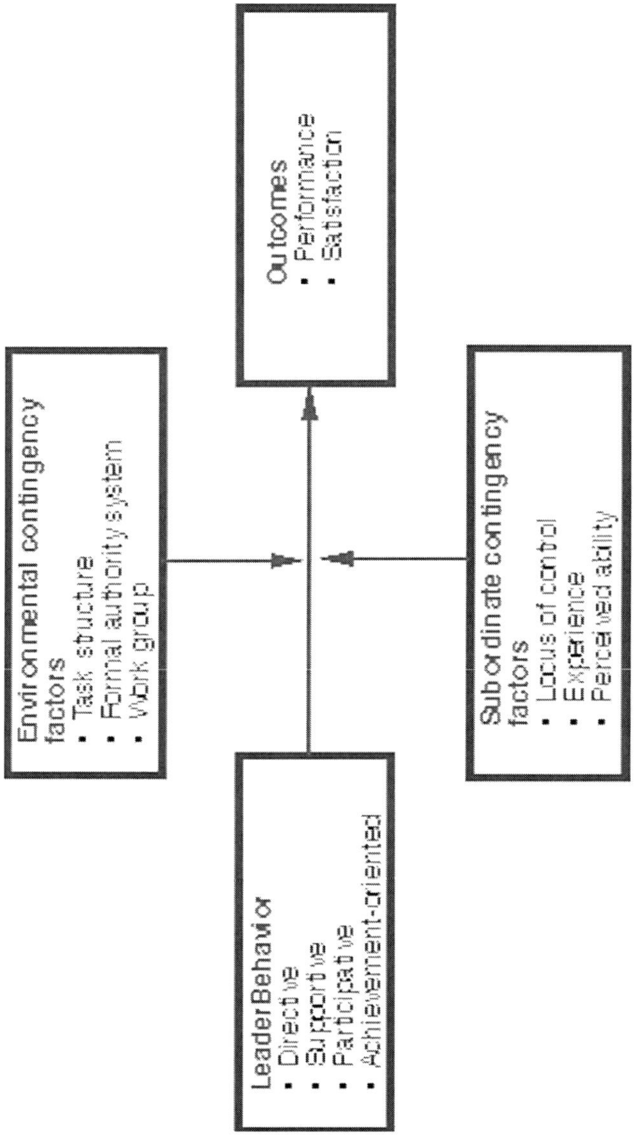

Figure 6 – Dubrin's Path Goal Method Source: House, R. J. (1971). A path-goal theory of leader effectiveness. Administrative Science Leadership Review, 16, 321-339

Figure 7 – Example: emergency Duany Housing in Haiti. (Duany 2010)

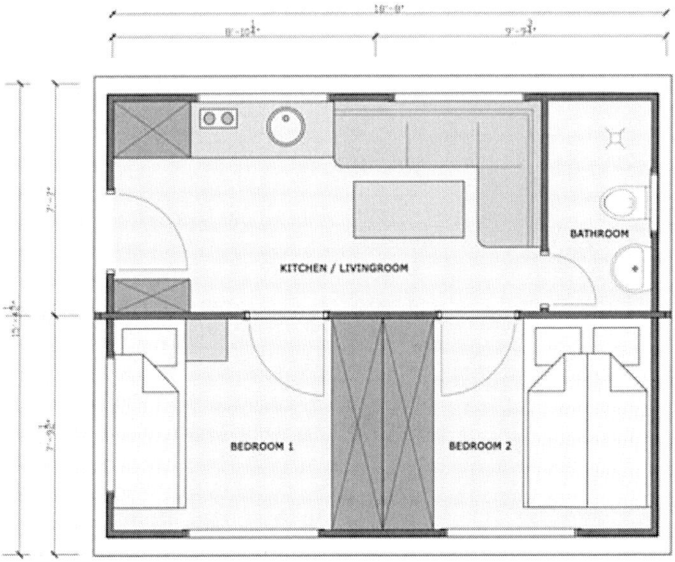

Figure 8 – Typical House Layout (Duany 2010)

Figure 9 - Example of traditional neighbourhood layout. Source: www.tndpartners.com/newurbanism/

Construction

Teams of volunteers and future community members will work together to build the project. Local products will be sourced where possible. Residents will contribute 10% to the cost of the housing to ensure a sense of ownership is obtained.

Governance, Maintenance and Monitoring

A community governance committee will be democratically appointed by the residents, similar to a company board elected by shareholders. Authoritarian forms of leadership will be avoided[7]. (Appendix 4).

Final Development

When a suitable purchase offer is obtained for the site (expected to be approximately 5 to 10 years from purchase date) it will be sold to a developer with the condition that the first houses are constructed for the original residents. As shown in Appendix 4, the developer will have an interest in providing quality housing as it will set the standard and price for the remaining development product.

4. Funding

Once seed-funded by donations, the project will become self-funding from the land sale profits as a result of betterment. Full details of funding sources and criteria are included in Appendix 4.

5. Costs

The detail of the estimated costs which will apply to the project will be divided and explained in three parts:

- Land purchasing
- Construction
- Maintenance

[7] "... persistent Authoritarianism is identified as one of the political reasons for continuing inequality in the report... Despite some progress towards democracy in Latin America, the culture of recurring authoritarianism has often stunted the effects of social and political reforms. This is probably because authoritarian rule is by its very nature elitist. (Page 167, Von Badden, 2010)

5.1 Land Purchasing

Funds will be raised via donations for the initial land purchases estimated at approximately US$1 million dollars for some 30 hectares of mostly un-serviced land with minor improvements approximately 12 kilometers south east of the Centre of Asunción. (Refer to Figure 5)

Table 3: Land Purchasing Cost

US$/Ha	Size (Ha)	Total Land Cost $US
33,333	30	1,000,000

5.2 Construction

The initial target of this project is to build 100 transitional houses; each house will include a small kitchenette, a bathroom, and two bedrooms. A typical house is 18 m2 (6m×3m) in size. It will be made of wood or aluminum and is designed to be low-cost, easy to construct, and re-locatable. It is planned to be built in two days by a team of eight to ten volunteers, working alongside the recipient family. Construction materials will preferably be locally sourced. 90 % of the construction cost will be supported by 'Un Techo' while 10 % of the cost will be received from the recipient family. The breakdown of the construction cost will show as in the following tables. Examples of the dwellings are shown in the Figures below.

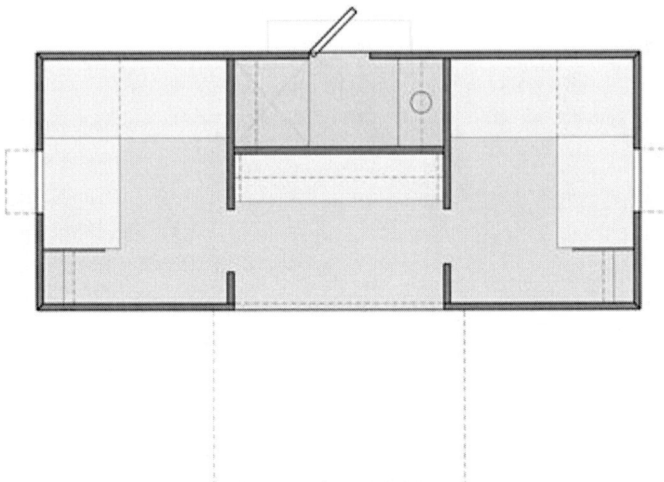

Figure 10 - Example floor plan of Duany House. Source: Andres Duany Haiti Project 2008

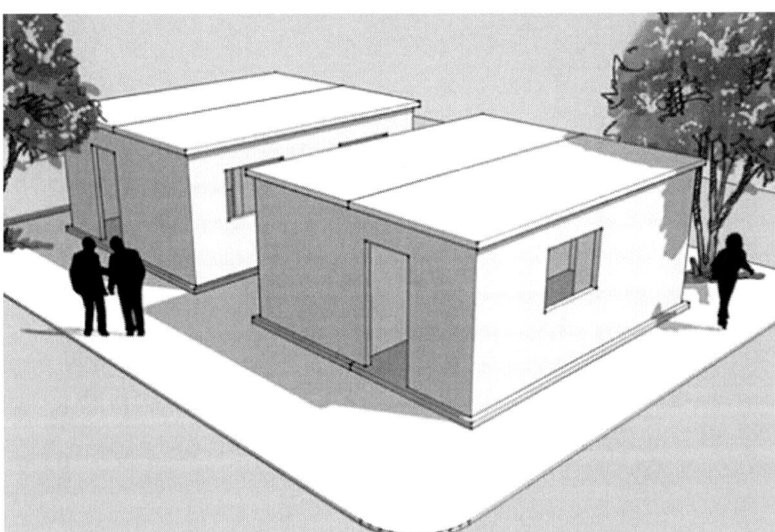

Figure 11 - Example of Duany Housing. Source: Duany (2008)

Table 4: Construction Cost in US Dollars

Item	Description	Cost($)
1	Foundation	800
2	Vertical structure	800
3	Floor and roof structure	400
4	Roofing and waterproofing	600
Shell cost		2600
5	Interior partitions, door and glazing	300
6	Floor, wall and ceiling finishes	200
Interiors cost		500
7	Plumbing system	300
8	Electric lighting and power	200
Mechanical and electrical cost		500
Total building construction cost		3600
9	Water supply	1000
10	Site paving, structures and landscaping	200
11	Utility on site	1000
Site construction cost		2200
Total building and site construction cost		5800
12	Overhead	200
Planned construction cost		6000

Source: Extrapolated from Rawlinsons Data (2010)

Labour for the construction teams will be entirely supplied by volunteers. As a result, labour costs will not be included in the construction cost.

The design contingency cost and the allowance for rising cost should be considered (Butcher, Demmers, 2003). The design contingency costs are normally around 5% to 15% of the total construction cost. The allowance of rising cost could be determined by an inflation rate.

Table 5: Construction cost with contingency cost and the allowance for rising cost

Item	Cost (US$)
Construction cost per house	6,000
Design Contingency	600
Allowance for rising cost	180
Total per House	6,780
Total cost for 100 houses	678,000
90% Funded by Un Techo	610,200
Plus Land Cost	1,000,000
Total Project Cost	**1, 610,200**

5.3 Maintenance

The community will be run by the residents under a democratically elected company board model. This board or committee will collect an annual percentage fee based on land value, which will be used to build and maintain infrastructure such as parks, roads and sewers. In some cases the community may also generate power on-site, and collect and distribute water, among other services. This self-run system may be more reliable than some municipal systems. Holding costs such as tax and rates payable to the Government are expected to be discounted due to the social advantages and the self-run nature of the project.

6. Section: Benefits

Benefits to Un Techo will be based around an improved model that is self-funding and places poor people in better locations, with secure tenure, job access, services and facilities. There will also be considerable financial benefits as a result of funds obtained from property value increases or betterment.

The social benefits will be improved education, health, and income for the residents. This will then give them the opportunity to enter the Middle Class and assist in achieving long term improvements in the country.

Benefits to contributors are many and include a better profile for the company, in addition to many other tangible and intangible benefits for both the company and staff members participating in the program by either raising funds or volunteering to assist with the construction of the houses. Benefits will be measured by both quantitative and qualitative methods.

Table A. Benefits to Un Techo

Item	Measurable	Achievable	Time Based
Additional Funding Source from land sale profit as a result of 'Betterment' Day (1995).	Easily measured when viewing organizations accounts. Expected at ~$2.5 million on first project	Depends on property market	~5 years.
Ability to transition poor people from emergency housing to middle class housing. Emergency houses can be re-used at next site.	Can be seen when new housing is constructed.	Depends on developer purchasing the property at a reasonable price.	10 years.
Ability to expand program internationally using a basic model that can be adapted.	Success can be gauged by number of international projects.	Yes.	Ongoing measurement over next 10 years.
Better location of Social Housing close to jobs and services.	Review can be done of quality of new housing locations. Employment statistics can also be reviewed.	Yes	5 years
Security of tenure for residents.	Review can be done of percentage of housing with secure title. Improvements to buildings will be a sign of this confidence.	Yes. Purchasing the land will allow Un Techo to control tenure.	5 years
Better location will assist in achieving better social outcomes such as higher employment rates for residents	Employment rate can be checked by annual census.	The site is located close to the hospital and university employment centres.	After 2 years higher employment rates should be achieved.
Increased credibility of Un Techo via success will lead to more donations.	Review donation intake annually.	Yes	Annual increases expected.

Table B. Social Benefits

Item	Measurable	Achievable	Time Based
Birth Weight Rises	Easily measured using median graphs. Take record of recent births for those families accepted on the program.	Improved living conditions and nutrition will raise birth weights	Ongoing measurement over next 10 years.
Crime Rate Drops	Easily measured if police publish statistics across area.	Improved living conditions will decrease desperate crime.	Ongoing measurement over next 10 years.
School Attendance Rises	Easily measured from school roles. Record average for children on program; then compare annually.	Will be achieved if families are in stable accommodation.	Ongoing measurement over next 10 years.
Tertiary Qualifications Increase	Would start at 0; then take records in following years.	Yes	Long term (Next Generation)
Happiness Increase	Annual questionnaire to residents.	Yes	Immediate
Savings Increase. This is a sign of being middle class.	Anonymous questionnaire to residents.	Yes	Annual check.
Employment Rate Increase	Annual questionnaire.	Yes	Annual check.
Average Income Increase	Annual questionnaire	Yes	Annual check.
Nutrition Improves	Annual study	Yes	Annual check.
Health Improves	Annual health checks records. (Visit by Un Techo nurse)	Yes	Annual check.
Better educated population elects better leaders and keeps them accountable.	We would expect to see a decline in corruption and a merit based appointment system.	Yes	Ongoing measurement over next 10 years.
Strong middle class rebuilds the economy	Various economic indicators can be reviewed		Ongoing measurement over next 10 years.

Table C. Benefits to the Company Sponsors

Item	Measurable	Achievable	Time Based
Company profile raised attracting more local customers	Questionnaire to clients about perceptions. Sales figures analysis.	Yes	Immediate and growing
Branding Associated with Charity Work	Questionnaire to clients about perceptions.	"	Immediate and growing
Self-fulfillment of employees from feeling they are making a difference	Questionnaire to employees.	"	Immediate and growing
Team building improves teams by working together to raise funds or doing construction and then monitoring the family adopted.	Questionnaire to employees	"	Immediate and growing
Customer loyalty improves due to greater credibility	Questionnaire to clients about perceptions	"	Immediate and growing
Social responsibility and sustainability objectives achieved (Good corporate citizen)	Questionnaire to clients about perceptions	"	Immediate and growing
More government contracts and concessions.	Total up contract values before and after participation.	"	Immediate and growing

7. Section: Qualifications

The project model requires a number of levels of leadership from varying styles. These have been matched to the team members.

Team Member	Qualification	Task Responsible For	Comment (Relevance to Leadership & Neg)
Carlos Escruceria	Master of Infrastructure Management	Infrastructure Design	Ability to work as a technical engineering officer in design and coordination of infrastructure servicing developments.
	Graduate Certification in Built Environment and Engineering	Project Management	Ability to lead professional, multi-disciplinary consulting teams.
	Exchange Student Experience	Cross cultural liaison	Ability to communicate with and understand multi-national consulting teams
	English and Spanish Proficiency	Translation	Ability to communicate with both local people in Spanish and consultants who may only speak English.

Ahmad Suliman Al Dahash	Industrial and Civil Engineer	House design and production	
	Production Management	Cost control	Ensuring the budget constraints are recognized and producing a design that can achieve this. Ensuring that tasks are completed on time and budget.
	Arabic and English Speaker	-	Ability to communicate with and understand multi-national consulting teams
Pailin Lomsinsab	Electrical Engineer		Possesses the technical skills to work as part of a collaborative design team.
	Thai and English Speaker		Ability to communicate with and understand multi-national consulting teams
Saree Koysiripong	Process Engineer experience (Major: industrial engineering)	Process control	Possesses the technical skills to work as part of a collaborative design team.
	Asset Control experience	Cost control	Ability to control work teams and ensure budget constraints are met.

	Thai and English Speaker		Ability to communicate with and understand multi-national consulting teams
Matt Doolan	Town Planner	Design of town layout	Possesses the technical skills to work as part of a collaborative design team.
	Development Consultant	Selection of project sites	Can give advice to Un Techo Ambassador regarding site selection.
	Project Manager	Project Management	Ability to lead multi-disciplinary development teams.
	English and Intermediate Spanish Speaker	Communication with local officials and community	Ability to understand needs and communicate in a culturally sensitive way.
	Australian Army Officer	Liaison with Defence Organizations	Ability to communicate with Defence Hierarchy.
	Chermside Soccer Club President	Developing Local Sports Clubs	Ability to raise and run local sports clubs.

8. The Challenge

Traditionally poor people have been marginalized on some of the worst land in cities and rural areas. This has added to their many disadvantages and served to trap them in cycles of poverty. The proposed model for social housing in Paraguay seeks to locate poor people on quality land. This will give them access to the services and facilities they need to have a chance to improve their lives through employment and education. The concept of harnessing 'betterment' is the key original feature of this social housing proposal.

The proposed social housing program forms just one part of a total strategy to build the Middle Class and cultivate its leaders. These new leaders supported by their Middle Class base will install policies to address inequality and eliminate poverty.

Note

Whilst this particular proposal has focused on land banking in urban areas it should be noted that the model could be adapted to rural areas. This would be based on the assumption that certain productive rural land could rise in value if commodity prices rise, and fertile land shortages develop, in the coming years. This could solve the employment issue with many of the residents being employed on the site in agricultural activities. The condition would be that appropriate services and facilities (such as schools and health care) would be available to an acceptable degree.

References

Arnold, D., G. (2010). Working Condition: Safety and Sweatshops. Journal of Business Ethics, 13:978-0-19-530795-5, doi: 10.1093/oxfordhb/9780195307955.001.0001.

Bryman, A. (1992), Charisma & Leadership in Organizations, Sage, London, in Hunt, J.G. and Peterson, M.F. (1997), "Two scholars' views of some nooks and crannies in cross-cultural leadership", The Leadership Quarterly, Vol. 8 No. 4, pp. 343-54.

Butcher, N., & Demmers, L. (2003).Cost estimating simplified. Retrieved from http://www. librisdesign.org/

Clive Hamilton, Christian Downie and Yi-Hua Lu (2007) The State of the Australian Middle Class, The Australia Institute Food and Agriculture Organisation *http://www.fao.org/ag/agn/nutrition/par-e.stm*

Day, P. (1995) Land, Australian Academic Press, Australia

Duany flat pack Haiti cabins. (2010). Retrieved from http://www.jetsongreen.com/2010/02/andres-duany-flat-pack-prefabs-haiti.html

Dubrin, Andrew, Carol Dalglish and Peter Miller. 2006. LEADERSHIP. 2nd ASIA-PACIFIC. 42 McDougall Street, Milton Qld 4064: John Wiley & Sons Australia, Ltd.

Economy Statistics > Gross National Income (per capita) (most recent) by country. 2010. Retrieved from http://www.nationmaster.com/red/graph/eco_gro_nat_inc_percap-gross-national-income-per-capita&b_printable=1

Estanislao, G (2004) Paraguay: Social Development Issues for Poverty Alleviation.

Fazio, María Victoria. "Poverty and Inequality in Paraguay: Methodological Issues and a Literature Review." version: January 17, 2005. Retrieved from http://www.depeco.econo.unlp.edu.ar/cedlas/monitoreo/.../review_paraguay.pdf

George, Henry. (1879) Progress and Poverty E.P. Dutton and Company

Glassman, R (1995) The Middle Class and Democracy in Socio-Historical Perspective E.J. Brill, Leiden, The Netherlands.

Global Property Guide 2010 http://www.globalpropertyguide.com/Latin-America/Paraguay

Goleman, Daniel (2002) The New Leaders, Clays Limited, London

House, R. J. (1971). A path-goal theory of leader effectiveness. Administrative Science Leadership Review, 16, 321-339

Marió, Estanislao Gacitúa, Annika Silva-Leander and Miguel Carter. " Paraguay: Social Development Issues for Poverty Alleviation." Retrieved from http://www.siteresources.worldbank.org/.../Paraguay_Country_Social_Analysis.pdf

Paraguay – Poverty and wealth. 2000. Retrieved from http://www.nationsencyclopedia.com/economies/Americas/Paraguay-POVERTY-AND-WEALTH.html

Paraguay – Crisis Poverty and Social Impact Analysis (PSIA). March 2010.content.undp.org/go/cms-service/stream/asset?asset_id=2430631

Paton, R., Mordaunt, J. and Cornforth, C. (2007), "Beyond nonprofit management education leadership development in a time of blurred boundaries and distributed learning", Nonprofit and Voluntary Sector Quarterly, Vol. 36 No. 4, pp. 148S-62S

Rawlinsons construction cost guide 2010 (18th ed.). (2010). Perth, Rawlhouse Publishing.

Sayles, L.R. (1999), "Managerial behaviour and a journey through time", The Leadership Quarterly, Vol. 10 No 1, pp. 7-11.

Un Techo para mi País, Youth building a Latin America without poverty. Retrieved from: http://www.untechoparamipais.org/english/pagina-principal/u-s-volunteers/

Von Badden, Erick (2010) Income Disparities In Latin America, School of

Doctoral Studies of the EU, Brussels.
Weber, Max, 'The Protestant Work Ethic and the Spirit of Capitalism'
Wolff, M., Young, S., Beck, B., Maurana, C.A., Murphy, M., Holifield, J. & Aitch, C. (2004) Leadership in a Public Housing Community. Journal of Health Communication 9(2). Doi:10.1080/10810730490425286.

Appendix 1

Statistics of Poverty and Inequality

Paraguay is a landlocked country in South America. It is surrounded, by Brazil, Argentina, and Bolivia. Its population is estimated at 6.4 million. The primary religion is Roman Catholic (some 90% of the population). Ethic breakdown is mestizo (mixed white and Amerindian) 85.6%, white 9.3% (of which German 4.4%, Latin American 3.4%), Amerindian 1.8%, Black 1%, other 2.3%. Source: *http://www.worldstatesmen.org*

Similar to many other Latin American countries, poverty, health, and education are some of the main issues that challenge development. According to 'Nation Master' the annual income per person is very low, ranking 98th in the world (out of 114 countries measured). In addition to this, opportunities for social mobility are generally low.

"Paraguay has an extreme gap between the small upper class and the large lower class, and there has historically been virtually no social mobility. Paraguay has the most unequal distribution of land in the region. Less than 10 percent of the population owned and controlled over 75 percent of the nation's land".
http://www.nationsencyclopedia.com/economies/Americas/Paraguay-POVERTY-AND-WEALTH.html

Appendix 2

Building the Middle Class

The Middle Class in Paraguay is not sufficiently large to support a successful democratic and economic system. As Glassman states…,

> *"In fact, the problem of economic equality becomes critical, given the capitalist base of society...extremes of wealth, generating rich and poor, become destabilising to democracy."* (Page 16)

and he goes on to say that…

> *"...Aristotle argued that the expansion of the size of the middle class was the best practical solution to the dual problems of democratic political stability and economic equality."* (Page 36)

This relates directly to the leadership issue; as Goleman in 'The New Leaders' explains, sociologist Max Weber had argued that **institutions thrive not because of one chance leader but because certain types of leadership are cultivated and supported by the system.** This is a twofold problem in Paraguay as it is unlikely that an appropriate leader will emerge out of the current system and, even if they do, they will not be supported by the current system and therefore cannot achieve success. Additionally, the institutions in Paraguay[1] are historically elitist and exclude the lower classes. If a new Middle Class can emerge, it can develop its own new institutions to rival these existing institutions. These institutions (Community Committees, Sports Clubs, and Local Fire, Military and Police) will become the cultivation zones for new leaders to train and network.

Paraguay could not be classed as a successful democracy due to its relatively small Middle Class. As a result corrupt[2], popularist leaders can utilize…

> *"...conscious intelligence and intersubjective communication to manipulate, confuse, coopt, terrorize or charismatise the groups into granting a non-rational consent, or an acquiescence to power."* (Page 10, Glassman).

With a larger Middle Class, leaders will emerge from this Class and be supported by this Class. These leaders can then install policies, such as appropriate levels of Land Taxation and Inheritance Taxation, which will redistribute wealth from the Upper Classes and create a more egalitarian society[3]. This will provide a long-term and non-violent solution to poverty.

---

[1] Sometimes referred to as the pillars of society, which often includes the Government, Military, Religion, and Commerce.

[2] "…a majority of the elections was fraudulent with the elite exercising disproportionate direct and indirect power to protect their interests". (Page 167, Von Badden, 2010)

[3] "The elite, as always, devised policies favoring the status quo, which were heavily tilted in their favor. These policies also prevented the majority from unfettered access to equal economic opportunities and ensured the survival of

In addition it will preserve the Upper Class capital and leadership assets (and foreign investment) within the country during the progressive implementation stages of the new policies.

Appendix 3

Social Change Methodology

It should first be explained that the proposal for social housing provision fits within a broader Social Change Model. This model involves three basic steps. The subject proposal for social housing fits within the first step and is highlighted in yellow below.

The first step is to build a sufficiently large Middle Class through the range of programs listed above. In the second step the Middle Class will then begin to produce leaders who will be cultivated in a range of Local Community based organizations, institutions and initiatives. (Examples could be the local soccer coach, local political representative, or local Fire Chief).These institutions will rival the traditional institutions of the elite. The new local leaders will survive and grow as they will now have a sufficient Middle Class support base. This support base is particularly critical when considering the need for numerical support in a democratic election system. Even if the first leaders fail others will continue to rise behind them as the tide turns toward a Middle Class-led society. Middle Class leaders, once in positions of power, will install policies which will maintain and build the Middle Class by re-distributing wealth (eg. Introduction of land and inheritance tax[4]) and progressively raising more Lower Class citizens into the Middle Classes.

There are a number of ways to build a Middle Class. The most obvious solution is to take members of the Lower Classes and place them in the Middle Class. Taking members of Upper Class and moving them down into the Middle Class will have little impact on total numbers and could be highly disruptive and potential unethical. Other alternatives hypothetically could include exporting or annihilating the Lower Classes (Genocide) or importing an additional Middle Class population (Colonization). Both of

---

inequitable distribution of income and other assets. "(Page 167, Von Badden, 2010)

[4] Paraguay has no Land Tax or Inheritance Tax. The main relevant taxes are income and capital gains tax at approximately 15%. (Global Property Guide 2010)

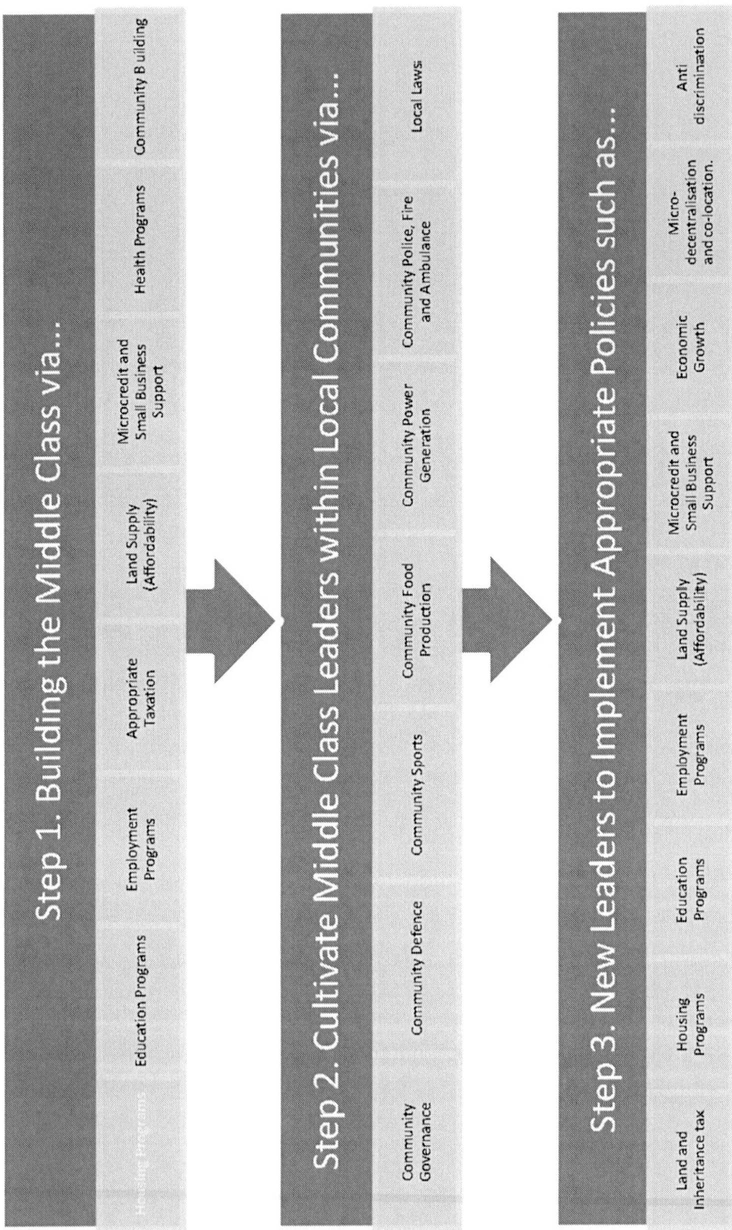

Step 1. Building the Middle Class via...
Housing Programs
Education Programs
Employment Programs
Appropriate Taxation
Land Supply (Affordability)
Microcredit and Small Business Support
Health Programs
Community Building
Step 2. Cultivate Middle Class Leaders within Local Communities via...
Community Governance
Community Defence
Community Sports
Community Food Production
Community Power Generation
Community Police, Fire and Ambulance
Local Laws
Step 3. New Leaders to Implement Appropriate Policies such as...
Land and Inheritance tax
Housing Programs
Education Programs
Employment Programs
Land Supply (Affordability)
Microcredit and Small Business Support
Economic Growth
Micro-decentralisation and co-location.
Anti discrimination

these last options are not considered appropriate. Other factors to consider include the prevention of Middle Class loss/export to developed nations (sometimes referred to as the brain drain). The figure below shows the current social class distribution which needs to be corrected.

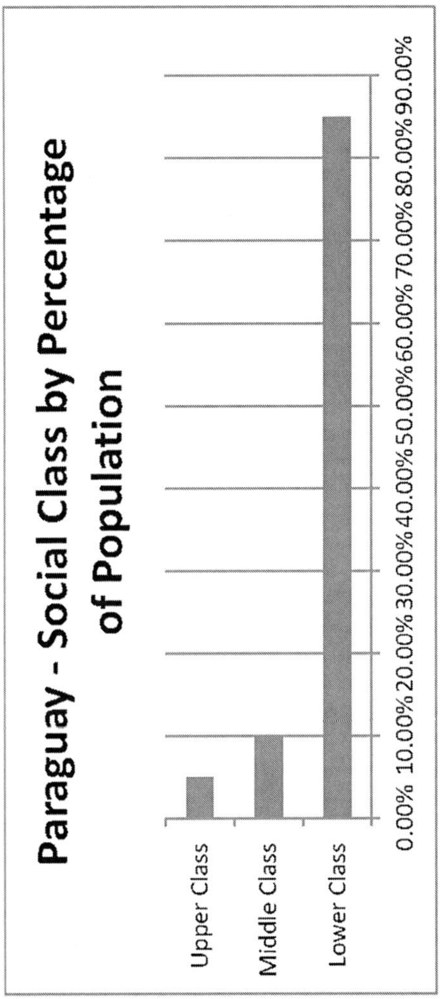

Figure 12 – Social class size in Paraguay – Source: Figures adapted from Nation Master Statistics (2010)

Appendix 4

Key Stakeholders Communication and Negotiation Plan

Land Selection

An example site has been selected in the middle to outer suburban areas of Asunción, capital of Paraguay. However, it could be a site in any significant city in a developing nation, providing it is in the path of future urban growth. The site must be free of contamination, be mostly free of flood, landslides and other constraints that are normally associated with the 'bad lands' siting of current slum areas. The site is to meet a criterion which includes close proximity to services, facilities, such as employment, schools, and public transport. The land must be relatively undeveloped and large enough to hold an initial 100 dwellings at relatively low densities (~1 dwelling per 3000m2). The example site chosen for this proposal is some 30 hectares of land in the south east area of Asunción in the suburb of 'Fernando de la Mora' (Refer to Figure 1 and 2 below). A suggested starting amount is US$1 million for land purchase.

Community Selection

A community will be selected using a set criterion. Individuals can apply or be recommended to the program by someone such as a local social worker. The emphasis will be equally on selecting people already located in the general area, with greatest need, and showing some potential to continue on the path to the Middle Class. Special consideration will be given to those who show an ability to start or run their own business. The intention is to encourage a sense of connection to the community with an identity described here as the 'noble tribe'[5]. The layout will foster this sense of place by including a centre with town square, small shops, community hall, sporting field and other community facilities. Sports in particular are to be encouraged to allow cross class / ethnic interaction which will reduce these barriers to social mobility.

It is anticipated that the basic layout used in the initial stages will be further developed in the ultimate development stage by the end developer.

Lower Classes with some small assistance can run highly efficient micro-businesses (e.g. Computer repairs, deliveries, small shops etc…). This would be encouraged and some commercial space could be allocated on or

[5] Related to concepts of the noble savage popular during the age of enlightenment.

near a central town square on the project site. These small business operators often have a vested interest in building a good community as it is in their financial interest to do so.

Governance, Maintenance and Monitoring
A community governance committee will be democratically appointed by the residents similar to a company board elected by shareholders. Authoritarian forms of leadership will be avoided[6]. This committee will ensure the housing area is well maintained and serviced. The committee will also assist in the establishment of clubs, schools, health care and other services. Special assistance will be given to encourage the creation of the first sporting, service and social groups.

A number of data-collecting methods, both quantitative and qualitative, will be employed to measure the success of the project throughout its life.

Individuals will not be permitted to sell their land or buildings in the first stage. If they do need to exit the program, a new family will be selected by 'Un Techo'.

A local flag will be designed by the community to encourage a sense of place and 'esprit de corps' to develop. 'Un Techo' will also run a monitoring and follow up to encourage and support residents to maintain and build their community.

Final Development
When a suitable offer is obtained for the site (expected to be approximately 5 to 10 years from purchase) it will be sold to a developer with the condition that the first houses are constructed for the original residents. The developer will have an interest in providing quality housing, as it will set the standard and price for the remaining development product. A development approval may be obtained by 'Un Techo' prior to sale if this is deemed to add further value to the property. Final development ratios will be some 1 dwelling per 300m2, or 1000 dwellings on the subject 30 hectare site. The residents will contribute to the cost of these houses. The original houses will be relocated to the next housing project

[6] "... persistent Authoritarianism is identified as one of the political reasons for continuing inequality in the report... Despite some progress towards democracy in Latin America, the culture of recurring authoritarianism has often stunted the effects of social and political reforms. This is probably because authoritarian rule is by its very nature elitist." (Page 167, Von Badden, 2010)

site. The intention is that eventually full ownership of the land and building/s will be passed to the individual residents. A 'win – win' approach to land sale would be adopted where both 'Un Techo' profits from the sale and the final developer can complete a successful profitable development.

Funding

Social housing construction cannot be financed entirely by the residents due to their low income. 'Un Techo' will be required to cover some 90% of the total construction cost. All land costs will be covered also by 'Un Techo'.

The majority of the funds will be raised by 'Un Techo' via donations. A focus of fund-sourcing will be on multi-national companies operating in Paraguay, which are extracting resources and profits. The program can link in directly with the social and environmental goals of these companies such as Shell's 'good neighbour'[7] policy. By contributing to the 'Un Techo' program these organizations can meet social goals without having to manage the projects themselves.

A significant portion of the funding will also be sourced from individuals and groups of individuals through the specific purchase of dwellings. They can assist in the building process and participate in the selection and monitoring of the recipient family.

---

[7] Shell states, "We aim to be good neighbours in the communities where we operate. This means more than running our operations cleanly and safely: it also means working with people close to our operations to address their concerns and help them benefit from our activities." Source: www.shell.com.au/home/content/aus/environment_society/society/

SAPREJ-12 Brief Report: 1st Conference and Training Programme on Sustainable Alternatives for Poverty Reduction and Ecological Justice

Julia Bradshaw

Diocese of Europe, Church of England
E-mail: juliaworld21@yahoo.co.uk

Abstract: *In the first SAPREJ conference it was agreed that Christians have a responsibility to help the poor of our planet.*

The fantastic location for this conference was the Greek Orthodox Academy in Kolymbari, Crete. It was a wonderful, if not tiring, experience. The enthusiasm and energy of the people involved from all corners of the globe was encouraging.

In true Greek fashion the proceedings started over an hour late, but at least this time there was a valid excuse; there had been a general strike on Wednesday making it difficult for people to arrive on registration day, so arrivals continued throughout the weekend.

We stayed in Greece for the opening lecture and this quote from the HAH Ecumenical Patriarch, Bartholomew, The Green Patriarch and his 'Care for Creation':

> "God created; we with greed and pride have destroyed. How do we repair? How do we sustain our repairs? The distribution of wealth needs to be fairer to enable repair."

Greece has a great need, like many other countries in the world, to reduce its poverty. The Greek Orthodox Church stresses that it works continually to help with this.

Schools are being involved so that the children grow up with sustainable ideas, hopefully bringing them into their home environments. On the first day we had many children attend, to show what they are doing and how they are being taught to achieve this.

Our next generation

Greece understands what it should be doing, but it is going to be a long hard journey to educate the people to follow the ecological needs of the planet.

The whole world needs to work together to sustain the environment.

Water, Air and Land must be treated with love and understanding. Peace would be the ideal solution with no conflict and everyone working together.

We visited 2 olive presses locally; one pictured here produces the oil traditionally, using stones to crush the olives. They both produce organic oil by using only pre-tested olives from accredited organic farms.

Nothing is wasted in the process. Leaves are used as compost and the pulp, stones and skin are used in the manufacture of cosmetics, creams and cleaning materials. Maybe olives should be used as the symbol of sustaining life rather than bread?

Stone Press

The Holy Convent of Chrysopigi was our next venue. The sisters there also produce their own olive oil and sell this along with candles and soap which they make. They involve the community by opening their doors to the public, but schools are especially welcome to educate them on environmental issues. How to grow crops organically, how to make compost, how to use waste products and to teach that all species must be saved, as they all have a part to play to sustain the planet.

The first lecture of day two was from Jan Willem Sneep of Holland, who talked of the need for nature conservation and ecological ethics. There are 7 billion people on the planet; 925 million are hungry! 41% do not have access to clean drinking water. Many die from water related diseases.

We need better management of water systems, to learn to cultivate crops without destroying the natural habitat, and to maintain our coastal and marine life on which 3 million of us depend for life.

Holy Convent Chrysopigi

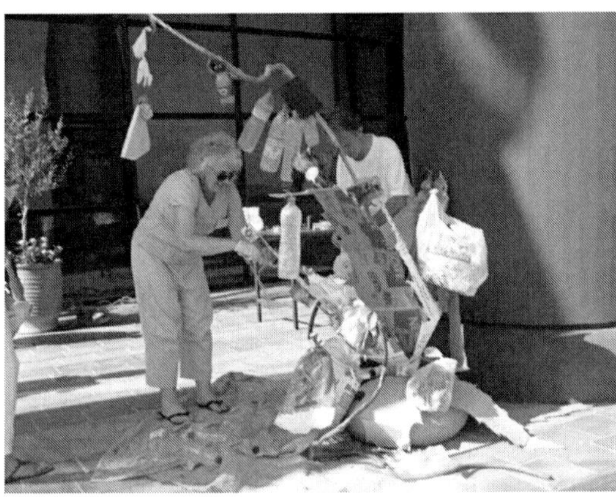

Building a rubbish sculpture

We did have breaks from the lecture theatre to spend time on the beach collecting rubbish. To show some of the amount of rubbish, most of which is not bio-degradable, a 'sculpture' was made.

Plastic is a big threat to the environment and there are various means to dispose of it and to re-generate it into many things. Encouragement is still needed in many parts of the world to re-cycle and not to use plastic carrier bags. Litter is still a very big problem globally.

Carlos Larrea, from Ecuador, spoke about why we should be using alternatives to oil. To keep oil underground is a challenge towards sustainability. Having always believed oil solved financial problems for countries that have oil deposits, I have now been persuaded otherwise. Ecuador has oil and this helped the country for about 10 years; they still have 47% poverty and over 10,000 people have emigrated. There is more oil in the country but they would have to destroy much of the rainforest to obtain this. This would have a great impact on the wildlife found there. Native tribes living in the forest that have never had contact with the outside world would have their homes and existence destroyed. Their government is working to keep at least 50% of fossil fuels underground, which will help with global warming as well as protecting the rainforest and all that is sustained by it.

Many countries are working on sustainable products, housing, light sources and clean water, and much more besides. It was interesting to listen to Dr. D'Souza and hear what is going on in the remote villages in India. They have encouraged villagers to develop their own co-operatives. For example, one village makes solar-powered lights. They sell these to surrounding areas. The women are taught to make and maintain these. The power lasts 12 hours, and every day they can be recharged at a central station. As there is no national grid electricity supply to some of these places this has given them a sustainable light source. In another village the electricity was hardly ever working. They decided to seek help. With the power of running water they were able to create enough electricity to power the village and have some to sell. The people trained to do this look after the system voluntarily and, as they cannot work in their fields, the other villagers do this for them in return. The biggest problem India has is a road and communication system that does not connect the whole country. This makes the distribution of food etc. extremely difficult. It will be many years before the whole country has access to clean water and electricity.

It was encouraging to listen to all these people and what is happening in their countries. There is still a long way to go. There is much poverty in the world. Do we have a responsibility as Christians to help the poor? Yes, we do. The poor need the rich just as much as the rich need the poor. As human beings we need each other. If more people 'with', helped more people 'without', then this would go a long way to having no hungry people. We also need to help and encourage, where possible, people around the world to be self-sufficient by using sustainable alternatives.

View across the hills to the sea. Crete.

7 billion people need a sustainable planet, and time is running out. All countries and governments need to be encouraged to achieve what was discussed in Rio 20 years ago.

We as Christians need to be actively involved on whatever scale we can to help with this, to preserve our countryside and planet by using sustainable alternatives. Hopefully, if we work together, this is an achievable target.

SAPREJ-12 MAILING LIST

Mailing List of SAPREJ-12 Participants

SURNAME	NAME	AFFILIATION	EMAIL
ABOOD	JAWAD	BABYLON TECH. INSTITUTE	R_ALGASBAWI@YAHOO.COM
ABRAMIDES	ELIAS CRISOSTOMOS	GREEK ORTHODOX METROPOLIS OF BUENOS AIRES	ELIASABRAMIDES@FIBERTEL.COM.AR
AKINDEJI	FALAKI	UNIVERSITY OF ILORIN	AKINDEJIAYO@YAHOO.COM
AMIRTHAM	ARUM	SOLAR COMPANY	ARUN.AMIRTHAM@SWISS-MANGO.COM
AMIRTHAM	ANJUN.	SOLAR COMPANY	ARUN.AMIRTHAM@SWISS-MANGO.COM
ANDRIANOS	LUCAS	ORTHODOX ACADEMY OF CRETE	LAOAC@OTENET.GR
ATTFIELD	ROBIN	CARDIFF UNIVERSITY	ATTFIELDR@CARDIFF.AC.UK
ATTFIELD	LEELA	CARDIFF UNIVERSITY	LEELADUTT@HOTMAIL.COM
BRADSHAW	JOHN	CHURCH OF ENGLAND	JULIAWORLD21@YAHOO.CO.UK
BRADSHAW	JULIA	CHURCH OF ENGLAND	JULIAWORLD21@YAHOO.CO.UK
BUDA	DANIEL	WCC CLIMATE CHANGE	DANIEL.BUDA@WCC-COE.ORG
WAIRIMU MUMBI	CATHERINE	GOSPEL RESTORATION CHRISTIAN CHURCH	GRCCBEVANS@YAHOO.COM
DOOLAN	MATT	QUEENSLAND UNIVERSITY OF TECHNOLOGY	MATT@DEVELOPTHIS.COM.AU

DOOLAN	WIFE	QUEENSLAND UNIVERSITY OF TECHNOLOGY	MATT@DEVELOPTHIS.COM.AU
EDWARDS	JULIA	PACIFIC CONFERENCE OF CHURCHES	EDWARDS@PCC.ORG.FJ
EVANS	MACHARIA KINENE	GOSPEL RESTORATION CHRISTIAN CHURCH	GRCCBEVANS@YAHOO.COM
GINTING	LONGGENA	UNITED EVANGELICAL MISSION	UEM.CLIMATE.ASIA@GMAIL.COM
GOGA D'SOUZA	NAFISA	LAYA RESOURCE CENTER	NAFISA.DSOUZA@GMAIL.COM
GRAPE	HENRIK	CHURCH OF SWEEDEN	HENRIK.GRAPE@SVENSKAKYRKAN.SE
HASIM	KHALID	BABYLON TECH. INSTITUTE	R_ALGASBAWI@YAHOO.COM
HENS	LUC	VLAAMSE INSTELLING VOOR TECHNOLOGISCH ONDERZOEK (VITO)	LUCHENS51@GMAIL.COM
JOHN	ROSEMARY	FLOWERS OF CRETE	ROSEMARY.JOHN@HOTMAIL.CO.UK
JONES	JULIA	FLOWERS OF CRETE	JULIA@FLOWERSOFCRETE.INFO
KAKOULAKI	MARIA	INSTITUTE FOR 21ST CENTURY AGORAS	MARIAKAKOULAKI@HOTMAIL.COM
KALACHANIS	KONSTANTINOS	MEDICAL SCHOOL OF NATIONAL AND KAPODISTRIAN UNIVERSITY OF ATHENS	*KALAHANIS@HOTMAIL.COM*
KENANIDIS	KONSTANTINOS	ORTHODOX ACADEMY OF CRETE	GDOAC@OTENET.GR
KENNEDY	JOY	UNITED CHURCH OF CANADA	*JKENNEDY@UNITED-CHURCH.CA*
KERBER	GUILLERMO	WCC CLIMATE CHANGE	GUILLERMO.KERBER_MAS@WCC-COE.ORG
KOTAMPASI	CHRYSANTHI	MEDICAL SCHOOL OF NATIONAL AND KAPODISTRIAN UNIVERSITY OF ATHENS	CK_RISIN@YAHOO.COM
KOULOUMPIS	VIKTOR	UNIVERSITY OF CUMBRIA	VIKTOR.KOULOUMPIS@CUMBRIA.AC.UK
LARREA	CARLOS	UNIVERSITY OF ECUADOR	CLARREA_2000@YAHOO.COM
LEITES	MARCELO	WORLD STUDENT CHRISTIAN FEDERATION	FUMEC@FUMEC-ALC.ORG

MAARUFU	SELEMANI FULGENCE	SOUTHERN AFRICA DEVELOPMENT FOUNDATION	SOUTHERNAFRICADEVELOPMENTF@YAHOO.COM
MADETE	RICHARD	UNITED EVANGELICAL MISSION	UEM.CLIMATE.AFRICA@GMAIL.COM
MARGARET	NJERI GATETE	GOSPEL RESTORATION CHRISTIAN CHURCH	GRCCBEVANS@YAHOO.COM
MATIATOS	IOANNIS	GREEK PARLIAMENT, MEDICAL SCHOOL OF NATIONAL AND KAPODISTRIAN UNIVERSITY OF ATHENS	IMATIAT@GMAIL.COM
MLYNEK	BENJAMIN	ECUMENICAL YOUTH COUNCIL IN EUROPE	BENJAMIN.MLYNEK@GMAIL.COM
MOSES	MBUGUA NGUGI	GOSPEL RESTORATION CHRISTIAN CHURCH	GRCCBEVANS@YAHOO.COM
MOTTE	JOCHEN	UNITED EVANGELICAL MISSION	JPIG@VEMISSION.ORG
MYRIAM	WAMBUI NJENGA	GOSPEL RESTORATION CHRISTIAN CHURCH	GRCCBEVANS@YAHOO.COM
NABEEL	H AL-SAATI	BABYLON TECH. INSTITUTE	R_ALGASBAWI@YAHOO.COM
NIKOLOPOULOY-STAMATI	POLYXENI	MEDICAL SCHOOL OF NATIONAL AND KAPODISTRIAN UNIVERSITY OF ATHENS	ASPIS@ATH.FORTHNET.GR
RATHGEBER	THEODOR	UNITED EVANGELICAL MISSION	TRATHGEBER@GMX.NET
SIPILAINEN	ILKKA	THE EVANGELICAL LUTHERAN CHURCH OF FINLAND	ILKKA.SIPILAINEN@EVL.FI
SKURATOVSKAYA	LARISA	WCC CLIMATE CHANGE	LASKUR@HOTMAIL.COM
SNEEP	JAN WILLEM	PLANTA EUROPA	J.W.SNEEP@MINELENI.NL
SOMPLATSKY-JARMAN	CAROL	PRESBYTERIAN CHURCH USA	BILL.SOMPLATSKY-JARMAN@PCUSA.ORG
SOMPLATSKY-JARMAN	WILLIAM	PRESBYTERIAN CHURCH USA	CAROLLSJ10@GMAIL.COM
STAVROULAKIS	GEORGIOS	TECHNICAL UNIVERSITY OF CRETE	GESTAVR@DPEM.TUC.GR
TOROITICH	ISAIAH KIPYEGON	ACT ALLIANCE	ISAIAH.TOROITICH@ACTALLIANCE.ORG

TRUBETSKOVA	IRINA	NEWHAMPSHIRE UNIVERSITY	IRINA@CISUNIX.UNH.EDU
TURNER	LINDA	BETHANY NEWTON UNITED CHURCH VANCOUVER	L.E.TURNER@SHAW.CA
PHILLIS	YANNIS	TECHNICAL UNIVERSITY OF CRETE	PHILLIS@DPEM.TUC.GR
WEBER	EBERHARD	UNIVERSITY OF THE SOUTH PACIFIC	EBERHARD.WEBER@USP.AC.FJ
TREVIÑO-CISNEROS	REYNALDO	21ST CENTURY AGORAS.	REYNALDOTR@GMAIL.COM
ATHANASOPOULOS	CONSTANTINE	UNIVERSITY OF CAMBRIDGE	CATHANAS@HOL.GR
FISHER	KONSTANTINOS	CITIZENS OF THE WORLD	KONHANIA@YAHOO.GR
WEST	MICHAEL	WCC PUBLISHER	MIKE.WEST@ WCC-COE.ORG
CHRISTAKIS	ALEKOS	INSTITUTE FOR 21ST CENTURY AGORAS	CWAALECO@AOL.COM
BENKING	HEINER	INSTITUTE FOR 21ST CENTURY AGORAS	BENKINGHEINER@GMAIL.COM
BABINA	KAHREL	INTERNATIONAL UNIVERSITY OF JAPAN	BABINAKHAREL540@HOTMAIL.COM

SAPREJ-12 Photo Memories

Thank you

RIGHTS TO HEALTH FOR POVERTY
ERADICATION AND ECOLOGICAL JUSTICE

Terra Creta
ESTATE

Photo Credits: Marcello Leites, Jan Willem Sneep, Julia Edwards, Carlos Larrea and Lucas Andrianos.

THE EDITORS

Dr. Lucas Andrianos (Andriantiatsaholiniaina) was born in Madagascar in 1969. After his graduation in Agronomy (University of Antananarivo, 1993) and his Master's Degree in Hydrology (University Libre of Brussels,1995), he received his PhD. from the Technical University of Crete, specializing in Environmental Sciences and Sustainable Development, in 2001. He started his university career at Simon Fraser University in Vancouver, BC, Canada. Since 2006 he has worked at the Foundation for Research and Technology (FORTH) and at the Orthodox Academy of Crete (OAC) in Kolympari as scientific officer and Head of the Institute of Theology and Ecology, where he organizes seminars, conferences and lectures on ecological theology, fuzzy methodology, hydrology, economic anthropology, sustainable development, environmental ethics, and structural greed. He is the founder and editor of a series of environmental books on "Ecological Theology and Environmental Ethics", "Conservation and Sustainable Use of Wild Plant Diversity" and "Sustainable Alternatives for Poverty Reduction and Ecological Justice". He is an active member of various academic and environmental organizations in the Americas, Asia, Africa and Europe. He is currently a member of Planta Europa Steering Committee, and collaborates with the World Council of Churches as sustainability consultant for the Greed Line study — "Poverty, Wealth and Ecology" diakonia. He is married to Georgia Glampedaki from Crete and they live in Chania with their three beautiful children: Selina, Giannis and Andriana.

Dr Jan-Willem Sneep was born in The Netherlands in 1948. He is a biologist and studied biology (ethology and botany) at the University of Utrecht. Since 1980 Jan-Willem has worked at the Ministry of Agriculture, Nature and Food Quality / Ministry of Economic Affairs in the Hague, and managed several functions in the field of national and international nature policy. In the 1990s Jan-Willem developed a number of species policies in the Netherlands (for example, species policy plans for badger, otter, butterflies, wall vegetation etc.). He was the Dutch representative at many international Conventions and Agreements such as the Bern Convention (Council of Europe), Convention on Biological Diversity (CBD), Convention on the Conservation of Migratory Species (CMS/Bonn

Convention), African-Eurasian Water-bird Agreement (AEWA), Agreement on the Conservation of Populations of European Bats (EUROBATS), and Agreement on the Conservation of Small Cetaceans of the Baltic and North Seas (ASCOBANS). In 1994 he received the (first) Golden Butterfly Award. Until July 2005, Jan-Willem was Head of the International Division of the Department for Nature of the Ministry of Agriculture, Nature and Food Quality. He developed the International Policy Programme for Biodiversity of the Netherlands 2002-2006, and also organized several international conferences in the field of nature conservation and sustainable biodiversity. In the period 2003-2007 he was Chair of the Committee of Experts for the development of the Pan-European Ecological Network (PEEN) of the Council of Europe. From 2004 to 2014 he was Chair of the Planta Europa Steering Committee. At present he is Director of the Planta Europa Foundation. Due to his important international plant conservation work he received the golden Planta Europa Jean Paul Galand Award in 2014.

Dr. Guillermo Kerber was born in Uruguay in 1957. Presently he coordinates the work on Care for Creation and Climate Justice at the World Council of Churches, based in Geneva, Switzerland. He holds academic degrees in philosophy and theology (ITUMS — PUG, Montevideo, Uruguay) and a doctorate in Sciences of Religion (UMESP, São Paulo, Brazil). The title of his thesis, published in Portuguese, is "The place of ecology in Latin American theology". In Uruguay, his home country, he was Professor of Ethics both at the National and Catholic Universities in Montevideo. He has authored various books and articles on ecology, climate change, human rights and international affairs from an ethical and theological perspective, published in Africa, Asia, Europe and the Americas. He sits on the board of the Global Call for Climate Action, Oeku – Church and environment, and the International Network for Human Rights.

Prof. Robin Attfield was born in 1941, and studied classics and philosophy at Christ Church Oxford, being awarded first class honours in Literae Humaniores in 1964. Moving to Regent's Park College (Oxford) he studied Theology and in 1966 was awarded first class honours in that subject. That year he married the author Leela Dutt, and they now have three children and six living grand-children. His doctorate, on Secularisation among Early Modern Philosophers, was begun at Manchester University and completed at Cardiff University, where he became a Lecturer in Philosophy in 1968. He also taught at the University of Ife, Nigeria in

1972-3 and at the University of Nairobi, Kenya in 1995. He joined the Religious Society of Friends (Quakers) in 1971. During the 1970s he became one of the pioneers of environmental philosophy. He was appointed a Professor of Philosophy in 1991, retiring from full-time employment in 2009. With Barry Wilkins he edited 'International Justice and the Third World' in 1992. With Andrew Belsey he organised the Royal Institute of Philosophy conference at Cardiff University in 1993 on 'Philosophy and the Natural Environment', and they edited a proceedings volume with the same title in 1994. During the 2000s he took part in the Environmental Ethics working party of UNESCO. In 2008 he was awarded a DLitt of Cardiff University for contributions to environmental philosophy. His books include 'God and The Secular' (1976 and 1993), 'The Ethics of Environmental Concern' (1983 and 1991), 'A Theory of Value and Obligation' (1987), 'Environmental Philosophy: Principles and Prospects' (1994), 'Value, Obligation and Meta-Ethics' (1995), 'The Ethics of the Global Environment' (1999 and 2015), the textbook 'Environmental Ethics' (2003 and 2014), 'Creation, Evolution and Meaning', (2006), and the textbook 'Ethics: An Overview' (2012). The collection of key essays in environmental ethics 'The Ethics of the Environment' that he edited was published in 2008.